Teaching and Learning of Energy in K-12 Education

Robert F. Chen • Arthur Eisenkraft • David Fortus
Joseph Krajcik • Knut Neumann • Jeffrey Nordine
Allison Scheff
Editors

Teaching and Learning of Energy in K-12 Education

Springer

Editors
Robert F. Chen
Environmental, Earth and Ocean Sciences
University of Massachusetts Boston
Boston, MA, USA

Arthur Eisenkraft
Education and Human Development
University of Massachusetts Boston
Boston, MA, USA

David Fortus
Weizmann Institute of Science
Rehovot, Israel

Joseph Krajcik
College of Education
Michigan State University
East Lansing, MI, USA

Knut Neumann
Leibniz-Institute for Science
and Mathematics Education (IPN)
Kiel, Germany

Jeffrey Nordine
San Antonio Children's Museum
San Antonio, TX, USA

Allison Scheff
Massachusetts Department
of Higher Education
Boston, MA, USA

Leibniz-Institute for Science
and Mathematics Education (IPN)
Kiel, Germany

ISBN 978-3-319-05016-4 ISBN 978-3-319-05017-1 (eBook)
DOI 10.1007/978-3-319-05017-1
Springer Cham Heidelberg New York Dordrecht London

Library of Congress Control Number: 2014935241

Printed on acid-free paper

Springer is part of Springer Science+Business Media (www.springer.com)

Acknowledgements

This book is a product of an international energy summit held at Michigan State University in December, 2013. The purpose of the conference was to get a better sense of what the research community knows about energy as a crosscutting concept and how to communicate this knowledge in kindergarten through secondary schools. Participants included scientists, science educators, science education researchers and teachers.

This book would not exist without the work of the many authors and the equal contribution of each editor.

We also wish to thank the National Science Foundation. This material is based upon work supported, in part, by the National Science Foundation under Grant NSF DUE-0928666. *Any opinions, findings, and conclusions or recommendations expressed in this material are those of the author(s) and do not necessarily reflect the views of the National Science Foundation.*

Contents

Chapter 1
Introduction: Why Focus on Energy Instruction?

Arthur Eisenkraft, Jeffrey Nordine, Robert F. Chen, David Fortus, Joseph Krajcik, Knut Neumann, and Allison Scheff

Energy is one of the most important ideas in all of science and is useful for predicting and explaining phenomena within every scientific discipline. Yet, there are substantive differences in how the energy concept is used across disciplines. While a particle physicist relies heavily on the idea that energy is conserved during interactions between subatomic particles, an ecologist is typically more concerned with the idea energy transfers across system boundaries.

A. Eisenkraft (✉)
Education and Human Development, University of Massachusetts Boston, Boston, MA, USA
e-mail: arthur.eisenkraft@umb.edu

J. Nordine
San Antonio Children's Museum, San Antonio, TX, USA

Leibniz-Institute for Science and Mathematics Education (IPN), Kiel, Germany
e-mail: jeffn@sakids.org; nordine@ipn.uni-kiel.de

R.F. Chen
Environmental, Earth and Ocean Sciences, University of Massachusetts Boston, Boston, MA, USA
e-mail: Bob.Chen@umb.edu

D. Fortus
Weizmann Institute of Science, Rehovot, Israel
e-mail: David.fortus@weizmann.ac.il

J. Krajcik
College of Education, Michigan State University, East Lansing, MI, USA
e-mail: krajcik@msu.edu

K. Neumann
Leibniz-Institute for Science and Mathematics Education (IPN), Kiel, Germany
e-mail: neumann@ipn.uni-kiel.de

A. Scheff
Massachusetts Department of Higher Education, Boston, MA, USA
e-mail: ascheff@bhe.mass.edu

R.F. Chen et al. (eds.), *Teaching and Learning of Energy in K-12 Education*,
DOI 10.1007/978-3-319-05017-1_1,

While the ecologist and the physicist are both aware that the energy of biological systems and of physical systems are fundamentally the same – that the only difference is the analytical methods used to track energy changes – students who are just learning about the energy concept may not be aware of this fundamental similarity. That is, students often do not connect the energy that they learn about in physics class with the energy that they learn about in biology, chemistry, or geoscience. After all, in physics class they talk about energy being "conserved", while in biology 90 % of energy is "lost" in transfers between trophic levels. In chemistry, energy is often described as being "stored" in chemical bonds, and in environmental science, they often discuss energy "flow" from natural resources to end users. In contrast, scientists know that these ways of talking about the role of energy are just a shorthand – a simplified way of speaking about energy that corresponds to the analytical lens we are using.

When a dietician tracks the energy requirements of the human body to help treat a diabetic patient, she needn't be concerned with the thermal energy increase in the surroundings as metabolism occurs. Talking about the body "using" energy from food to carry on life processes is typically sufficient. While she is aware of energy conservation, tracking transfers to the Earth's atmosphere doesn't help her to treat patients, so this portion of energy analysis is typically omitted. In fact, including this portion of energy analysis may ultimately detract from treatment because it can distract her patient from the main idea she is trying to convey.

Teachers face a complicated prospect when teaching students about energy. Like the dietician who is aware that energy is quantitatively conserved but chooses not to discuss this with patients for their own benefit, teachers must choose how to present energy in their discipline-centered classrooms such that the analysis does not unduly confuse students but that is still true to the nature of energy. With the release of the *Framework for K-12 Science Education* (National Research Council 2012) and the *Next Generation Science Standards* (Achieve Inc. 2013), teachers have a new challenge to teach energy as both a core disciplinary idea and a crosscutting concept. That is, teachers are now charged with not only teaching about energy as a disciplinary idea but also teaching explicitly about energy as an analytical framework that cuts across disciplines. While scientists typically do not make these crosscutting connections explicit in their day-to-day work, we are asking teachers to instruct their students in such a way that these connections are made clear. The bet we are placing through these new standards documents is that teaching energy as a crosscutting concept will help to prepare a new generation of scientists and engineers who are well equipped to think about the cross-disciplinary problems that are becoming increasingly important in our world. To respond to the challenge of teaching energy in new ways, teachers need guidance from the science and science education research communities on how to present energy in their classrooms.

In December 2012, 40 scientists, science educators, and teachers gathered for an energy summit to better understand the importance of the energy concept in school science and how to best promote student understanding of the energy concept. While much previous work has been done to understand students' conceptual difficulties with learning the energy concept and the instructional imperatives that emerge from these difficulties, we recognized new imperatives that were not addressed by existing

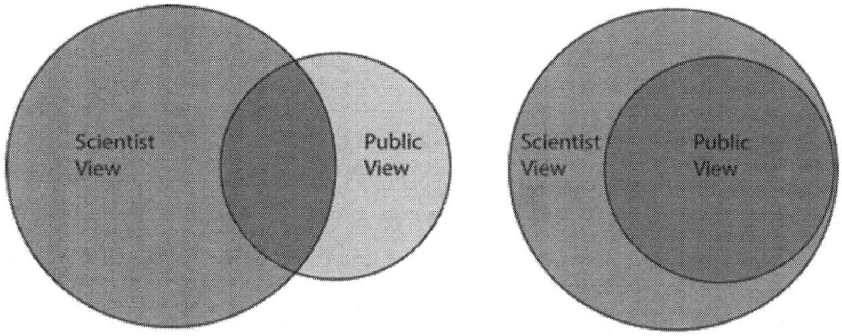

Fig. 1.1 Contrasting possibilities of how scientists' views of energy and the public's view may be related

research. Specifically, no consistent strategies for teaching the concept of energy exist to foster the development of a comprehensive understanding that spans across disciplines, as an empirically validated learning progression of energy that spans K-12 is missing.

As a group, we shared our insights and prior work on a set of three related questions. The first of our questions, "What should people know about the energy concept?" required a careful look at the scientists' view of energy and how this contrasts with the public's view of energy. Imagining a Venn diagram of the scientists' view of energy and the public's view of energy, how much overlap is required? (See Fig. 1.1) In the Venn diagram on the left, the scientists and the public have some overlap. We expect that the scientists will have additional knowledge that is not shared by the public owing to their expertise. The troublesome domain is that of public knowledge that is not shared by the scientists. We might hope that the public's domain of understanding would be completely within the scientists' domain (as in the Venn diagram on thc right), but we know that the public (and scientists on many occasions) will adopt a common usage of energy – e.g. vim and vigor – that has little to do with scientific understandings.

The term "energy" has many meanings within everyday contexts. There is no problem with words having multiple meanings if both parties are aware of the different uses and are clear on which meaning is being employed within a conversation. When the common usage makes it difficult for people to understand the scientific view of energy, we need strategies to draw sharp distinctions among the uses of the term "energy" and its meaning.

A fundamental challenge that exists within the energy concept is that a clear, crisp definition of the term seems to be out of our reach. While the principle of conservation of energy is remarkably simple for any isolated system (the total amount of energy never changes), a rigorous and self-consistent definition of that which is being conserved is remarkably difficult to state. Richard Feynman, in his famous Lectures on Physics captured the essence of the energy conservation law:

> There is a fact, or if you wish a law, governing all natural phenomena that are known to date. There is no exception to this law – it is exact so far as is known. The law is called the conservation of energy. It says that there is a certain quantity, which we call energy,

> that does not change in the manifold changes which nature undergoes. That is a most abstract idea, because it is a mathematical principle; it says that there is a numerical quantity, which does not change when something happens. It is not a description of a mechanism, or anything concrete; it is just a strange fact that we can calculate some number and when we finish watching nature go through her tricks and calculate the number again, it is the same (Feynman et al. 2011, pp. 4–3).

Feynman goes on to say, "It is important to realize that in physics today, we have no knowledge what energy is" (Feynman et al. 2011, pp. 4–3). From this abstract notion that "whatever energy is . . . it is conserved" scientists then introduced the related concepts of energy forms/types, transfer, transformations, and degradation/dissipation.

Professional scientists acquire a tacit understanding of energy and it is the job of the teacher to assist students in acquiring a similar understanding. This brings us to our summit's second question, "What are the challenges we are facing in teaching students about energy?" Why is student learning about energy different from learning about the cell or ionic bonding or buoyancy? What are the unique features of the concept of energy that pose difficulties?

Since energy is both a disciplinary core idea and a crosscutting concept as articulated in the *Frameworks* and NGSS, teachers need students to help understand how energy is a part of the living environment and the physical world. They have to learn about energy in the context of biology, chemistry, physics and the earth and environmental sciences. Simultaneously, they must recognize that the energy of living (e.g., their bodies and various organisms) and nonliving systems (e.g., chemical reactions, roller coasters, tectonic plates) are the same energy.

Students have to learn about energy even though "we have no knowledge of what energy is." We must provide students with opportunities to explore energy even though they cannot touch it or see it. Though students cannot touch energy, see energy, or even measure it directly, every student has experienced the feeling of "having a lot of energy" or being tired and feeling "low on energy". Though we cannot define energy, we all feel an intuitive connection to the idea through our everyday lives. In school, students measure temperature and calculate thermal energy or measure speed and calculate kinetic energy, but these activities help very little in helping students gain insight into the energy that they "feel" and "use" in their everyday lives.

Students experience unique challenges in learning about energy because it is a fundamentally abstract idea, yet has precise scientific uses. Further, students use the term energy in their everyday lives well before learning about it in school and come to develop an intuition about it that may or may not map onto a scientific view of energy. The challenges students face when learning about energy are substantial, and much research has been done to try to understand these challenges. Yet, with a strengthened emphasis on helping students to not only understand energy in scientific and everyday context but also across scientific domains, much work remains in understanding the challenges that students face when learning about the energy concept.

As the summit attendees tried to reach consensus on these first two questions ("What should people know about the energy concept?" and "What are the challenges we are facing in teaching students about energy?"), we then confronted the last of our questions, "What can be done to meet the challenges?" How can the topic of energy be approached across the curriculum and across all grades? Compared to the first two questions, the last question is much more difficult. Precious few studies have been done to investigate promising approaches to energy instruction and even less research has been done to explore how K-12 students can be taught to build an integrated understanding of energy (i.e., an understanding that cuts across contexts and is organized around the most broadly applicable principles) over the course of many years. Yet, summit participants were invited precisely because they have been making progress on this front – either through ongoing projects or their own instruction, or both.

Of course, the ultimate goal of the summit and this book is to impact classroom instruction by providing teachers with a clear direction to take in their own energy instruction via consensus recommendations for instruction and identification of promising research directions. As you will see in the subsequent chapters, consensus is difficult to achieve for this complex and indispensible concept that so many disciplines use in different ways. But there is hope. By asking an international and diverse group of scientists, science educators, and teachers to share their own work relative to the three summit questions and to critique each other's ideas in light of their own experience, we were able to make progress toward identifying key energy understandings, naming important challenges faced by students, and sharing promising instructional approaches.

In the remainder of this introductory chapter, we will share how we came to realize a need for the summit and describe the structure of the summit to illuminate how we shared our work and discussed it from a variety of important perspectives. Then, we will introduce the structure of this book and how the chapters within it have been grouped to illustrate some of the common ground – and areas of ongoing debate – that emerged during the energy summit.

1.1 Realizing the Need for a Summit

All attendees have a personal story that brought them to the summit. As one of the organizers of the conference, the University of Massachusetts Boston has been researching energy as a crosscutting concept through our Boston Energy in Science Teaching (BEST) grant funded by the National Science Foundation.

Boston Energy in Science Teaching (BEST) is a partnership between the University of Massachusetts Boston, Boston Public Schools, Northeastern University, and Roxbury Community College. Through this grant, the partnership is looking at how the teaching and learning of energy as a crosscutting concept can impact classroom instruction, student achievement and engagement, teacher content knowledge, and faculty research. This project, which began its investigation prior to the publishing of the NRC Framework and Next Generation Science Standards, has had three full years to investigate where energy is taught in the classroom, what connections can be made between curricula, what type of professional development can help teachers begin to teach with an energy lens, and what it looks like to teach with an energy focus in the classroom. BEST has provided an opportunity for the partnership to contribute to the research summit and lead the teacher summit.

The BEST grant emerged from an interdisciplinary course that we had taught to K-12 teachers for 3 years prior to this grant as part of the NSF sponsored Boston Science Partnership (a Math Science Partnership). The course instructors were professors in biology, physics and environmental chemistry who were present at each class meeting. The experience of teaching this course has enriched their understandings of energy across disciplines and helped them recognize the varied ways in which energy is treated in each discipline.

When the teachers evaluated the course, one comment struck a nerve. A few teachers, not a majority but a few teachers nonetheless, remarked that they wished the professors would have prepared more so that they would not argue in front of us. The teachers making this evaluation had not appreciated the opportunity to see knowledgeable scientists trying to better understand each other, but rather saw the dialog as a problem. These teachers wanted to know the right answer to tell their students and in expressing this desire exposed the impoverished view of science in their classrooms.

What were the professors "arguing" about in the classroom? When discussing the conduction of heat, the physics professor would present a simple equation showing how heat flow across two dissimilar materials was related to the difference between T1 and T2, the contact area, and the thermal conductivity of the materials. The chemical oceanographer saw that the transfer of heat from the warm surface ocean to the colder deepwater as a similar problem but noted that oceanographers model this apparent conduction with the same equation, but use "eddy diffusivity" rather than conductivity. The biology professor claimed that this equation really did not tell the whole story. For an animal with fur, it is difficult to determine T1 and T2, the contact area, and, in fact, the animal could change her metabolism to change T1. All of these perspectives are scientifically correct, but illustrate the diversity of applications for this concept.

The Energy Course was deemed a success by our pre and post test measures and the compilation of student comments, but the concerns of the teachers mentioned above were a catalyst for wanting to further explore energy as a crosscutting concept. We worked with teachers on K-12 vertical articulation of the energy concept and created a second course that focused specifically on how energy is taught in the classroom at each grade level.

Similarly, at UMass-Boston, we began with the premise that there are very few science colloquia where you couldn't raise your hand at the close of the talk and appropriately ask, "What are the energy considerations of your work?" We further realized that almost every science course at the University includes energy and yet we find that our undergraduate students do not realize that ATP in biology, activation energy in chemistry and kinetic energy in physics are all the same energy.

Researchers across the world are having similar experiences as the UMass-Boston group as they pursue the understanding of the energy concept and how it should be taught in the schools. Recognizing the commonalities across our research, the organizers of the conference from the University of Massachusetts Boston, Weizmann Institute of Science (Israel), Michigan State University, Leibniz-Institute for Science and Mathematics Education (IPN), Kiel (Germany) and Trinity University (Texas), began developing a funding proposal and mapping the outcomes of our proposed summit. The proposal was funded by the National Science Foundation as a supplementary grant to the larger and ongoing Boston Energy in Science Teaching (BEST) grant.

1.2 Structure of the Summit

1.2.1 Goals and Participants

Along with addressing the three questions discussed earlier in this chapter, there were three goals for the summit: (1) to synthesize current research on the conceptual understanding of energy, (2) to identify directions for future research on the teaching and learning of energy, and (3) to foster international collaborations among science education researchers.

Key to making progress relative to these three goals was getting the right people to the table. While logistics and funding often prevent assembling an ideal group that includes all relevant players – and this summit was no exception – we attempted to gather a group of scientists, science educators, and teachers who had been conducting their own work in this field and who could represent an important perspective while still keeping the group of participants small enough to have sustained and substantive conversations. In the end, we assembled a group of 40 participants who represented the major branches of science, possessed strong experience in science education, and reflected diversity in their country of origin and career stage. There are, of course, notable exceptions to our participants (for example, we were not able to arrange a scholar from Africa to attend). Still, the assembled participants were representative of many different perspectives and backgrounds. The list of attendees and their affiliations are provided in Appendix A.

1.2.2 Surfacing and Discussing Ideas

The general structure of the summit was interactive. Eighteen of the summit participants were experts who have conducted prior research on the teaching and learning of energy, and these participants wrote a 15–20 page paper prior to the summit. The papers described their research, opinions, and responses to the three guiding questions on the teaching and learning of energy.

At the summit, submitted papers were discussed during small group discussion sessions. Using a "tuning protocol" (Blythe 2008), two sub-groups of participants simultaneously had structured conversations about the ideas presented in three papers (for a total of six papers per tuning protocol session). By engaging in this structured conversation, participants were supported in focusing the discussion around ideas presented in each paper and in providing the author with targeted suggestions for revisions of their papers, which ultimately appear in this book.

Each tuning protocol session was immediately followed by three simultaneous "report-out" sessions that used a "Jigsaw" format (Aronson and Patnoe 1997). These sessions grouped participants such that each tuning protocol sub-group was equally represented; this grouping helped ensure that participants looked for areas of overlap between the tuning protocol conversations.

The summit included three tuning protocol sessions and three report-out sessions (allowing for discussion and synthesis related to all 18 submitted papers). Each report-out session was focused on a different summit guiding question, and participants were tasked with identifying areas of overlap and disparity between the papers presented in the preceding tuning protocol conversations. Thus participants recorded areas of consensus and dispute relating to what students should know about the energy concept, what challenges students face in learning about energy, and promising instructional approaches.

Throughout each tuning protocol session and report-out session, participants were intentionally grouped based on their scientific background and research/teaching experience. This grouping allowed us to explore both disciplinary and cross-disciplinary perspectives as we discussed the three summit questions.

It is also important to note that the summit did not include plenary discussions or keynote speakers. This organizing team made this choice in an effort to keep the attention focused on the collaborations of the researchers presenting their work and to work towards an environment in which no one person's perspective was systematically elevated above another's.

1.2.3 Teacher Voices and a Second Summit for Teachers

At the close of the second day, the K-12 teachers hosted a panel discussion where they shared their reflections of the papers and the discussions of which they had been a part. The summit drew to a close on the third day. In the morning, groups synthesized the commonalities and disagreements from the summit discussions. In addition, each group discussed the structure for this book. This work also paved the way for plans for a teacher summit, which was held in July 2013. To ensure continuity and build upon the work done in the researcher summit, the teacher summit included all teacher participants from the researcher summit and a few scientists and science education researchers. Just as the researcher book has culminated in a book primarily intended for scientists and science education researchers, but that is useful for teachers as well, the teacher summit will culminate in a book for teacher practitioners that will be useful for scientists and science education researchers who are interested in how the recommendations from this book may play out in practice.

1.3 Organization of This Book

While all authors addressed each of the three guiding summit questions, many felt better positioned to comment substantively on a single question and focused their papers accordingly. After the summit, the organizing team grouped the papers (which had been revised based on feedback received during the summit) based on the question to which we felt they made the strongest contribution. In this process, we recognized a need to create a fourth category for papers, since some papers were quite strong in representing what existing research has to say about the teaching and learning of energy. Thus, the parts of this book are organized around four major questions.

- Part I: What should students know about energy?
- Part II: What does the research say about the teaching and learning about energy?
- Part III: What are the challenges about the teaching and learning about energy?
- Part IV: What opportunities/approaches exist for teaching and learning about energy?

Each part begins with a brief introduction and summary of the chapters in that part. Each summary then ends with conclusion statements, recommendations, and a few discussion questions. The part introduction is followed by the research papers that were presented, discussed, and revised by summit participants.

- Part I: What should students know about energy? This part includes two chapters by physicists (Helen Quinn and Ramon Lopez, both of USA) and one by a group of science educators (Jenny Dauer, Hannah Miller, and Charles Anderson, also of USA) who discuss energy in a biochemical context. Since the authors represent multiple disciplinary and instructional backgrounds, it is illuminating to note where their ideas both overlap and diverge.
- Part II: What does the research say about the teaching and learning about energy? This part includes a chapter by Reindeers Duit (Germany) summarizing the prior research on the teaching and learning of energy in grades K-12. This is followed by an analysis of the standards documents from nine countries with hints of a research based model of energy in chemical reactions by Lie Wang and Wang Weizhen (China). Cari Hermann-Abell and George DeBoer (USA) then describe their efforts in creating assessment questions to test for student understanding four key ideas – forms, transfer, transformation and conservation – and the

results of administering this exam to 24,000 students. The part concludes with a chapter by Bob Chen, Allison Scheff, Erica Fields, Pam Pelletier, and Russ Faux (USA) focusing on a concept mapping approach used in Boston Public Schools to identify elements of instruction that can coordinate discussions of energy across different grade levels.

- Part III: What are the challenges about the teaching and learning about energy? The first of five chapters in this part Hui Jin and Xin Wei (USA) explore how common, everyday uses of the term "energy" can become an obstacle to students learning the scientific view of energy. This is followed by their attempt at an energy learning progression that moves from the common language to the scientific uses of the term "energy." In the second chapter, Xiufeng Liu and Mihwa Park (USA) call for a broader exposure to energy in history classes and dealing with the political aspects of energy use. Robin Millar (England) gives his perspective on everyday use of "energy" and how this can be the launching point for learning how science treats energy. Nicos Papadouris and Costas Constatinou (Cyprus) articulate reasons why energy is such a difficult concept and opt for a philosophical approach that emphasizes energy as a crosscutting concept. This chapter concludes with a paper by Margot Vigeant, Michael Prince, Katharyn Nottis, and Ronald Miller (USA) that elaborates on problems associated with teaching energy concepts to engineering students and the inability of many students to understand the concepts even while correctly answering numerical problems.
- Part IV: What opportunities/approaches exist for teaching and learning about energy? The six chapters in this part describe research based curriculum efforts that can provide guidance on how we can effectively teach energy concepts. In the first chapter, Sara Lacy, Roger Tobin, Marianne Wiser, and Sally Crissman (USA) describe their efforts to introduce energy concepts to elementary school children and map out a learning progression for grades 3–5. Kristen Wendell (USA) evaluates an engineering program to see where energy concepts are present and where there may be missed opportunities to introduce additional energy concepts. Angelica Stacey, Karen Chang, Janice Coonrod, and Jennifer Claesgens (USA) explicitly show the dangers of introducing energy simplifications in chemistry and how these can lead to misconceptions that exacerbate other student learning. Melanie Cooper, Michael Klymkowsky, and Nicole Becker (USA) continue describing energy as it relates to chemistry at the college level and how their curriculum addresses the molecular, macroscopic and quantum mechanical approaches to understanding energy. The fifth chapter, written by Rui Wei (China), Lei Wang (China), and William Reed (USA) critique the different metaphors we use for energy and try to determine the benefits and hazards of our reliance on these metaphors in our teaching. The final chapter by Lane Seeley, Stamatis Vokos, and Jim Minstrell (USA) describe some professional development activities in which teachers acquire a more sophisticated view of energy.

These four parts are then followed by a conclusion and future directions that readers can consider as a means to continue this engaging and important work.

As an organizing team, we were thrilled with the outcomes of the summit and are excited to share with readers the contributions from each of the authors in this book. The papers that follow reflect not only the work of the authors, but also the thoughtful comments, insights, and suggestions for revisions from the scientists, science educators, and teachers from around the world who participated in the summit.

References

Achieve Inc. (2013). Next Generation Science Standards. Retrieved October 10, 2013 from, http://www.nextgenscience.org/next-generation-science-standards.

Aronson, E., & Patnoe, S. (1997). *The Jigsaw classroom* (2nd ed.). New York: Longman.

Blythe, T. (2008). *Looking together at student work* (2nd ed.). New York: Teachers College Press.

Feynman, R. P., Leighton, R. B., & Sands, M. (2011). *Feynman lectures on physics 1: Mainly mechanics, radiation, and heat*. New York: Basic Books.

National Research Council [NRC]. (2012). *A framework for K-12 science education*. Washington, DC: National Academies Press.

Part I
What Should Students Know About Energy?

Energy plays a central role in our everyday lives, as well as in all science disciplines. As a disciplinary core idea and a concept that cuts across all science disciplines, it is clear why energy is a critical idea for students to learn school and consequently why this concept receives particular consideration in policy documents, such as in various standards documents across the world. Energy is one of several "big ideas" of science that is also referred to as a "crosscutting concept" or an "enduring understanding" and is found in almost every year of schooling and across disciplines. However, what is it about energy that is critical for students to learn? Is it its importance as an accounting principle that determines what cannot occur? Is its cross-cutting nature what is important, since this allows it to become an integrator and perhaps unifier among the various science disciplines, allowing for a truly inter-disciplinary understanding of science? Is its economic, political, and societal relevance what students really need to understand?

This part contain three chapters, two by physicists and one by a group of science educators, each of which presents a different perspective on energy and what about it is important to understand to make sense of particular types of phenomena.

The first chapter, written by Helen Quinn, presents a particle physicist's musings about energy. Beginning with the lack of a worthy definition of energy at the macroscopic level and the different lingo that scientists in different disciplines have when discussing energy, she makes a case why energy seems so confusing. She argues that a coherent understanding of energy can be attained only when it is considered at the smallest scales and that therefore our aim should be for students to understand energy at these scales and to be able to use these understandings to make sense of energy at macroscopic scales as well. After using this small-scale understanding to describe macroscopic manifestations of energy (thermal, chemical, mechanical, electrical, and nuclear energy), the applicability of the mass-energy relation to all phenomena, and some of the conceptual difficulties associated with each type of energy, she describes four key ideas about energy that she thinks can and should be the basis of K-12 education about energy.

The second chapter, by Ramon Lopez, uses a complex phenomenon from his research field as a probe into student understanding and learning. He provides a brief overview of what happens when the solar wind impinges on a planet with a magnetic field, such as the Earth, focusing in particular on magnetic reconnection and magnetohydrodynamics, and emphasizing the central role energy transformations play in understanding these two topics. He then discusses some of the energy-related conceptual difficulties that many graduate physics students face when learning about these topics, conceptual difficulties that have their root in inadequate K-12 education about energy. From these conceptual difficulties he raises ideas about how energy should be taught at all levels, including K-12.

While not disagreeing with Helen Quinn who focuses on energy as an actual physical entity, Jenny Dauer, Hannah Miller, and Andy Anderson, in the third and final chapter in this part, present energy conservation as an analytical tool, "rules to be followed". Quinn writes of the need to identify matter with energy to reach a coherent understanding of energy; Dauer, Miller, and Anderson write of the necessity to help students distinguish between matter and energy. They describe an instructional scaffold (twist ties) that helps students focus on matter and energy as separately conserved entities.

Quinn, Lopez, and Anderson were all deeply involved in the development of the Framework for K-12 Science Education and the Next Generation Science Standards (NGSS.) The three different perspectives on energy in this part provide insights into how they played out in these policy documents. Is Quinn's perspective appropriate for K-12 students or must K-12 instruction begin by helping students view matter and energy as separate entities (even though they naturally see them as one) and only later combine both perspectives? Do the conceptual issues identified by Lopez really originate in a poor understanding of energy conservation and flow or can complex mathematics mask the physical meaning of equations and thus the problem is not a poor understanding of energy conservation but the under-developed ability to translate between mathematics and the physical entities represented by the mathematics? The lack of uniformity of these three perspectives stimulates further insights.

Chapter 2
A Physicist's Musings on Teaching About Energy

Helen R. Quinn

2.1 Introduction

Energy is at the same time a topic of high relevance for our everyday life and one of the deepest and most subtle ideas of science. When asked about examples of energy, some students list phenomena involving light, heat or electricity (e.g. Trumper 1990). Some may give examples such as energy stored in fuel (e.g. Lijnse 1990), food (e.g. Solomon 1983) or water (behind a dam) (e.g. Duit 1984). Adults may add terms like nuclear energy, solar energy, chemical energy or mechanical energy. Looking at such a list, it is very hard to see what all these diverse phenomena have in common, where they overlap and where they are distinguishable. Adding to this the fact that the way energy is described within different disciplines of science varies greatly – at times so much so that it is difficult to see connections between them – it is not surprising that energy is such a difficult concept for students to understand (e.g. Duit 1981; Driver and Warrington 1985; Liu and McKeough 2005; Neumann et al. 2013; see also Chap. 5 by Duit, this volume).

Students' need to know what energy is often leads to them being acquainted with a simple definition of energy (cf. Papadouris and Constantinou 2011). The teaching of simple definitions to students is based on the misconception that we learn words and concepts by being told their definitions. In fact we learn them by experiencing and applying them in multiple contexts (cf. Bransford et al. 2000). Any definition of energy at the macroscale level that would be general enough to be correct is either vague enough to be worthless, or contains a long list of "forms of energy" that seem so disparate that no concept can be abstracted from such a definition. This may lead to frustration in both teachers and students Perhaps it helps to discuss the fact that

H.R. Quinn (✉)
SLAC National Accelerator Laboratory, Stanford University, Stanford, CA, USA
e-mail: quinn@slac.stanford.edu

R.F. Chen et al. (eds.), *Teaching and Learning of Energy in K-12 Education*,
DOI 10.1007/978-3-319-05017-1_2, © Springer International Publishing Switzerland 2014

science did not arrive at the concept of energy by defining it, but rather by exploring it (see for example Coopersmith 2010), and that this the path that learners must take too, in order to understand it.

Different forms of energy are measured in different units because they were discovered and categorized at different times. Conservation of energy, first applied only for special and idealized cases (conservative systems), emerged as a more general principle as the relationships and transfers between the different types of energy and the conversion factors between their measures were recognized, and the deeper mathematics behind equations of motion explored. As always in teaching science, we need to untangle the ideas from their history, and decide when recapitulating the historical development of the idea is helpful to students and when it simply immerses them in confusions that they do not need to repeat to get reach a conceptual understanding of the topic being taught as it is understood today. In teaching about energy it is also important to make connections between the concepts related to energy used in different disciplinary contexts, as well as the everyday meanings of the word.

Physicists talk about kinetic and potential energy, using gravitational potential energy for most of the examples of energy transfer they introduce at the high school level, or perhaps elastic potential energy in a spring. Electrical potential differences are introduced in different units and used only to talk about electric circuits. What do they have to do with potential energy? Power is introduced with its own units; the fact that it is a rate of energy flow is not transparent. Energy concepts related to electric and magnetic fields are not discussed till advanced undergraduate courses. Mass-energy equivalence through $E = mc^2$ may be introduced in high school physics in the context of special relativity or nuclear processes, but the true generality of this relationship is seldom stressed. Physicists have adopted a convention that the term heat can only be used for energy transfers between systems, whereas for almost everyone else heat means thermal energy, whether or not it is being transferred. The deep inter-relationship between energy and forces is seldom introduced until advanced undergraduate courses, but the capacity of forces to transfer energy is stressed in introductory physics introducing the added concept of work, which is sometimes presented as a way to define energy (the capacity to do work) which is not particularly enlightening. Chemists talk about bond energy. Nuclear physicists use the term binding energy. Biologists and earth scientists talk about chemical energy, or food and fuel as sources or stores of energy. Engineers talk about electrical and mechanical energy and about energy conversion. Where in all this terminology is a student to develop a coherent concept of energy?

2.2 The Particle Physicist's View of Energy

As I am a particle physicist, the view of energy and matter at the smallest scales informs my thinking. I discuss it here, not because I think we can teach this view as the starting point for understanding energy, but because I think discussing this level of understanding energy allows us to think about what to teach, and when, in order for students to be moving over time towards a deeper and more consistent understanding of energy (i.e. it will prepare the ground for developing learning progression of energy).

At the level of quantum physics, or even advanced classical mechanics, we find that to define energy is to write the laws of nature. If we can define how the energy of a system depends on the relative positions and motions, and on the charges and masses, of the particles within a system, then we can predict (at least probabilistically) how that system will behave. The quantity (technically the Hamiltonian or the related quantity known as the Lagrangian) that describes and defines energy in a system is what determines the laws of physics (i.e. the equations of motion) for that system.

At the atomic or subatomic scale, energy has two basic components, it is either kinetic energy or energy stored in the interaction fields (electromagnetic, gravitational or subnuclear) between the particles. Electromagnetic radiation provides a tricky bridge between the two, because it can be described either as massless particles (photons) which nonetheless carry kinetic energy, or as time-changing and travelling electromagnetic fields carrying energy across space. Both descriptions say it carries energy from place to place, and which is most appropriate to use depends on the situation.

The energy of any system is built up out of these fundamental forms of energy, the motion and interaction energies of the fundamental particles it contains, just as matter is built up from those particles. At different scales it is convenient to describe both the structure of the system and the energy it contains in different ways. However, in the end, I think that, just as we cannot understand many properties of matter without atomic and sub-atomic understanding, we cannot clearly understand many of the commonly used terms for forms of energy until we break them down again and into the underlying particles and their interactions.

The fact that total energy is conserved is a fundamental theorem at this scale, closely linked by the magic of mathematics (Noether's theorem) to the fact physics does not depend on the time, location, or frame of reference. If we write a theory of matter and its interactions for which the function that describes energy has these desirable (and observed) invariance properties, it predicts conservation of energy and momentum among its consequences. However the mathematics that underlies these statements takes us well beyond high school mathematics, so the law of conservation of energy must be presented as a rule which has little empirical support. It is truly difficult to measure all forms and flows of energy, and so any demonstration of the law is at best approximate. While they may be able to see it as a limiting case, that is as true for an idealized system, students have no way to know that it how exact and general a law it is, except by being told it.

Perhaps the most widely recognized and least understood formula in all of science is $E = mc^2$. Most people, including Ph.D. level chemists and biologists, think it is something that only applies in nuclear physics. Instead it is a deep statement that says the quantity we call mass and the quantity we call energy are in fact indistinguishable. (The c^2 in the relationship is just an expression of the fact that we measure them in very different units.) The relationship tells us that, as viewed in the rest frame of the center of mass of any system, what we define and measure as the mass of a system is not just the sum of the masses of the particles that it contains. It includes all forms of energy within it. From outside the system, without probing inside it in some way, there is no measurement that can tell whether the system has a large mass because it contains some high mass objects, or because it contains less-massive but rapidly moving objects.

Indeed as we go to the most fundamental theories we find that most of the mass of protons and neutrons, which means most of the mass of any matter made from atoms, arises from the kinetic energy and interaction energy of the quarks within the protons and neutrons. The sum of the masses of the quarks is only a small fraction of the proton or neutron mass. (Even the quark masses appear as interaction energy. They are due to the interaction of the quarks with the omnipresent Higgs field.) Thus the notion that mass is anything other than an accounting of all energy within a system (when the center of mass of the system is at rest) disappears. Furthermore, for a moving object or system, the division of the energy of a moving particle into two parts, mass-energy (mc^2) and kinetic energy ($1/2\ mv^2$) turns out to be a low speed approximation to the more complete statement of Einstein's formula, which can be written as $E = mc^2/(1 - v^2/c^2)^{1/2}$. In this relationship mass-energy and kinetic energy for a moving system are not separable, but are inextricably intertwined.

While the equivalence of mass and energy is essential to gaining a fundamental understanding of energy, and of conservation of energy, it is irrelevant for most practical purposes, and certainly in the most of school science. In all but nuclear physics situations we do not need to discuss it. We simply leave out mass-energy in all our calculations of energy, because it is a large quantity that, if we are careful about the rest of the accounting, we can treat as a constant. This has an important consequence. Once we have excluded some energy we can never talk about total energy; we can examine only examine changes in energy. However if we are going to discuss conservation of energy as a system changes, we need to be sure we maintain a consistent definition for the energy we have excluded from the accounting.

Kinetic energy for a moving and unchanging object is relatively easy to describe, what is much harder for students to conceptualize is all the various forms of potential energy. In particle theories these all come down to energy stored in fields, relative to that in some reference situation. Theories of fundamental physics are built on a mathematical model in which the interactions between particles are mediated by fields. These fields are essential for modeling the mechanism of forces between distant objects and for modeling interaction energy, and the related concept of potential energy. The key idea is that these fields exist and vary across space, contain energy, and can transfer energy between distant objects. While they are invisible, their presence can be measured by their effect on a test charge or magnet, or in the case of gravitational fields, a test mass, placed in the field. The concept of a force field requires careful qualitative development. It can be introduced well before students are prepared to treat such fields mathematically. Even if students have a vague and science-fiction-based idea of an invisible force field (e.g. Adrian and Fuller 1997) this can be used as a starting point.. The concept can be refined and shaped as students experience phenomena, such as the effect of a magnet on iron filings, or "static electricity," that can be described and explained in terms of fields.

Without the concept of the fields, the interaction energy between the objects is not attached to anything and does not have any location that can be included in the students' mental models of phenomena. In this situation observations that masses speed up as they fall, and that magnets move things without touching them, appear to contradict the notion of conservation of energy. Students tend to conceptualize energy as a thing (e.g. Duit 1987). Physicists conceptualize it as a quantity that can be associated with things, and transferred from one thing to another, but which itself is not a substance[1]. Of course, force fields are not substances either, put they do have a detectable physical reality, that perhaps makes them more readily conceptualized than energy itself. This needs study. How can the concept of interaction energy as energy stored in the space between the interacting objects best be modeled for students? What experiences and activities help students develop this concept? At what stage can potential energy be conceptualized as a difference in interaction energy compared to a reference situation? When does the concept of a force field help, when is it just another meaningless set of words?

2.3 Descriptions of Various Types of Energy

I now examine many of the everyday terms used to describe energy. They overlap and are not generally well defined. It is useful to clarify what they represent and when they are useful. In most cases, as far as I can see, it is not useful to try to define them more precisely – when precision is needed we can achieve it without most of these terms.

2.3.1 Thermal Energy

Many students do not distinguish between heat and (thermal) energy (e.g. Kesidou and Duit 1993 or Maskill and Pedrosa de Jesus 1997). In strict physics definitions this is not acceptable, physics uses the term heat only for energy transfers, and not for energy present in a system. One reason for this is that, as discussed above, total energy present is not a useful concept in most situations, and furthermore it can be difficult to decide what part of that total energy in a given situation should be labeled as thermal energy.

---

[1] "... in physics today we have no notion of what energy is. We do not have a picture that energy comes in little blobs of a definite amount. It is not that way." (Feynman et al. 2011, pp. 4–1).

Indeed until one has a clear particulate understanding of matter, thermal energy cannot even be described. At the particulate level, it is often described as the energy of random translational motion of particles within a system; that is, as kinetic energy. However this description is only true for an ideal (non-interacting) monatomic gas. Whenever we have molecules or solid matter present, thermal motion also includes rotational motions of the molecules and vibrational motion of the atoms in a molecule. If we look more closely we see that the potential energy of interaction of atoms within the material is changing all the time as the atoms vibrate. Energy is constantly being transferred between the atomic motion and the potential energy between the atoms as the molecule stretches and contracts. A little thought makes it clear that if these changes in potential energy were not included in the definition of thermal energy, thermal energy would fluctuate as a molecule vibrates. That would be a most inconvenient definition. So, except in the ideal gas of non-interacting atoms, thermal energy must include some potential or interaction energy as well as kinetic energy.

As soon as we introduce interaction energy we are into the morass of defining energy relative to some fixed condition. Any set of interacting masses and charges has a total energy that depends on the relative positions and motions of the charges and masses, but we seldom need to know or care what that total energy is, in fact we only need to know how it changes when the positions and motions change. In principle we define absolute zero temperature (0 K) to be the temperature at which there is no thermal energy, but since we cannot actually get anything to that temperature that is more a theoretical statement than a practical one. For practical purposes we can relate changes in temperature to changes in thermal energy per unit volume, or per mass of material. With the exception of the ideal gas case, this relationship cannot be easily predicted but rather is extracted from measurements, and it is different for different substances.

The fact that it takes different amounts of heat to achieve the same change of temperature for the same mass of two different substances makes it clear that temperature cannot be measure of energy, or even of energy per unit mass. Students initially conceive of heat and temperature as much the same thing (Kesidou and Duit 1993), after all both have to do with getting hot! Learning to distinguish them and to understand their true relationship is an essential step in reaching a clear view of thermal energy. Many textbooks discuss the relationship only for an ideal gas, which elucidates only a part of the complex relationship.

The concept of heating as an increase in thermal motion clearly breaks down when we consider what happens as matter transitions from solid to liquid, where the energy of interaction between its constituent particles changes significantly. Ice at zero degrees has less energy than the same amount of water at zero degrees, as can be seen by the fact that it takes energy to melt the ice. The water molecules in ice are bound together into a solid. The energy needed to unbind them (that is to break the inter-molecular bonds) is called the latent heat of melting. This is amount of energy we must add to melt a given quantity of ice. This makes it a bit tricky to compare "thermal" energy of ice with that of water. The added energy has broken the bonds that formed the ice crystal. Likewise a change in interaction energy takes place as matter goes from liquid to gas, again energy is added without a change of temperature to achieve the change of state. This energy is called the latent heat of evaporation. So should we call those changes of state changes of thermal energy?

We simply do not need to try to answer to that question. It is a choice, just as defining what part of the energy we remove from the problem by calling it the mass of the system is a choice. Just like total energy as a whole, total thermal energy is not generally a useful concept. (Indeed to completely define the mass of a system you have to define not only its configuration, but also its temperature, because thermal energy too contributes to mass-energy. All of this is generally irrelevant for the problem at hand for K-12 students.) The idea of thermal energy is useful for talking about changes in a system, and where energy goes when it leaves a system, but not for calculating absolute quantities of energy.

2.3.2 Chemical Energy

In any chemical process the set of atoms present does not change, so the mass-energy of the atoms present is constant and thus irrelevant for any energy changes that do occur. Any chemical process takes a set of molecules and converts them to a different set, with different bonds between the atoms. With this in mind we understand why chemists focus on differences in total bond energies to explain energy released or captured in a chemical reaction.

All bond energies are negative because the stable molecule has less energy than the separated atoms. This can also be understood by looking at the electromagnetic fields due to the charged substructure of the atoms, and how the total energy stored in these fields can be reduced by bringing atoms together and "sharing" some of the electrons between them. Actually calculating such changes in energy from first principles is a complex quantum chemistry problem. The language of chemical bonds and bond energies is a useful shorthand to describe the results of such a calculation, or of measurements of energy differences. However it is completely wrong to talk about energy stored in a chemical bond – every chemical bond is a shortage of energy. So what do we mean by chemical energy?

Generally we mean some energy that has been, or could be, released in a chemical process. The energy captured or released in any chemical interaction is the difference between the sum of the bond energies before and after the reaction. Released energy typically manifests itself as increased thermal energy. The energy captured in the inverse process can come from thermal energy or from other forms such a sound energy or radiation. If energy is released, it is because the molecules after the reaction are more tightly bound than those before the reaction – the resulting molecules between them have a greater shortage of energy than the starting ones (compared always to the separated atoms). Thus the term chemical energy is, like thermal energy, not easily defined in any absolute way. All we care about are the changes. It is not meaningful to talk about total amounts of chemical energy, but it is meaningful to talk about the amount of energy released or captured in a particular chemical process.

In everyday language we say a battery converts chemical energy into electrical energy, or that food or fuel contains chemical energy (Calories), but neither of these statements is particularly precise. There is certainly stored energy in a battery, energy that can be released through a chemical process that occurs when the terminals are connected to a circuit. For rechargeable batteries energy can added to the battery by driving that process in the reverse direction with an external electric power source. Since we charged up the battery using an electric current should we now call the stored energy electrical energy, or have we converted it to chemical energy? Does it matter what we call it?

Biologists talk about food or biomass as having chemical energy, or as a source or a reservoir of energy, for example when discussing food webs or photosynthesis. We all do the same when we talk about fuel as a source of energy, or about the number of calories we eat, (A calorie is a unit of energy, defined as the amount of energy needed to heat 1 g of water by 1 °C, the ones we eat are actually Calories, that is to say kilocalories). In fact the food or fuel only provide energy by reacting with oxygen. Saying that the food provides the energy ignores the critical role oxygen plays in the energy balance of the chemical processes of combustion, respiration and photosynthesis. Oxygen (in the form O_2) is removed from the atmosphere (or from the ocean, lake or river) in reactions that release energy, and is added to them in photosynthesis, a process that captures energy from the sun to drive the reverse chemical reaction.

In calculating the energy changes in the processes that turn O_2 plus hydrocarbons into CO_2 plus water (combustion, respiration), or the reverse process (photosynthesis), the changes in the oxygen bonds are an important part of the energy balance. So technically it is incorrect to say the energy either comes from or is stored solely in the food or fuel. However, from a practical point of view, in an oxygen rich environment, the availability of food or fuel controls the availability of energy, so the language, while imprecise for understanding energy, is useful for understanding a food web or the societal needs for fuel. (Of course in oxygen- poor environments organisms rely on different set of chemical processes to release energy for their needs, but the principle that it takes a chemical process, not just one of the reactants, to provide the energy is the same.)

If the inter-dependence and competition between species in a food web can be understood by young students as interdependence in obtaining food, will it help them to discuss it in terms of energy when they really have little idea of what energy is? The stress on energy arises, I think, from the importance of the energy from sunlight for the development of biomass from air and water. Beyond that the food web model says little about energy, at least at the level it is presented to students. Much of the energy flow is at best implicit in the food web model.

For students to connect ideas about energy across the disciplines, in particular between chemistry and biology, it may help if the "food is energy" language were avoided. Can we discuss the food web as a biomass flow rather than an energy flow in the system? Can we say biomass or food provides organisms with access to energy for life functions, rather than that is or provides energy? Even very young students are likely to know they need to breathe as well as eat, when do we connect the need for oxygen with the need for energy? When do we introduce the idea of chemical changes as processes that can release or capture energy? That can be demonstrated as a phenomenon well before the atomic level chemistry is accessible to students.

What is puzzling, and indeed tricky, about the energy captured in photosynthesis is where the captured energy is stored. It is in both of the product materials, and is in fact part of their mass-energy. But as we are leaving mass energy out of the problem, all we can say is that the products of photosynthesis have more stored energy than the reactants did because the product molecules have less negative total bond energies. Calling the difference chemical potential energy is fine, because we see the energy can be released again in the reverse chemical interaction during cellular respiration, but saying any one substance has stored chemical energy eventually leads to confusion. Every chemical bond is a lack of energy. Those negative energies are a puzzle to most students, even at the high school level (e.g. Boo 1998). Certainly they are not needed to understand ecosystems. The shorthand of saying the energy is stored in the biomass simplifies the discussion of the ecosystem. However at some point it may begin to confuse the students. Probably somewhere in the middle school grades, discussion about differences in usage becomes important, acknowledging that, from the point of view of chemistry, the biological terminology about energy is imprecise. The differences in usage must be discussed in order for students to link their thinking about energy in chemical change to their thinking about energy in ecosystems and living organisms.

It is important to recognize that Ph.D. level biologists talk of biomass as energy in an ecosystem, and Ph.D. level chemists think conservation of mass is exact in chemical processes because they do not ever think in terms of the masses of molecules. These conventions are deeply embedded in the language of these disciplines and we cannot change them by changing how we teach at the K-12 level. However what we can do is be aware of the barrier to understanding that these differences across fields can create for students and help diminish that barrier by being explicit about these differences.

This conclusion, that one must discuss the fact that words have multiple meanings, and are used differently in different situations, is one of the major realizations for me in thinking about teaching energy across disciplines. A word may have very particular restricted usage and definition in a certain area of science, but we cannot say that is the only correct definition of the word. It has other meanings in everyday usage, and still others in other areas of science! Part of learning to "talk" science is learning to understand when the restricted definition is being applied, and when the word is being used in a related but less strictly defined fashion. All students can benefit from a discussion of language such as this, but it has particular value for those students whose home language is not the language of instruction.

2.3.3 Mechanical and Electrical Energy

Like chemical energy, mechanical energy and electrical energy are imprecisely defined though commonly used terms. Consider an operating machine with an electric motor (say an electric toothbrush) that is driven by a battery. Does it have mechanical energy, electrical energy, or chemical energy? Perhaps we can agree that it has some of each, but can we define how much of each? Rarely do we care! The brush moves, electric currents flow and the battery runs down through a process of chemical change. We could just as well say the system has motion energy and potential energy, we do not need to define the terms mechanical, chemical and electrical energy to describe it.

Any machine operates with some energy source, often either a chemical process or an electrical one, carries out some motions, and in the end stops, with some objects possibly moved to new locations and different stored energy. The term mechanical energy generally refers to the energy of the moving parts of the machine, but may include elastic or electromagnetic potential energy (such as that of a stretched spring) or even gravitational potential energy that plays a role in the cycles of that particular machine. In my opinion we really never need the term mechanical energy in a science class. Eliminating it is easier than defining it. Eliminating it does not mean ignoring it. As with all everyday terms that overlap with technical terms, students need to discuss the imprecise nature of everyday language in order to understand why scientists introduce and carefully define the new terminology, in this case the terminology of kinetic and potential energy.

Electrical energy arriving via the power grid seems to be one of the biggest mysteries for students (c.f. Stocklmayer and Treagust 1996; also see Bodzin 2011). When energy moves from the power plant to your house over the power grid, given that the grid is alternating current, electrons do not flow from one place to the other, they simply move back and forth in the wires. The kinetic energy of their motion is tiny. However because electrons carry electric charge, when they move the electric and magnetic fields around them change. These changing fields and their effect on matter or magnets are what heat your toaster, light your electric light, ring your doorbell, or drive the electric motors in your blender or can-opener. So we say that these devices are driven by electrical energy. Keeping track of where that energy resides when your appliances are turned off is a bit messy. Eventually it is transferred from the system that drives the generators at the power plant to the system you are using, and you pay for the amount that flows through your meter, without concerning yourself about where it was the moment before you flipped your switch. Modelling these systems and the fact that energy is transferred between them via the power grid is more useful than trying to model where the energy resides at any instant.

Students hear, learn and use all of these imprecise terms; for everyday uses they are quite adequate. The question for teaching about energy is whether and when it is important to define them or eliminate them – when does striving for precision add clarity, when does it just confuse? Clearly a transition to thinking about energy in terms of motion and interactions at the particulate level cannot precede the same transitions in thinking about matter. But can we use a little care and avoid reinforcing the misconceptions or contradictions of everyday language around energy? Can we discuss everyday terms without seeking to artificially define them to try to make them more scientific?

2.3.4 Conservation of Mass?

Historically and practically it is important to chemists to emphasize that mass is conserved in chemical processes. Well before anybody understood the variety of elements or the nature of their atoms, chemists had observed this fact. In trying to understand any process knowing that something is not changing is a very important step because it severely delimits possibilities. Even alchemists did not try to transmute light substances into gold, they knew that was impossible! With a modern atomic view we can see that the law of conservation of mass and the law of constant proportions in chemical processes can both be understood as consequences of the law of conservation of atoms in chemical processes. These empirically-discovered laws preceded, and helped lead to, our understanding of atoms. Furthermore we can readily measure masses of reactants and products but we cannot so readily observe atoms, so conservation of mass remains important as a phenomenon that students can observe.

However the statement that mass is conserved in all chemical processes contradicts the relationship $E = mc^2$ from physics. Conservation of energy and conservation of mass cannot both be exact in chemical processes. Kinetic energy changes in such a process. If mass does not change then some energy has appeared from nowhere. How can we resolve the discrepancy? Only by giving up conservation of mass as a principle.

To get a consistent view across disciplines, it is necessary to conclude that the mass of a molecule is actually a tiny bit less than the sum of the masses of the atoms it contains, by exactly the binding energy of the molecule divided by c-squared. Differences in binding energies are accompanied by differences in mass-energy, and thus in mass. However, the difference between the mass of a molecule and the sum of the masses of the atoms it contains is such a tiny fraction of the mass of the molecule that it is not measurable by any chemical balance. Furthermore the large difference in scale between the mass of the atoms and this mass difference makes it very inconvenient to discuss both in the same the units. Obviously, since atoms are conserved, the sum of the masses of the atoms is constant in any chemical process. Chemists therefore say mass is conserved and talk only about energy differences, that is differences in binding energy. They never actually discuss the mass of the molecules, or if they do, they treat it as being the same as the sum of the masses of the atoms, which it is to the accuracy of their measurements.

Even Ph.D. level chemists may be shocked by the idea $E = mc^2$ applies to molecules in this way, but eventually agree, that, while not measurable by their methods, this may be true in principle. From the chemist's perspective this is a totally irrelevant fact. From the perspective of gaining a common understanding of phenomena of different types I think it is critical. At the point when students are learning the meaning of $E = mc^2$ in physics, this issue needs to be discussed.

2.3.5 Energy Flows (Convection, Conduction and Radiation)

Energy moves from place to place in three generic ways, through movement of matter, through energy transport within matter without bulk movement of the matter (conduction), and through radiation. These mechanisms cannot be described with any precision before students have a particulate view of matter.

Whenever a local source heats a region the thermal energy so produced tends to be spread around by more than one of these mechanisms. Which one is the dominant effect depends on the situation. Students are often asked to say (or told) which of the three occurs in a sample situation, even though the situation, viewed in detail, actually involves more than one. Take for example a room warmed by a radiant space-heater –does convection or radiation dominate? – that probably depends on where in the room you are standing. Yet students are given this as an example of radiation. I think this kind of oversimplification confuses rather than clarifies. It would be much better to allow students to have a nuanced discussion to decide which type dominates than to present these as mutually exclusive options.

Obviously any moving object carries energy from place to place as it moves, since motion itself is a form of energy. In fluids energy can be moved around by a flow of hot fluid from one place to another within the fluid. When this occurs as a cycle driven by a heat source and gravity, and perhaps also by earth's rotation, we call the flow a convection current. Locally heated fluid rises because it is less dense than unheated fluid above it. Cooler fluid flows in from the sides to replace it, only to be heated in turn by the heat source, and thus to rise, setting up a flow pattern. In a spinning earth, its oceans or its atmosphere, earth's rotation also contributes to the patterns of the flows. The patterns of the winds, and of ocean currents, as well as the flow of fluidized rock deep within earth's crust are all important in earth's systems. Understanding and modeling these flows of matter and of energy are an important part of the earth sciences. Thus in earth sciences physical, chemical and even nuclear processes deep in the earth's core play complex and intertwined roles in understanding and modeling matter and energy flows. How and in what detail these phenomena can be treated depends on the order in which students are presented with the different disciplinary ideas, but whatever the order, if teachers do not make linkages across the disciplines and untangle different conventions for talking about energy within the disciplines, the students can not be expected to do so.

Radiation, the third type of energy flow, seems perhaps the most mysterious to students particularly when it is not visible (e.g. Libarkin et al. 2011). Any object is constantly radiating and absorbing electromagnetic radiation to and from the surrounding environment. If the object is hot enough we can see this radiation as a red glow, or hotter yet a "white hot" glow, but even objects that do not glow in the visible part of the spectrum are emitting radiation, just at longer wavelengths than those we see. Plants and animals glow in the infrared, as can be seen using infrared sensitive detectors or film. Night vision goggles take advantage of this effect.

The fact that light transports energy can be connected to the fact that we feel it as warmth when it is absorbed in our skin, but the relationship between visible light and other electromagnetic wavelengths is not obvious to young children and cannot be made so until they are well adapted to abstract models for scales that they cannot see. Models of matter at the scale of atoms and their substructure need to precede and inform models of how matter can produce and absorb electromagnetic radiation, and models for that radiation as it travels across space.

The term radiation carries a negative notion for many students because some radiation is both invisible and dangerous to our health (e.g. Millar 1994). This includes short wavelength electromagnetic radiation, where each photon carries enough energy to ionize atoms in our bodies. Most of the ionizing radiation from the sun is absorbed when it ionizes atoms in the upper atmosphere, but some ultraviolet penetrates to earth's surface and can cause sunburn and possibly skin cancer to those over-exposed to it. X-rays are even shorter wavelength and more dangerous ionizing electromagnetic radiation, and gamma rays are even more extreme.

Radioactivity introduces a different confusion around the word radiation. Some nuclear decays indeed produce ionizing electromagnetic radiation (gamma radiation). Other nuclear decay processes produce fast moving particles such as helium nuclei (alpha radiation), neutrons, or electrons (beta radiation). These are all matter particles, but when produced by nuclear transitions they are generically and confusingly referred to as nuclear radiation, and the source nuclei as radioactive. This terminology predates any understanding of the nature of the produced particles but persists in both everyday and nuclear physics usage today. Indeed, these energetic particles too can cause tissue damage and ionization, so from a medical perspective they are likewise described as radiation and assigned dose limits for safety. However from the point of view of trying to clarify different ways that energy is transmitted, these are massive moving particles, and the term radiation means electromagnetic radiation. How confusing is that? Again the contradictory terminologies cannot be avoided, so must be discussed.

2.3.6 Nuclear Energy

Nuclear energy is yet another poorly defined term. It is often used to mean electrical energy produced by a nuclear power plant. We could define it to mean energy released due to either nuclear fusion of nuclear fission processes. This energy first appears as motion of product particles or radiation, and then, in the power plant example, gets used as a way to heat water to drive a steam turbine to produce electric power.

All nuclear processes depend on one or other of the two nuclear interactions, strong and weak interactions. One characteristic of these processes that makes them notable is that the changes in stored energy are large enough that the changes in mass are a much larger fraction of the mass present than in a chemical reaction. Hence it is in nuclear processes that the equation $E = mc^2$ is usually introduced as an explanation of where the energy released came from – it came from a reduction in mass. But equally it can be described as coming from changes in interaction energy within the nucleus (e.g. in alpha decay), or even within the nucleons (in beta decay). Since this interaction energy is measurably included in our definition of the mass-energy, and hence the mass, of the nucleus or nucleon involved we are forced to say that mass changes in this case.

2.4 Key Energy Concepts for K-12 Science Education

I now turn to discussing the four key ideas about energy that I think can be taught to K-12 students. As in all science, one big part of this teaching must be to clarify and stress the distinction between technical usage of words and everyday usage of words. Indeed as discussed above, it turns out that to make connections across disciplines, you also have to understand that the term energy is used differently in different science disciplines, and so you also need to understand those differences as you try to understand the language (or rather languages) of science.

2.4.1 Only Changes in Energy Matter (Who Cares How Much You Have if Most of It Is Not Negotiable)

While energy is not a substance, it has one thing in common with matter as viewed at the K-12 level; both are conserved quantities – stuff we can neither make nor destroy. When we talk about energy transfers or energy flow, it can lead students to conceptualize energy as a material thing (Warren 1983), which it is not. Perhaps it would help to compare it to net worth, which can be held or transferred in many ways (of which currency is only one) and for which it is important to keep track of its coming and going through a system of book-keeping. The net worth of a school district includes the value of its physical plant, the schools and (usually) the land they stand on, but in deciding the budget for the coming year, most of that is fixed and not negotiable, so the total net worth of the school district does not matter, what matters is its projected income for the year, and its plan for spending. Keeping track of energy is like keeping track of a budget in that way. (Perhaps we could go even further with the analogy and think of kinetic energy as cash, and potential energy as money in the bank.)

When we discuss energy in any situation we are actually only concerned with changes in energy – how much it transferred between objects or systems, how much is captured or released during any change in the system. The absolute energy of the system never matters, unless we are trying to build the system, to create its massive matter, from energy alone. That only occurs in particle physics collisions where we collide and annihilate matter and antimatter at high energies and produce new particles and antiparticles with different masses. In any other situation we start with some matter, and it undergoes some processes, but the mass of the matter is not changed by a significant amount, except when nuclear processes occur. In all non-nuclear cases it is convenient to treat the mass-energy quite separately from other forms of energy, and to leave it out of the book-keeping for energy altogether. So, while the principle that mass is energy is general, and to my mind critical to full understanding, in most cases it makes sense to treat mass and energy as separate concepts. Mass-energy is not included in the definitions of energy for a train or a car. If we were to include it, it would be a large constant energy in any process. Then, in order to look at any other forms of energy we would be calculating differences that are tiny fractions of the whole. That is always inconvenient. Much better to take the large constant mass-energy out of the problem and deal only with changes.

Obviously, once we are leaving out one of the aspects of energy, we can never talk about the total energy of a system, only about changes in its energy. Even then, the way we describe and account for these changes depends on the scale at which we are describing the system. It also depends on the choice we make in order to define the mass of the system, that is the part of the energy that we want to remove from the equations. There is always arbitrariness to this choice. We must choose some reference situation, which we define to have zero potential energy. Whatever interaction energy, indeed whatever energy of any type, is present in this reference situation is to be included in the mass of the system and removed from the energy accounting problem. One consequence is that potential energy will sometimes be a negative quantity in our equations. This can be very confusing to students (Stephanik and Shaffer 2011). But what does negative energy actually mean? It simply means the system has less interaction energy than the reference system which we arbitrarily chose to define as the zero potential energy case.

As an example of this arbitrariness let us think about the system that consists of a mass hanging from a spring, bouncing up and down in the gravitational field at the surface of the earth. To study energy changes during the motion of this system we must consider how three things change – the kinetic energy of the mass, the gravitational potential energy of the mass, and the energy stored in the spring. The energy of the spring we separate into two parts, a constant mass-energy which we want to remove from consideration, and interaction energy differences relative to that, which we call the elastic potential energy. We have to pick a reference length of the spring to define its mass-energy and thus fix the zero value the elastic potential energy terms. We also have to pick a location for the mass at which we define its gravitational potential energy to be zero. There is no "right" choice.

Suppose we choose our reference position for both parts of the calculation to be the one where the spring has its relaxed, unweighted length. In that case the elastic potential energy will always be positive, because when it is stretched or compressed relative to this length the spring has added energy. However the gravitational potential energy will be sometimes be negative and sometimes positive relative to this position, depending on whether the mass is below or above it. Alternately, we could choose the lowest point of the motion as the reference point, the point that we pull the mass down to before we let it go to bounce up and down. Then gravitational potential energy will always be positive but elastic potential energy will be negative relative to this situation, because the spring is less stretched anywhere else in its motion. There are other possible choices. Each choice changes the equations we write, but not the basic underlying fact that changes in one type of energy are balanced by changes in the other two types, or by the loss of energy to the surrounding environment, which eventually will bring the mass to rest at a position that is different from the one where we let it go. We can even separate the two definitions, and define the elastic potential energy relative to one location and the gravitational potential energy relative to another – that may be confusing and certainly takes is careful book-keeping, but it is not wrong. In no case is it meaningful to talk about the total energy of the spring-plus-mass system, because it is interacting with the earth. Gravitational potential energy is part of a larger system spring-plus-mass-plus-earth.

It does not matter which choice we make – each has advantages and disadvantages. We just have to be clear about our choices and keep them consistent throughout our treatment of the problem. In principle, in each choice the spring-plus-mass system has a different mass-energy with the hanging mass held at rest at our reference point. Indeed, the larger system earth-plus-spring-plus-mass also has a different mass-energy at each choice of reference point. For either system the differences in mass between the different choices are too tiny to measure with any mass-measurement that we could make (and anyway for a system as artificial as "earth-plus spring-plus-mass" the mass of the system is never separable from lots of other mass and energy in the world around it). But we do need to recognize that for each different choice we make for defining the zero of any type of potential energy, we have decided to drop different parts of the energy out of our equations. While the changes in mass are a tiny fraction of the mass present, the changes in both gravitational and elastic potential energy are important for our problem, so we must carefully define the reference situation in order to write our equations for energy.

The idea that a reference system is needed to define what part of the energy we remove from our problem as a constant, and what part we treat as potential energy in our problem is seldom clearly introduced. Students are told that the zero of potential energy is arbitrary and can be chosen as they wish, but not that this is because they have in effect defined whatever energy is present in the reference situation as part of the mass of some object or system, to get it out of the way, and to avoid the irrelevant and highly complex question of total energy.

2.4.2 *Any Change in Energy Is Balanced by Some Other Change in Energy (You Can't Make or Destroy Energy, Only Move It Around)*

With total energy removed from consideration, conservation of energy becomes the statement that any change in energy is balanced by some other change in energy. In any system energy can be transferred between the components of the system, or between motion and interaction energy within the system, or it can be transferred into or out of the system. Keeping track of energy requires keeping track of all these things.

In everyday language we talk about producing and using energy –what we actually mean by producing energy is producing fuel that we can move around and burn to release energy when and where we choose, or generating electricity, which also serves to allow us to move energy around to use when or where we need it. To "use energy" means to use the fuel or electricity to provide energy to do whatever it is we want to do (move, keep warm, produce light). Once we have "used" the energy it is not gone. Energy always ends up dispersed into the surrounding environment as light, sound and heat and in waste materials. Diffuse energy in the environment is generally hard to capture and re-use, so we think of it as "gone" or "lost", but from a strict energy accounting point of view it is still there. Students can be confused by the contradiction between the common admonition that they should strive to conserve energy and the physics principle that it is always conserved. The differences in the meaning of the word conservation in these two cases merit some explicit discussion.

Obviously if matter flows into or out of the system it can carry energy. Even if there is no matter flow, energy can enter or leave the system as heat –either by radiation, or by conduction if there is any contact between the system being studied and anything else (for example the air around it). Physicists talk about the ideal notion of an "isolated system" but no system can actually be isolated in a way that prevents it from radiating energy, or absorbing radiation. The energy that leaves a system as heat is hard to measure. In general students cannot verify conservation of energy through their own measurements. There are a few examples, such as a collision of two different size pucks on an air table, where conservation of energy, together with conservation of momentum, can be used to predict outcomes, to the level of accuracy of the measurements. Even in these cases a student might observe that the collision made a sound, that friction, while reduced, is not zero, and that there is some drag on the objects from the atmosphere. So the conservation of energy in this situation is at best approximate. It took me many years of physics study before I realized that conservation of energy was a fundamental principle, not an idealization that would be dropped once I got to a deeper level of understanding. I do not know at what stage of a student's education it is worthwhile and meaningful to stress this difference.

Whenever we define a system that is not in fact isolated, there is another way that energy can be transferred into or out of it. That is through forces due to objects that we defined as external to the system, forces that act upon the system and change its state. Physicists define a quantity called "work", though it has little to do with our everyday concept of work. However the calculation of work done on a system allows us to keep track of the changes in energy due to forces acting on it. While very useful for physics, the physics concept of work is not natural to students (Gilbert and Osborne 1980) and is not much used in other disciplines. Physicists sometimes offer "the capacity to do work" as a definition of energy. In my opinion this definition is entirely useless for gaining a conceptual understanding of energy, and only meaningful in very restricted and idealized situations. Indeed it is logically circular, as the definition of work was arrived at by asking how much a force changes the energy of the object it is acting upon.

The concept of work is only one side of the relationship of forces to energy, the side that encodes the fact that forces acting on an object can change its energy. The other side of the coin is that any pair of (equal and opposite) forces acting between a pair of objects are an indication that there is interaction energy between them. This interaction energy would be reduced if each object moved in the direction of the force on it due to that interaction. Objects fall, or roll down hill because that reduces the gravitational interaction energy between the mass of the object and the mass of the earth. Like charges repel each other because moving them apart reduces the energy stored in electric fields between and around them, and unlike charges attract because the electric field between and around them, and thus the energy stored in that field, is reduced as they move closer together.

These phenomena that are usually learned as rules without explanation, yet they have explanations in terms of force fields and/or interaction energies. Potential energy (whether gravitational, electromagnetic or even nuclear) and negative chemical bond energy are other such phenomenon. Chemical bonds, elastic and tensile forces within matter, and contact forces between matter objects all depend on the charged substructure of atoms and the electromagnetic fields, and hence forces, between them. I would very much like to see some studies of whether (and at what stage) introducing the concept of fields can help students develop models that allow them to better model, interpret, apply and relate energy and force phenomena, and better understand chemical bonds and properties of bulk matter.

2.4.3 Energy Availability Governs What Can Happen (You Can't Do Anything Without Energy)

So if we always leave out some energy, and we cannot verify conservation of energy in any system students could observe, why do we talk about conservation of energy at all? The answer I think lies in the fact that it has important everyday consequences: access to energy controls and delimits what a system can do. Understanding energy flow and redistribution throughout a system is often a key to understanding the functioning of the system as a whole.

Conservation principles are useful precisely because they delimit possibilities. The fact that atoms are conserved in chemical processes allows us to do atom book-keeping to track matter through chemical reactions. It greatly reduces the set of possible processes, for example compared to those that could happen if mass were conserved but atoms could change type freely. Likewise the knowledge that energy is conserved restricts possibilities.

One consequence is that, in order for any system (whether natural or designed) to move anything it needs a way to capture or collect the energy needed for that task, at least temporarily. Generally at the end of any cycle that energy has been distributed into the surrounding environment as thermal energy and in waste materials, so the system needs a continued input of energy in order to continue operating. Thus to understand any system it is valuable to investigate how matter and energy is captured or provided to it, what it is needed for within the system, and how it is redistributed as a system functions. (This idea is highlighted as one of the "cross-cutting concepts" in the Framework for K-12 Science Education, for details see NRC 2012).

Another application of energy conservation seems trivial at the macro-scale but understanding has important consequences for understanding smaller scale examples. If two objects stick together then the combined object has less energy than the two objects separately – energy must be provided to pull them apart again. Thus we can explain why they do not fall apart as a consequence of conservation of energy: they cannot fall apart because they simply do not have enough energy to do so, just as ball sitting at the bottom of a hill cannot spontaneously roll up the hill. This idea seems obvious when we think about ripping apart Velcro or pulling up sticky tape, but can become a mystery to students when it applies at the atomic scale (Boo and Watson 2001). Perhaps emphasizing the parallel would help. A chemical bond is a lack of energy, any stable molecule has less energy than the set of atoms that it contains would have if they were widely separated. Chemists call this difference the bond energy. Analogies that liken the bond to a rubber band are confusing because the rubber band itself adds energy to, and becomes a part of, the system it holds together, even as it creates a combined system that is stable and cannot be taken apart without adding some more energy. A chemical bond is not an object, it is an interaction between objects, and one that lowers their combined energy compared to the situation when they are separated. The chemical bond is like the interweaving of the hooks in the Velcro, not like the Velcro itself.

Negative chemical bond energy is an example of negative potential energy. In both cases we are talking about differences in interaction energy relative to some reference situation. However, the parallel is rarely made, and the reference system is seldom mentioned. (It is neither the starting nor the ending set of molecules in a chemical process, but the hypothetical case of a collection of widely separated atoms.) Even students who understand the notion of negative potential energy in a physics example may become confused when they meet the binding energies in a chemistry class if neither the language of reference situation, nor that of potential energy, is introduced there. Conversely the student who may have grasped the chemical idea of negative bond energy is not necessarily encouraged to see that as an example of negative potential energy when they get to a physics class. It seems to me that these are the kinds of connections can help students integrate knowledge across disciplines. Why are they so rarely made?

2.4.4 Energy Tends to Spread Itself Around as Much as Possible

The final energy principle is perhaps the most mysterious when stated in its technical form – entropy tends to increase. The basic concept here (for the K-12 level) is that particle thermal motions and collisions, and thermal radiation tend to disperse energy throughout any system, and move it between systems. Energy concentrated in a small region of a system is unstable, because processes within the system tend to spread the energy throughout it, and to radiate it away from the system. Objects or regions that are hotter than their surrounding environment lose energy to that environment. Conversely cooler regions get heated. Without energy inputs systems evolve towards a condition of equal temperature throughout, which is a condition of maximally distributed thermal energy. Not only thermal energy, but interaction energies also tend to minimize local concentrations, as rocks fall, and the charged particles within matter move to find positions where the forces on them are balanced against each other. Indeed any large concentrations of stored energy can be dangerous if released rapidly, water behind a dam does incredible damage if the dam breaks, and batteries with high energy-density can catch fire or even explode.

Any process in a machine or living system always ends up heating the surrounding environment and thereby losing some energy. It is a fundamental law of physics that one cannot build a perpetual motion machine –one that goes on running forever with no input of energy – because of this effect. (The formal proof of this statement is not accessible to high school students, indeed many college physics majors struggle to comprehend it; despite that I think it is an important idea for students to learn and consider.) The unavoidable dissipation of energy means that machines need ongoing inputs of energy, and makes production of transportable energy- that is producing or extracting fuel (for combustion reactions with oxygen that release energy) or production of electric power – a major task in industrialized societies.

2.5 When and How Can Students Learn About Energy?

The challenge in all these detailed statements about how to describe energy comes down to the fact that everyday usage gives no way to unify diverse phenomena around energy, or even to define energy. The unifying ideas and technical definitions are all at the atomic or sub-atomic level. Only a limited and idealized set of cases can be treated quantitatively at the macroscopic level. This means that many of the concepts around energy cannot be made precise until students have a firm grasp of particulate models of matter. However, even at the high school or college level, there is no commonality in the way energy is discussed across disciplines, and the differences in definitions and language can leave students struggling to make connections.

Add to this the fact that the everyday usage of terms such as "having energy" or "feeling energetic" to describe the way a person feels or acts is quite a distinct concept from the technical meaning of the word energy, while at the same time it shares some aspects, for example the idea that more motion means more energy. This is the entry point into thinking about energy for young children, and it must be taken into account.

Everyday words that overlap but are not the same as technical words are not errors in usage, but they can lead to misconceptions about the technical meaning unless the differences are acknowledged and discussed as the technical usage is introduced. "Potential" energy has a different problem with everyday meanings. When we say "potential" in everyday language we mean something that might be, but does not yet exist – such as a potential partnership. Potential energy is actual energy stored in some interaction between objects, negative potential energy is a lack of such energy compared to a reference situation –in neither case is potential energy a possibility of energy yet to be realized. So the term potential energy brings its own confusions. Add to that the arbitrariness of the reference situation from which we calculate differences in interaction energies to determine the potential energy, and we see why students struggle to grasp the ideas around potential energy.

In the elementary grades student ideas about energy are necessarily going to be general rather than quantitative. Students experience energy-related phenomena – motion, heat, light, and sound, melting, evaporation, temperature changes. When should the language of energy be introduced? When is it needed? When does it clarify and when does it confuse? How do we help students connect ideas about energy across all the science disciplines? I do not have answers to these questions, but I am convinced that answering them requires both classroom research and a discourse across the disciplines as to how best to teach these ideas at various levels, including at the college level. That is why I am happy to contribute my thoughts to this volume.

Acknowledgement I acknowledge helpful comments from David Hammer and Joseph Krajcik that have influenced the final form of this article. I also acknowledge the careful editorial work of David Fortus, Knut Neumann, and Allison Scheff who have much improved the clarity and added some needed references to a literature that I do not know well.

References

Adrian, B., & Fuller, R. (1997). *A qualitative investigation of college students' conceptions of electric fields*. Paper presented at joint meeting of the Arkansas-Oklahoma-Kansas and Nebraska AAPT Sections, Manhattan, KS. Retrieved June 17, 2013, from http://physics.unl.edu/~rpeg/Electric_fields.pdf.

Bodzin, A. M. (2011, April 3–6). *What do eighth grade students know about energy resources?* Paper presented at the NARST 2011 annual international conference in Orlando, FL, USA.

Boo, H. K. (1998). Students' understandings of chemical bonds and the energetics of chemical reactions. *Journal of Research in Science Teaching, 35*, 569–581. doi:10.1002/(SICI)1098-2736(199805)35:5<569::AID-TEA6>3.0.CO;2-N.

Boo, H. K., & Watson, J. R. (2001). Progression in high school students' (aged 16–18) conceptualizations about chemical reactions in solution. *Science Education, 85*, 568–585.

Bransford, J. D., Brown, A. L., & Cocking, R. R. (2000). *How people learn: Brain, mind, experience, and school: Expanded Edition*. Washington, DC: National Academies Press.

Coopersmith, M. (2010). *Energy – The subtle concept*. Oxford: Oxford University Press.

Driver, R., & Warrington, L. (1985). Students' use of the principle of energy conservation in problem situations. *Physics Education, 20*, 171–176.

Duit, R. (1981). Students' notions about the energy concept – Before and after physics instruction. In W. Jung, H. Pfundt, & C. von Rhoeneck (Eds.), *Proceedings of the international workshop on 'problems concerning students' representation of Physics and Chemistry knowledge* (pp. 268–319). Ludwigsburg: Paedagogische Hochschule.

Duit, R. (1984). Learning the energy concept in school – Empirical results from the Philippines and West Germany. *Physics Education, 19*, 59–66.

Duit, R. (1987). Should energy be introduced as something quasi-material? *International Journal of Science Education, 9*, 139–145.

Feynman, R. P., Leighton, R. B., & Sands, M. (2011). *Feynman lectures on physics 1: Mainly mechanics, radiation, and heat*. New York: Basic Books.

Gilbert, J. K., & Osborne, R. J. (1980). Some problems of learning science. *School Science Review, 61*, 664–674.

Kesidou, S., & Duit, R. (1993). Students' conceptions of the second law of thermodynamics – An interpretive study. *Journal of Research in Science Teaching, 30*, 85–106. doi:10.1002/tea.3660300107.

Libarkin, J. C., Asghar, A., Crockett, C., & Sadler, P. (2011). Invisible misconceptions: Student understanding of ultraviolet and infrared radiation. *Astronomy Education Review, 10*, 010105–1. doi:10.3847/AER2011022.

Lijnse, P. L. (1990). Energy between the life-world of pupils and the world of physics. *Science Education, 74*(5), 571–583.

Liu, X., & McKeough, A. (2005). Developmental growth in students' concept of energy: Analysis of selected items from the TIMSS database. *Journal of Research in Science Teaching, 42*(5), 493–517.

Maskill, R., & Pedrosa de Jesus, H. (1997). Pupils' questions, alternative frameworks and the design of science teaching. *International Journal of Science Education, 19*(7), 781–799.

Millar, R. (1994). Students' understanding of key ideas radioactivity and ionizing radiation. *Public Understanding of Science, 3*, 53–70.

National Research Council [NRC]. (2012). *A framework for K-12 science education: Practices, crosscutting concepts, and core ideas*. Washington, DC: The National Academies Press.

Neumann, K., Viering, T., Boone, W. J., & Fischer, H. E. (2013). Towards a learning progression of energy. *Journal of Research in Science Teaching, 50*, 162–188. doi:10.1002/tea.21061.

Papadouris, N., & Constantinou, C. P. (2011). A philosophically informed teaching proposal on the topic of energy for students aged 11–14. *Science & Education, 20*(10), 961–979.

Solomon, J. (1983). Messy, contradictory and obstinately persistent: A study of children's out-of-school ideas about energy. *School Science Review, 65*(231), 225–230.

Stephanik, B. M., & Shaffer, P. S. (2011). Examining student ability to interpret and use potential energy diagrams for classical systems. arXiv:1109.2074 [physics.ed-ph].

Stocklmayer, S. M., & Treagust, D. F. (1996). Images of electricity: How do novices and experts model electric current? *International Journal of Science Education, 18*(2), 163–178.

Trumper, R. (1990). Being constructive: An alternative approach to the teaching of the energy concept. Part I. *International Journal of Science Education, 12*, 343–354.

Warren, J. W. (1983). Energy and its carriers: A critical analysis. *Physics Education, 18*, 209–212.

Chapter 3
A Space Physicist's Perspective on Energy Transformations and Some Implications for Teaching About Energy Conservation at All Levels

Ramon E. Lopez

3.1 Introduction

Our sun is a magnetically active, variable star, and those variations can have a significant effect on our planet. Space physics, also known as Heliophysics (the NASA term), is the study of the sun, its variations, and its interactions with the space environments of the Earth and other bodies in the solar system (Kivelson and Russell 1995). The sun produces a continuous outward flow of plasma (comprised primarily of protons and electrons), and associated magnetic field, called the solar wind. At Earth orbit the typical solar wind flow speed is 400 km/s, the typical density is 5 cm^{-3}, and the typical magnetic field is 5 nT. Thus the dominant form of energy in the solar wind is in the flow, while the magnetic field energy density is about 1/64 the flow energy. Random, thermal motions of the solar wind particles typically contains 60 % the energy in the magnetic field.

When the solar wind impinges on a body in the solar system, a complex set of interactions result. These interactions are fundamentally electromagnetic, and plasma physics is the branch of physics that describes the kinds of interactions that occur between the solar wind and the magnetic fields and plasma environments of planets, comets, and other bodies in the solar system. When the solar wind approaches a planet with a significant magnetic field (like the Earth), a cavity is formed in the solar wind, the boundary of which is determined by the balance of solar wind pressure and planetary magnetic field pressure. Such an object is called a magnetosphere (see Fig. 3.1).

Occasionally, the sun rapidly releases magnetic energy in the solar atmosphere, giving rise to solar flares, which are the most energetic events in the solar system. The relaxation of solar magnetic fields to a lower energy state also can produce large clouds of plasma and magnetic fields called Coronal Mass Ejections (CMEs) that

R.E. Lopez (✉)
Department of Physics, University of Texas at Arlington, Arlington, TX 76019, USA
e-mail: relopez@uta.edu

R.F. Chen et al. (eds.), *Teaching and Learning of Energy in K-12 Education*,
DOI 10.1007/978-3-319-05017-1_3,

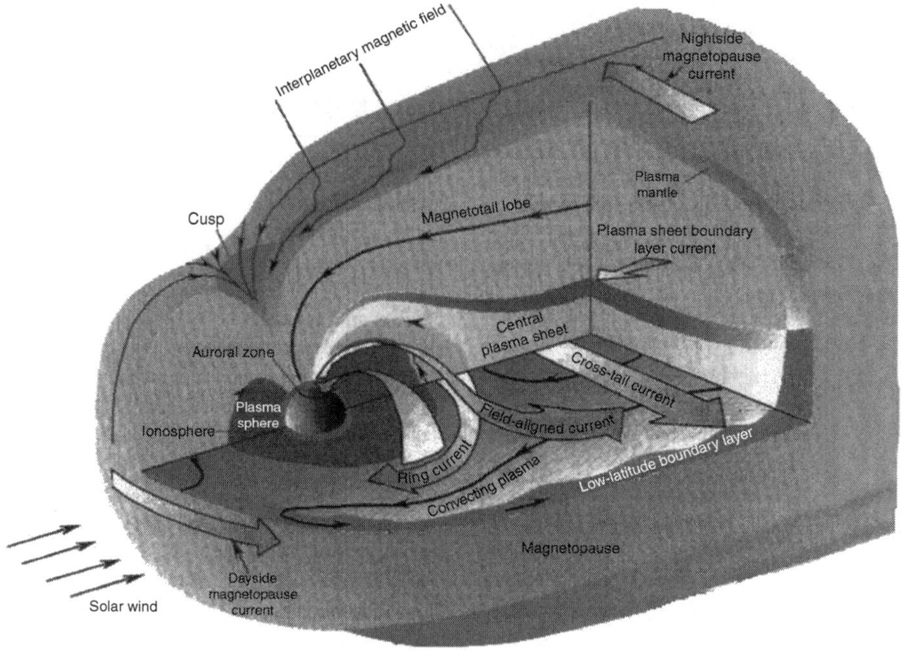

Fig. 3.1 A schematic depiction of Earth's magnetosphere

race through the solar system. CMEs in turn can transfer some of their energy to the magnetospheres of objects in their path.

The interaction of the solar wind with the Earth's magnetic field is of particular interest, especially when phenomena like CMEs are involved. That interaction can transfer significant amounts of energy to the geospace system, producing environmental changes known as space weather (e.g., Carlowicz and Lopez 2002). Space weather can have a severe impact on the space- and ground-based technologies upon which we depend. Particles in space (protons and electrons) energized to high energies (hundreds of MeV), either by the CME in interplanetary space, or by processes in the magnetosphere, can damage spacecraft electronics or pose a danger to astronauts. The variable electric currents that flow through the ionosphere during these magnetic storms can induce currents on the ground that can damage our electric power grid. In fact, a National Academy of Sciences report pointed out that a superstorm (like the kind that hit Earth in 1859, a worst case scenario) could severely damage the US electrical grid and cause trillions of dollars in economic loss (*Severe Space Weather Events–Understanding Societal and Economic Impacts* 2008).

A key feature of heliophysics research is to trace the flow of energy through the system. The recently released Decadal Survey of Solar and Space Physics (*Solar and Space Physics: A Science for a Technological Society* 2012b) picks up this theme in the second paragraph of the Introduction. It states: "*The energy Earth receives from the Sun determines its environment. This energy, primarily visible light, but also including ultraviolet and X-ray radiation, establishes the temperature, structure, and composition of Earth's uppermost atmosphere and ionosphere. The Sun also has a corpuscular output — the magnetized solar wind and energetic particles — that expands throughout interplanetary space, interacting with the Earth and affecting humans in numerous ways...*" Throughout that report, which will guide NASA, NSF and other funding agencies over the next decade, the issue of the transport of energy through the system and its conversion through various forms is of paramount importance and is a major theme of the science goals.

Given that energy transformations are at the heart of Heliophysics, how do graduate students understand with these concepts when they first encounter them? What conceptual framework about energy do they bring to bear on the problem? In the remainder of this paper I discuss two specific topics in Heliophysics: magnetic reconnection, and the energy transport equation in magnetohydrodyamics. For each of these I will provide some qualitative background about the physics, but then examine the fact that I have seen a remarkable inability of graduate students to apply basic concepts of energy conservation, concepts they should have learned in high school, to understand just what the mathematical formalism means. For the most part, these are just observations that have been collected over time. However, in the case of the energy transport equation, I do have some evidence about student response to approaching the subject through an active engagement classroom model that was conducted as an experiment. From these examples I will make some recommendations regarding how energy, fields, and especially energy conservation and transfer, should be taught at all levels.

3.2 Magnetic Reconnection: Energy in Fields

The most important interplay and energy exchange between plasmas and the magnetic field is through a process called magnetic reconnection. In reconnection, stress in the magnetic field built up by the motion of plasma is released as the magnetic field rapidly relaxes to a lower energy configuration. A simple analogy is that of a twisted rubber band. To twist the rubber band mechanical work had to be done. The energy transferred to the rubber band creates a twisted configuration. If the rubber band is allowed to unwind it will transfer the energy in the twists to the environment. And if it snaps, it will untwist very rapidly, suddenly releasing the energy. This is essentially what happens in a solar flare.

Reconnection occurs through a topological change in the magnetic field such that the magnetic energy decreases. This energy appears in the heating and acceleration of the plasma. The basic reconnection picture is illustrated in Fig. 3.2. Magnetic fields of opposite polarity come together in the diffusion region. This allows magnetic field lines to reconnect (field lines are not actual objects, they are defined by the plasma on them), and the resulting magnetic tension from the new configuration accelerates the plasma out of the reconnection region. Again the analogy can be made to rubber bands. Consider the magnetic field to be two rubber

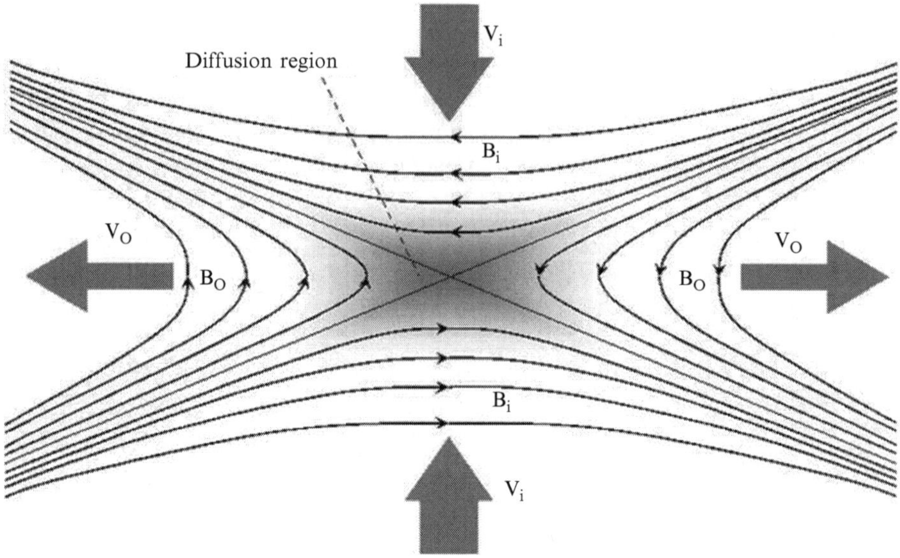

Fig. 3.2 A schematic depiction of reconnection, with plasma flowing into and out of the reconnection region. In the diffusion region, the magnetic field is no longer frozen into the plasma

bands stretched and anchored to the right and left of Fig. 3.2. If we cut the rubber bands in the middle and glue the loose ends to the top rubber bands to those of the bottom, the two reconnected rubber bands will snap back to the right and left.

In discussing reconnection one often hears the phrases "annihilation of magnetic energy", "dissipation of magnetic energy", or simply "energy release." What is really going on is a transition from one configuration, in which there is a lot of stress (magnetic energy) in the magnetic field, to one where that stress is reduced (much less magnetic energy), just as in the case of the rubber bands. And energy conservation dictates that the energy difference between the configurations appears in the plasma.

While the rubber band analogy is imperfect – rubber bands are physical objects and field lines are not, rubber bands can have loose ends and field lines cannot – it illustrates the basic point about energy transfer for magnetic fields to plasmas. Reconnection is ubiquitous in space plasmas. It is the process by which there are sudden releases of magnetic stress on the sun that we call solar flares. It is the most important process by which solar wind energy is transferred to the Earth system, and it is also at work in Earth's magnetosphere dissipating magnetic field energy extracted from the solar wind-magnetosphere interaction.

Students often have trouble with reconnection because they approach the energy considerations in a patchwork fashion. And one particular notion is generally missing from their toolkit is the notion of energy in fields. The idea that an invisible agent, a "field", is responsible for objects exerting forces on each other at a distance is a difficult abstraction (Törnkvist et al. 1993), and undergraduate students readily confuse the notion of fields and potentials or potential energy (Loverude 2005; Meltzer 2007). Without understanding that fields contain energy, and that this is related to the configuration of objects in the system, a whole host of phenomena (not just esoteric things like reconnection) become mysterious.

Consider two carts with magnets on the bumpers moving toward each other on an air track. If the magnets both have the same polarity pointing out, the carts may stop and reverse their motion without even touching. How can one explain this observation? In the language of forces one can invoke "like poles repel", but if one discusses energy it is clear that at a certain point there is no kinetic energy in the system when the carts stop. Where did it go? And then the carts start moving again. Where did the energy come from? In order to use a consistent energy conservation framework to interpret the observations one must include the idea that kinetic energy was transferred from carts into the magnetic field, then back to the carts from the field.

The NRC document *A Framework for K-12 Science Education: Practices, Crosscutting Concepts, and Core Ideas* (2012a), explicitly takes the position that a proper understanding of forces and energy requires students to have at least a qualitative understanding of fields and to recognize the role of fields in energy transformations. This is not to say that the concept of potential energy is to be abandoned. On the contrary, potential energy is very important when discussing the role of gravity, or elastic energy in things like springs or twisted rubber bands. However, over time students should build a conceptual framework around fields. In fact, elementary students already do develop some rudimentary notions about fields (Bradamante and Viennot 2007), and some of those notions appear to be important in resolving other issues, such as the creation of a coherent concept of a round earth (Vosniadou and Brewer 1992).

The development of the field idea, as laid out in the *Framework*, begins in elementary grades, where students explore the fact that objects (such as magnets) can exert forces on each other without touching each other. In middle grades students acquire the notion of a field as a real thing that transmits forces and energy between objects through space. High school students extend that understanding to the fact that energy can be stored in the field itself, and that this depends on the configuration of the system. So as one pushes two north poles closer together, the configuration of the magnets changes as does the energy in the field. This approach would allow students to form coherent models of energy transfer by non-contact forces by invoking fields, which must be recognized a physical entities. To develop the notion of a field as a thing, students can begin with concrete experiences, including activities such as mapping fields, then explore more abstract notions of a field through simulations (e.g., Wieman et al. 2008). And while quantitative calculations of field energy are not expected of students, a conceptual understanding of energy storage and transfer in fields (along the lines of the colliding carts example) is expected of all students.

But the notion of the energy in a field really makes sense in the context of the conservation of energy. After all, why invoke an invisible thing to store energy unless you have a robust conceptual model of energy conservation? And while the conservation of energy is a topic found in K-12 science everywhere, it is also a topic where even physics graduate students often have an unclear and fragmented understanding that inhibits applying the idea to understanding new physics.

3.3 The Energy Transport Equation in Magnetohydrodyamics: Energy Conservation and Transfer

While active engagement techniques have become widespread in undergraduate physics instruction, the use of such techniques has been more limited in graduate physics education (Lopez and Gross 2008). To test the impact of active learning in a graduate classroom setting, an experiment was conducted in which a group of graduate students agreed to sit in on a sample class. The students were then interviewed about the class by a physics education graduate student who had not participated in the class.

The topic of the sample class (which lasted 1.5 h) was the derivation of the energy transport equation in magnetohydrodynamics, a basic equation when considering the behavior of space physics systems. The participants were seven graduate students at UT Arlington, of whom three were US students and four were international students. Only two had had undergrad classes taught with active engagement techniques, and none had ever take plasma physics. At various points in the derivation the students were grouped in pairs (or three students) and asked to derive the next step, which was outlined. At the end of the derivation the energy transport equation was obtained (Fig. 3.3), where P is the plasma pressure, $\vec{V}$ is the plasma bulk velocity, $\vec{E}$ and $\vec{B}$ are the electric and magnetic fields, respectively, and γ is the ratio of specific heats (an adiabatic thermal equation of state was assumed).

The equation looks rather formidable, but it actually expresses a simple, yet difficult to acquire idea. After deriving the equation, the students were broken up into teams and told to pick a term in the equation and describe it in simple English. As that discussion proceeded, the students realized that the first term is the rate at which the sum of kinetic, thermal, and magnetic energy densities change per unit time in a volume. The second term is the rate at which the sum of kinetic, thermal, and electromagnetic energy is going out of the volume minus the amount going into the volume. So, while appearing formidable, the equation just says that the net

$$\frac{\partial}{\partial t}\left(\frac{\rho V^2}{2}+\frac{P}{\gamma-1}+\frac{B^2}{2\mu_0}\right)+\nabla\bullet\left(\frac{\rho V^2}{2}\vec{V}+\frac{\gamma P}{\gamma-1}\vec{V}+\frac{\vec{E}\times\vec{B}}{\mu_0}\right)=0$$

Fig. 3.3 Energy transport equation in ideal magnetohydrodynamics

amount of energy that goes into or out of a volume is equal to the change of the energy in the volume – energy conservation.

If more energy – in whatever form – leaves a volume than enters a volume then the second term is positive. For the sum of the two terms to be zero the first term must be negative, which means that the energy density in the volume decreased. Similarly, if more energy enters the volume than leaves the volume, the second term is negative and the energy density increases. What goes in goes out, except what stays there.

A week after the sample class, the students were interviewed, the interviews were videotaped, and then transcribed. Some student comments are as follows:

So he was relating the physical world to the math world and giving some explanation or some connection between the two ... So that is what I expect from a physics class ... the math and the physics and the two of them at the same time.

It was..it made the students active and it was two way.. Otherwise usually all the lectures are one way ... teacher teaches ... its just on the blackboard or from the notes and class ends. So it's all one way ... so here it was two way and I liked it.

Before I had this lecture, I mean of course we have talked about conservation ... but I never put ... I actually never put the two together ... it seemed like I made some connections that I've never made before ... just from the hour ... half hour lecture ...

While it is heartening to hear that students, most of whom had never experienced this kind of class, liked the active engagement, the third comment reported here is quite remarkable. All these students were Ph.D. students who had passed their qualifying exams. And here is a Ph.D. student saying that he never really quite got conservation of energy until that class!

The issue that undergraduate students have difficulty utilizing a coherent framework around the conservation of energy is not a surprise since it has been documented by numerous studies (e.g., Lawson and McDermott 1987; Arons 1999). One might be a bit surprised that this issue persists well into graduate study. But the difficulty that graduate students have in applying conservation of energy to a new area like space physics seem to be similar to issues that confront introductory students in university physics classrooms. We know that in other contexts with different forms of energy (such as the First Law of Thermodynamics with work done and internal energy), students have great difficulty in correctly applying conservation of energy (Loverude et al. 2002). Moreover, as Loverude et al. (2002) point out "*There was a strong tendency to treat the theorems as formulas, not as mathematical models of important physical principles.*"

A second issue that needs to be addressed is student understanding of what is meant by the system in question. Lack of understanding of what comprises the system has been identified as a significant factor in undergraduate students' inability to use the conservation of energy (Lindsey et al. 2012). And a full description of the system may require the consideration of the energy in the field, as discussed above.

Certainly there when friction is involved one must consider an internal energy term to account for the "loss" of energy (Arons 1999; Legge and Petrolito 2004).

In the same way the undergraduates struggle with solving roller coaster problems using energy, graduate space physics students struggle with understanding configurational changes in the magnetic field and what that means for the dynamics of the system. Simple ideas, like *what goes in goes out except what stays there*, need to be emphasized. Moreover, it seems that additional scaffolding is required for students to connect mathematical representations with the physics. That kind of scaffolding was explicit in the sample graduate class, but is often missing from instruction in general.

The difficulties advanced students encounter point to specific issues that can be addressed in K-12 instruction, and which are found in the *Framework*. Fundamentally, students need to develop a robust model for energy conservation and transfer. The use of multiple, concrete representations of energy (marbles representing photons, boxes whose volumes represent a number of joules, etc.), can help students build the notion of energy conservation (or conservation of any quantity for that matter). These representations should be flexible enough to allow students to include "new" forms of energy and see equivalences between them. In fact, there is evidence that elementary students are capable of just this kind of reasoning (Rodoff et al. 2010). And as soon as quantitative descriptions of energy conservation and transfer are addressed (in High School according to the *Framework*), explicit connections need to be made between diagrammatic, conceptual, and/or physical models of energy, and mathematical models of energy conservation and transfer so that students don't see formulas as just "plug and chug" recipes but as a quantitative means to express a fundamental feature of systems.

Such concrete representations must very explicitly define the system in order to address the fact that a fuzzy understanding of what constitutes a system is an impediment for college students (Lindsey et al. 2012). The issue of energy transformation needs to be stressed over and over, and tightly couple to energy conservation. Students should diagram energy flow through systems, and then connect those diagrams to algebraic expressions and other representations. Much of this is best accomplished in an active learning setting where students have to engage the concept in context and explain it in their own words. Students who develop robust ideas around energy conservation and transfer in K-12 education will be much better prepared to address these issues in undergraduate science education (which itself must change).

3.4 Conclusions

Graduate students in physics are remarkably unable to apply basic a energy conservation framework to issues in space physics, and considerable effort must go into training them to recognize the importance of energy in fields, how changes between different configurations means changes in energy, and how to connect mathematical formalisms to simple ideas like *what goes in goes out, except what stays there*. While the examples in the paper are drawn from space physics, almost certainly any area of physical science could point to analogous issues. The underlying problem is that student conceptual frameworks around energy are not as robust as we would like them to be, and (as in every area of science) there are problems in connecting representations of phenomena into useful mental models. These issues, while persisting into graduate school, extend throughout K-12 education.

The *Framework*, and the upcoming *Next Generation Science Standards*, represent an opportunity to create a much more robust understanding of energy in K-12 science education. The emphasis on fields in the Framework will allow for the creation of conceptual models of energy transfer and conservation that are not possible without considering fields, and which are accessible through direct physical investigation (especially in the case of magnetism). Given what is known about the difficulties that undergraduate and graduate students have with these topics, instruction needs to focus on creating concrete models of energy transfer and conversion in systems in which students are explicitly aware of the components of the system, including components that are invisible (fields). Such a foundation in K-12 education would help address many difficulties with the topic of energy and energy conservation that is encountered in college classrooms.

References

Arons, A. B. (1999). Development of energy concepts in introductory physics courses. *American Journal of Physics, 67*, 1063–1067.

Bradamante, F., & Viennot, L. (2007). Mapping gravitational and magnetic fields with children 9–11: Relevance, difficulties and prospects. *International Journal of Science Education, 29*(3), 349–372.

Carlowicz, M. J., & Lopez, R. E. (2002). *Storms from the sun*. Washington, DC: Joseph Henry Press.

Kivelson, M. G., & Russell, C. T. (Eds.). (1995). *Introduction to space physics*. Cambridge, MA: Cambridge University Press.

Lawson, R. A., & McDermott, L. C. (1987). Student understanding of the work-energy and impulse-momentum theorems. *American Journal of Physics, 55*(9), 811–817.

Legge, K. A., & Petrolito, J. (2004). The use of models in problems of energy conservation. *American Journal of Physics, 72*, 436–438.

Lindsey, B. A., Heron, P. R. L., & Shaffer, P. S. (2012). Student understanding of energy: Difficulties related to systems. *American Journal of Physics, 80*, 154–163.

Lopez, R. E., & Gross, N. A. (2008). Active learning for advanced students: The Center for Integrated Space Weather Modeling graduate summer school. *Advances in Space Research, 42*, 1864–1868. doi:10.1016/j.asr.2007.06.056.

Loverude, M. E. (2005). Student understanding of gravitational potential energy and the motion of bodies in a gravitational field. *AIP Conference Proceedings, 790*, 77–80. doi:http://dx.doi.org/10.1063/1.2084705.

Loverude, M. E., Kautz, C. H., & Heron, P. R. L. (2002). Student understanding of the first law of thermodynamics: Relating work to the adiabatic compression of an ideal gas. *American Journal of Physics, 70*, 137–148.

Meltzer, D. E. (2007). Analysis of shifts in students' reasoning regarding electric field and potential concepts. *AIP Conference Proceedings, 883*, 177–180. doi:http://dx.doi.org/10.1063/1.2508721.

National Research Council. (2008). *Severe space weather events – understanding societal and economic impacts: A workshop report*. Washington, DC: National Research Council.

National Research Council. (2012a). *A framework for K-12 science education: Practices, cross-cutting concepts, and core ideas*. Washington, DC: The National Academies Press.

National Research Council. (2012b). *Solar and space physics: A science for a technological society*. Washington, DC: The National Academies Press.

Rodoff, J., Goldberg, F., Hammer, D., & Fargason, S. (2010). The beginnings of energy in third graders' reasoning. *AIP Conference Proceedings, 1289*, 269–272. doi:http://dx.doi.org/10.1063/1.3515220.

Törnkvist, S., Pettersson, K.-A., & Tranströmer, G. (1993). Confusion by representation: On student's comprehension of the electric field concept. *American Journal of Physics, 61*, 335–338.

Vosniadou, S., & Brewer, W. F. (1992). Mental models of the Earth: A study of conceptual change in childhood. *Cognitive Psychology, 24*, 535–585.

Wieman, C. E., Perkins, K. K., & Adams, W. K. (2008). Oersted medal lecture 2007: Interactive simulations for teaching physics: What works, what doesn't, and why. *American Journal of Physics, 76*, 393–399.

Chapter 4
Conservation of Energy: An Analytical Tool for Student Accounts of Carbon-Transforming Processes

Jenny M. Dauer, Hannah K. Miller, and Charles W. (Andy) Anderson

4.1 Introduction

Energy is a key concept in the K-12 science disciplines, including biology, chemistry, physics, geology and astronomy. In our work we have focused on the fields of biology and environmental science, with a particular focus on carbon-transforming processes. These processes create organic materials (photosynthesis), transform organic materials (biosynthesis, digestion, fermentation), and oxidize organic materials (cellular respiration, combustion). They are the key mechanisms by which energy is transformed in living systems and in human energy systems. It is important for students to understand carbon-transforming processes for many reasons, most importantly that the cause of global climate change is the current worldwide imbalance among these processes.

Helping middle and high school students develop scientific understandings of the role of energy in these processes is especially challenging. Some of these challenges and a learning progression for energy in carbon-transforming processes are described in other publications (Jin and Anderson 2012; Chap. 9 by Jin and Wei, this volume). In this chapter we use these previous findings to underpin our focus on implications for curriculum and instruction: We explain appropriate goals for students' knowledge and practices about energy in carbon-transforming processes, suggest three key challenges in meeting those goals, and briefly describe some instructional supports we are developing to help teachers and students meet these challenges.

J.M. Dauer (✉)
University of Nebraska – Lincoln School of Natural Resources, Lincoln, NE
e-mail: jenny.dauer@unl.edu

H.K. Miller • C.W. (Andy) Anderson
Department of Teacher Education, Michigan State University, East Lansing, MI, USA

R.F. Chen et al. (eds.), *Teaching and Learning of Energy in K-12 Education*,
DOI 10.1007/978-3-319-05017-1_4,

4.2 A Key Goal: Using Energy Conservation as an Analytical Tool

Richard Feynman suggested that the key concept of energy across disciplines is that energy is conserved during processes:

> It is important to realize that in physics today, we have no knowledge of what energy is. We do not have a picture that energy comes in little blobs of a definite amount. It is not that way. However, there are formulas for calculating some numerical quantity and when we add it together it gives "28"—always the same number. It is an abstract thing in that it does not tell us the mechanisms or the reasons for the various formulas. (Feynman et al. 1963).

Feynman suggests that the conception of energy can be used to create measurements and formulas that allow for prediction, modeling and analysis, even if the mechanism or reason for the formula is unknown. According to this conception, energy should be used as a *tool for analysis,* and one key characteristic makes it a valuable tool: *Energy is conserved in physical and chemical changes.* This conception of energy is supported as a suitable goal for K-12 learning by the *Next Generation Science Standards* (Achieve, Inc 2013) and the *Framework for K-12 Science Education* (NRC 2012):

> One of the great achievements of science is the recognition that, in any system, certain conserved quantities can change only through transfers into or out of the system. Such laws of conservation provide limits on what can occur in a system, whether human-built or natural.... The supply of energy and of each needed chemical element restricts a system's operation—for example, without inputs of energy (sunlight) and matter (carbon dioxide and water), a plant cannot grow. Hence it is very informative to track the transfers of matter and energy within, into, or out of any system under study. (p. 94)

We note that this passage relies on essentially a nineteenth century definition of energy, focusing on chemical and physical changes and not mentioning relativistic and quantum conceptions of energy. We agree with this emphasis; in this chapter we focus on conservation of energy as a tool for analysis of carbon-transforming processes. We feel that this achievement arms a student with tools for interpretation and analysis of multiple situations, including important socio-ecological issues relating to global climate change. In particular, this description emphasizes three characteristics that are essential to a useful scientific model of energy. Energy is: different from matter, without mass, and conserved in physical and chemical processes and is therefore traceable through these processes. So these qualities define our key goals with respect to students' conceptions of energy. If students can understand a model of energy that includes these qualities and apply it successfully to carbon-transforming processes, then they will take a major step toward appreciating and using the power of energy as an analytical tool.

4.3 Challenges and Instructional Supports

The power of energy as a concept lies in its application across contexts and disciplines, yet different disciplines offer distinct challenges for teaching and learning. Here we focus primarily on biological systems at multiple scales, from carbon-transforming processes described at the atomic-molecular scale to energy flow through ecosystems and global environmental systems. Discussing energy in biological contexts (e.g. photosynthesis, cellular respiration, digestion) as opposed to energy in physical contexts (e.g., pendulums, projectiles, electrical circuits) exposes unique challenges to student learning. In physical contexts, energy indicators (for example, motion for kinetic or elevation for gravitational potential energy) are often easily observed, and quantification is often possible. In contrast, in biological contexts, which almost always involve chemical energy and heat transfer, energy indicators are more difficult to observe and quantify.

Biological systems pose another kind of challenge. Applying physical laws to real-world systems is daunting because of the complexity of the systems themselves. For example, the trajectory of a batted baseball depends on the initial velocity of the ball, wind speed and direction, the spin on the ball, air pressure, the texture of the ball's surface, and other factors. Learners cannot possibly account for all of these factors in explaining or predicting the ball's trajectory. Physical science classes typically deal with this complexity by simplifying the systems; rather than analyzing actual systems in the real world, students analyze idealized, simplified systems (for example balls batted in a vacuum, ideal gases and pure chemicals). This option is usually not available for the life and Earth sciences where learners study real plants, animals, and ecosystems in complex physical settings. But the need for simplification persists; learners are still unable to analyze living systems in their full complexity. So in order to make our analyses of living systems comprehensible to students, we sometimes must simplify the models and principles instead of simplifying systems.

Simplifying models and principles is a standard practice in all of science. All models simplify the real world, and scientific reasoning always involves choosing the appropriate simplifications for the problems at hand. When we are teaching, though, the "problems at hand" often involve student comprehension as well as systems in the material world. This leads to an important issue that we address in this paper: How can we develop appropriate simplifications of energy-related models and principles for young learners studying living systems? While many agree that a simplified model of energy in complex systems is necessary (Cooper and Klymkowsky 2013; Chap. 11 by Millar, this volume), few have proposed satisfying and specific solutions for simplification in biology instruction. We suggest that appropriate simplifications need to meet at least four criteria. First, they should be comprehensible to students, as indicated by empirical methods like our learning progression research (for example: Mohan et al. 2009; Jin and Anderson 2012). Second, understanding should be achievable within reasonable constraints on instructional time. Third, simplifications should help to position students to understand more sophisticated models and principles in their future learning. Fourth, they should be able to be applied consistently across the range of systems and processes and across spatial and temporal scales.

Through our learning progression work we have identified three core instructional challenges in teaching students to use energy as a tool for analyzing carbon-transforming processes:

1. understanding the purpose of the concept of energy
2. identifying forms of energy in living systems
3. tracing energy separately from matter

In the remainder of this chapter we summarize findings from our research and other research that describe the nature and dimensions of these challenges as we teach students about energy in carbon-transforming processes. We also discuss the implications of these research findings for our goals in teaching middle school and high school students—our judgments about achievable outcomes that students should be expected to learn. Finally, we propose instructional tools and strategies, as well as necessary simplifications of scientific conceptions of energy that we are using to address these challenges and to reach achievable learning outcomes.

In our work we have asked students in grades K-16 about many different carbon-transforming processes at a range of scales. The issues we describe arise consistently across all of these carbon-transforming processes and persist across age groups. In this chapter we will illustrate these issues and our instructional approaches with examples of student responses from two questions: an interview question asking about how trees grow, and a written question asking about what happens to energy when a mouse dies. The questions and examples of responses are included in Table 4.1. These student responses are part of a larger set of data (approximately 150 interviews and written responses to 1,100 student tests) collected from 2011 to 2012 in 6th–12th grade classrooms of 20 teachers in six states. Data collected during the 2012–2013 school year data will be analyzed to assess the effectiveness of the teaching tools and strategies described in this paper.

4.3.1 Understanding the Purpose of the Concept of Energy

Our research to develop a learning progression framework on carbon-transforming processes describes common 4th–12th grade student accounts that use energy as a resource that enables actors to make events happen (Mohan et al. 2009; Jin and Anderson 2012). Other research has documented how students think of energy as the cause of events more widely in physical and chemical contexts

Table 4.1 Sample student responses

Questions asked	Energy is causal	Energy is a tool for analysis Student doesn't have all the details Student attempts to trace energy Energy is an enduring entity
Interview questions including: Does a tree need energy? Where does a tree get energy? What happens to the energy when it is inside a tree?	INTERVIEWER: ... What is the difference between the things that give the tree energy and things that don't give the tree energy? STUDENT A: Because things that give the tree energy they are what make it grow so like the water and the nutrients and the sun and the carbon ... since they're like the food for the tree it is the tree's energy. And I think it has to do with the cells, like the cells need it for the tree to live INTERVIEWER: And do you think that there's energy in the trees, like bark and wood and leaves or any other parts of the trees? INTERVIEWER: Okay. And what about things that don't give the tree energy? You know what kinds of things that that would not include? STUDENT A: Well certain animals like those caterpillars that eat the tree down... INTERVIEWER: Okay. Why does a tree need energy? STUDENT A: So that it can live INTERVIEWER: So it can live? STUDENT A: Grow INTERVIEWER: And grow? Okay	INTERVIEWER: Does the tree need energy to grow do you think? STUDENT B: Yeah, it needs the light energy definitely. I don't know about the other ones, but definitely light energy they need to grow INTERVIEWER: And how does it use the energy to grow? STUDENT B: There's a chemical reaction of some sort. And I don't know—obviously, it's not like a fire, like, burning, but it's some sort of, like, combustion or something that's happening in the molecules inside the plant that is brought on probably by the light energy STUDENT B: I mean there's chemical bonds in pretty much every molecule... INTERVIEWER: So do you think the tree stores energy for later? STUDENT B: Yeah I think so cause it seems like in the winter it would be, like, there's less sun and stuff like that, so it would need to store up energy INTERVIEWER: Where do you think it does that? STUDENT B: Maybe the tree does it in the trunk. I don't know INTERVIEWER: Does it store it in molecules do you think? Or is it stored in some other way? STUDENT B: Probably, but I never thought about it. I guess molecules would make—well, if it's just in the chemical bonds, then yeah, I guess that would make sense INTERVIWER: Is there another way do you think it could store energy besides chemical bonds? STUDENT B: Not that I can think of

(continued)

Table 4.1 (continued)

Questions asked	Energy is causal	Energy is a tool for analysis Student doesn't have all the details Student attempts to trace energy Energy is an enduring entity
Written questions:	STUDENT C:	STUDENT D:
(A) What kinds of energy are stored in the living mouse? Where did they come from?	(A) The energy that the living mouse had stored is the food he had ate. He also might have slept and that made him wake up with energy	(A) Energy can neither be created or destroyed. I'm not sure what kind of energy the mouse has and where it came from
(B) What kinds of energy are stored in the dead mouse (if any)? How are they connected to the energy in the living mouse?	(B) There is no energy in the dead mouse. If there were any he would still be alive	(B) All of the energy is still there, but other organisms who are decomposing the mouse will help convert that energy into breaking down the dead mouse

(Trumper 1990, 1993; Watts 1983). Students often enter biology classrooms with these causal conceptions of energy, which can be compared to Aristotle's concept of energy characterized as being-at-work, and explaining why and how events happen (Jin and Anderson 2012).

For example, middle school students' accounts of how trees grow typically describe a tree as an actor with a purpose in life—to grow. Anything that enables or causes a tree to grow and be healthy is a source of energy for the tree, including water, air, nutrients and sunlight. Examples of students using the concept of energy as causal or a source of vitality are in Table 4.1, where students talk about energy needed for plants growing and about what happens to energy after a mouse dies. In the plants growing example, "Student A" associates energy with a cause that results in growth and vitality, which allows the student to incorrectly include water, nutrients, sunlight and carbon all as providers of energy for a tree, whereas agents which do not cause growth or vitality, like caterpillars, do not provide energy. In the mouse dying example, for "Student C," energy is anything that allows the mouse to thrive, including sleep, and when a mouse dies—energy is gone.

The complexity of biological systems permits students to either make or maintain incorrect associations between energy and vitality, animation or growth. These powerful associations made between energy and vitality is often not important in physical contexts. For example, the kinetic energy of a billiard ball can be both a *cause* of transferred motion (the cause for another billiard ball to be struck and move), and also a *tool for analysis* (tracing kinetic energy from one billiard ball to another). In living systems, however, student conceptions of energy (such as Students A and C who associate energy with purpose, need or vitality, Table 4.2) often do not align with scientific conceptions of energy, and can be particularly distracting when students are explaining biological phenomena (Fig. 4.1).

Table 4.2 The three questions

Question	Rules to follow	Connecting atoms with evidence
The location and movement question: Where are atoms moving? Where are atoms moving from? Where are atoms going to?	**Atoms last forever** in combustion and living systems All materials (solids, liquids, and gases) are made of atoms	When materials change mass, atoms are moving When materials move, atoms are moving
The carbon question: What is happening to carbon atoms? What molecules are carbon atoms in before the process? How are the atoms rearranged into new molecules?	Carbon atoms are bound to other atoms in molecules **Atoms can be rearranged to make new molecules**	The air has carbon atoms in CO_2 Organic materials are made of molecules with carbon atoms Foods Fuels Living and dead plants and animals
The Energy Question: What is happening to chemical energy? What forms of energy are involved? How is energy changing from one form to another?	**Energy lasts forever** in combustion and living systems C–C and C–H bonds have more stored chemical energy than C–O and H–O bonds	We can observe indicators of different forms of energy Organic materials with chemical energy Light Heat energy Motion

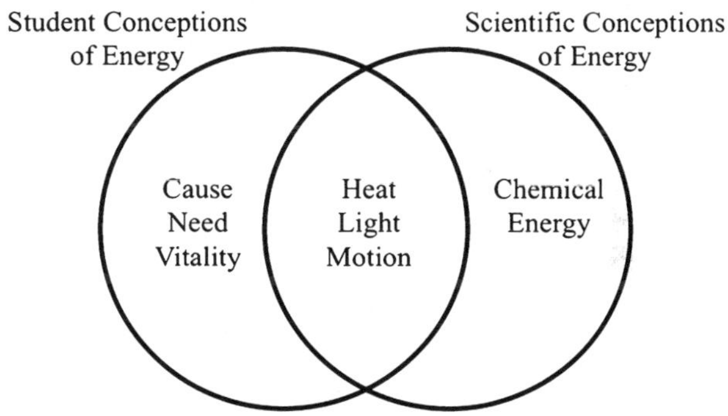

Fig. 4.1 Student conceptions of energy compared to scientific conceptions of energy

Our conclusion is that learning to use energy as a tool for analysis in physical contexts such as pendulums and roller coasters does not address the core learning barriers that students encounter in using energy to analyze living systems. Common middle school definitions of energy such as "the ability to do work" or "the ability to cause a change" can reinforce the idea that causes of events are the same as energy sources, which is especially problematic in living systems. Our instructional approaches address these problems directly.

Thus our goal is to help students move from accounts where energy is an ephemeral cause, into accounts that treat energy as a tool for analysis. In their new accounts, energy must be an enduring entity in a system and traceable through processes. This entails (a) developing a sense of necessity about energy conservation in living systems—the same amount of energy must be present at the beginning and end of a process—and (b) helping students to use quasi-quantitative representations of energy in carbon-transforming processes.

4.3.1.1 Developing a Sense of Necessity About Energy Conservation

Student explanations of carbon-transforming process are complete only when they have accounted for energy before, during and after processes. In our curriculum materials, we treat the principles of conservation of energy and conservation of matter as rules to be followed rather than relying on students to discover these ideas empirically or to construct them from first principles in the classroom. This is consistent with our conceptualization of energy as a tool for analysis; through metacognitive prompts, students use this rule as a tool for analysis of their own accounts of observations, relying on the authority of the laws of conservation as a checkpoint for their own ideas. These rules make it possible for students to self-assess if their accounts of matter and energy in processes and all systems are constrained by the same rules throughout the curriculum.

The rules for students to follow are embedded in our "Three Questions" learning framework (Table 4.2). We teach students that adequate scientific accounts of carbon-transforming processes must include answers to all "Three Questions." The first two questions focus on movement of matter and changes in matter; we focus here on the "Energy Question." As Table 4.1 shows, each question comes with associated "Rules to Follow" including our expression of conservation of energy: "Energy lasts forever in combustion and living systems." (The "Evidence to Look For" relates to indicators of forms of energy, which is discussed below.)

The "Rules to Follow" are particularly important for engendering a sense of necessity about energy conservation. Students are required to apply these rules as they give explanations of carbon-transforming processes and when they interpret evidence from classroom investigations about each carbon-transforming process. During an investigation of mealworm respiration, for example, students readily recognize that the mealworms are moving, which involves energy. The idea that energy "comes from" the food the mealworms ate is also consistent with students' causal notions about energy. The notion that energy must have been present all along in the food (in the form of chemical energy) and that chemical energy in the food was transformed into the energy of their motion rather than simply causing their motion is new to most students. Even newer is the idea that the energy of the mealworms' motion must still exist in some form after the mealworms stop moving. Thus the "Rules to Follow" are the foundations of instructional strategies to instill a sense of necessity of conservation of energy in the students' accounts. The conception of energy can become a tool for analysis of students' own observations and explanations, requiring them to conserve and trace energy through a process.

4.3.1.2 Quasi-quantitative Representations of Energy

Our basic goal is to enable students to account for energy qualitatively in terms of energy forms and transformations; scientific quantification of energy is impractical in these complex systems. Students should be able to treat energy as an enduring entity, accounting for its forms and transformations during a biological process. However, simply asking students to name forms of energy involved in processes has been insufficient to achieve our goals; we have concluded that accounting for energy and its transformations requires both physical representations of energy and rules for tracing energy.

Students who are using physical representations of energy and rules for conservation of energy are able to move towards explanations that treat energy as an enduring entity and students begin to trace energy through processes. College classes that used a physical accounting system for energy showed substantial progress toward principle-based reasoning about conservation of energy (Rice et al. 2014). We have developed similar strategies for K-12 students. We use physical representations of energy (twist ties, Fig. 4.2) during molecular modeling exercises. Representing energy in this way allows students to develop accounts of energy as an enduring entity that it is separate from matter. Twist ties can be referred to as "units of energy" which allows for accounting for energy before and after a process, without quantifying different forms energy.

When students trace energy through complex biological systems, there are simplifications that help students build a coherent story. For example, in our instruction we simplify the Second Law of Thermodynamics by not mentioning the concept of entropy, though we do emphasize degradation of energy in living systems. Energy is not like matter in that it cannot be recycled, and instead flows through living systems, ultimately being lost as heat. So, all processes change energy from more useful to less useful forms, especially low-grade heat. Degradation of energy through an organism or ecosystem can be illustrated for a student by using twist ties that represent sunlight or chemical energy before a process and then heat after a process.

Examples of student responses that show these practices are in the right-hand column of Table 4.2. These are students who have developed a sense of necessity about conservation of energy *even though they do not understand the details of the processes*. In the plant growing response, "Student B" is able to identify the sun as a source of energy, and to trace energy to materials in the trunk of the tree, but is not able to discuss specifics of process of photosynthesis or biosynthesis. In the mouse dying example, "Student D" is not sure about what kind of energy exists in a living mouse, but knows that it will not disappear after a mouse has died, and if it goes somewhere that it is likely to be used by decomposers that are decaying the

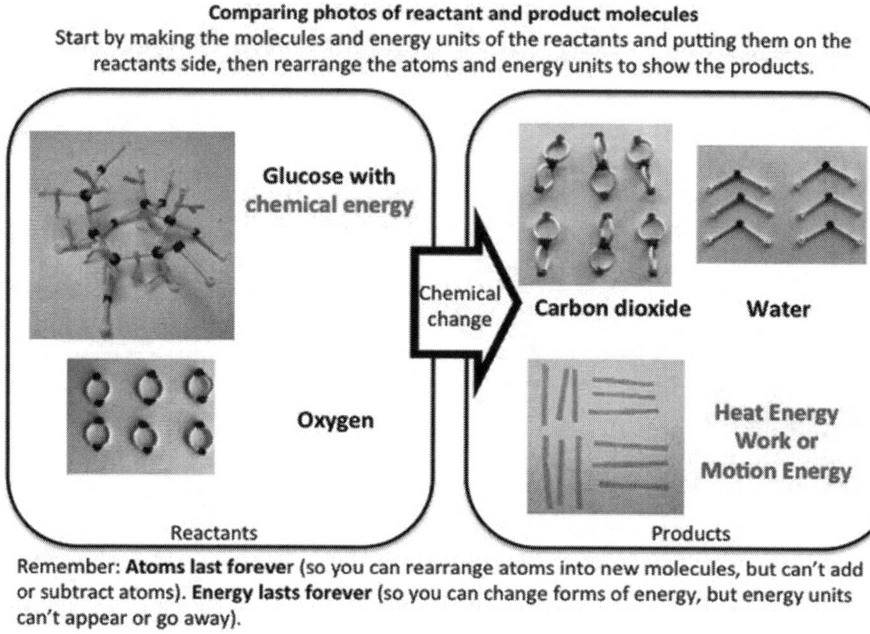

Fig. 4.2 Twist ties and molecular models during an exercise on cellular respiration

mouse. This illustrates that students are able to begin tracing energy and generating worthwhile unanswered questions about a process when the concept of energy is used as a tool for analysis, and when students are given rules to apply. We think that this type of reasoning is both an achievable goal and a useful way for student learning to progress.

4.3.2 Identifying Forms of Energy in Living Systems

Principles such as conservation of energy are useless if students cannot correctly identify the forms of energy in a process (Lee and Liu 2010). Yet many researchers have documented students' non-scientific associations with energy (Nordine et al. 2011; Watts 1983; Trumper 1990) that include energy as human-related, depository, activity-related, as an ingredient, product, function or fluid-like substance. As we have documented in our research (Jin and Anderson 2012), living systems are specifically challenging because students make powerful and incorrect associations between energy and cause, vitality, or growth. For example, students answering the question about a mouse dying often include sleep and exercise as sources of energy for a living mouse, and we commonly talk about energy in similar colloquial ways in our own lives ("my energy is dragging so I'll get more caffeine"). Thus, instruction must help students distinguish between their many colloquial conceptions of energy and a scientific view of how energy is manifested only in specific forms.

In our instruction, we address these problems by making several simplifications. These simplifications include (1) limiting our discussion of energy to specific forms of energy that are important in carbon-transforming processes, (2) describing chemical energy as something that is associated with C–C and C–H bonds, (3) defining "heat" and "work/motion" as forms of energy, and (4) simplifying the second Law of Thermodynamics.

1. First, we limit our treatment of energy to four specific types of energy: chemical energy, light energy, work or motion energy, and heat or thermal energy (ignoring gravitational and other forms of energy), each of which is simplified in some way. We limit the discussion of energy to these four forms as students trace energy through carbon-transforming processes.
2. Second, we describe chemical energy as "stored" in high-energy molecules with C–C and C–H bonds and "released" when these molecules are oxidized and those bonds are replaced with lower-energy C–O and H–O bonds. In doing this we follow common practice in biogeochemistry, acknowledging the critical role of oxygen in transforming chemical energy while recognizing that in the Earth's atmosphere organic materials are usually the limiting reactant. Thus substances in equilibrium with the atmosphere do not have available chemical energy, while substances out of equilibrium with the atmosphere, including organic substances, have available chemical energy. The amount of energy available from a particular substance is equal to its heat of combustion—a more complicated measure of energy than we can use with students who have not studied chemistry. However, counting the number of reduced carbon and hydrogen (C–C and C–H) bonds in an organic molecule provides a reasonable approximation of amount of energy available from its oxidation.
3. We also use simplified ideas of work, kinetic energy, and heat. We conflate work and motion energy; we do not distinguish between work as a process of energy transfer and kinetic energy. In the same vein, we do not clearly define "work," using the word to designate a suite of complex metabolic processes involving transport and biosynthesis as well as organismal motion. For heat energy, we do not distinguish between heat as an energy transfer process and thermal energy, or between "heat energy" and infrared electromagnetic radiation into space.
4. As described above, we simplify the Second Law of Thermodynamics by describing waste heat as a product of carbon-transforming processes without mentioning entropy.

With the help of these simplifications, students can trace energy through all of the carbon-transforming processes that we study, using energy labels for twist ties that are limited to the four forms of energy identified above. We feel that the benefits of this approach exceed the costs. In particular, students can learn to avoid the multitude of non-scientific meanings for energy that they bring from their everyday discourse, and they can begin to trace energy through carbon-transforming processes in a rigorous way. We can see this in the example of "Student B" in Table 4.2 where the student identifies light energy as the source of energy for the tree, and then eventually reasons, with some assistance, that the sunlight energy can be traced to chemical energy associates with the bonds of the molecules that make up the tree. Because the student was limited to notions of energy (light, heat, motion and chemical energy), tracing energy through transformations becomes easier to deduce.

4.3.3 Tracing Energy Separately from Matter

To use energy successfully as a tool for analysis, students must learn to treat energy as an enduring entity and to trace energy through transformations in living systems. Our research suggests that current instructional practices enable very few students (less than 3 %) to achieve this practice consistently in their accounts of carbon-transforming processes (Jin and Anderson 2012). As described previously, less advanced students trace sequences of cause and effect rather than attempting to trace energy as an enduring entity or are unable to distinguish scientific from colloquial meanings of energy. For more advanced students who attempt to trace energy through processes, another substantial barrier remains: these students often conflate forms of matter with forms of energy (e.g., "glucose is energy;" "plants transform sunlight into food;" "the man lost weight by transforming his fat into energy when he exercised.").

We have documented the problem of students conflating matter and energy in our previously published work (Mohan et al. 2009; Jin and Anderson 2012), identifying it specifically with Level 3, the next-to-highest level in our learning progression. Conflation of matter and energy is particularly problematic when learning about biological systems when there are chemical changes such as cellular respiration that transform solids and liquids into gases. In these cases, students have difficulty conserving and tracing energy because they fail to conserve matter. Students who readily assert that gases have mass still have trouble believing that gases have *enough* mass to account for substantial mass changes in living systems (Mohan et al. 2009). So it is easier for students to believe that "fat is transformed into energy when a person exercises and loses weight" than that "a man who exercised and lost 20 pounds breathed out most of that mass in carbon dioxide and water vapor." Much of our instruction aims to have students account for matter and energy as separate and enduring entities.

We have described some of the key elements of our instructional strategies for addressing student conflation of matter and energy, including the "Three Questions" framework, which requires separate tracing of matter and energy, the use of molecular models with separate representations of energy (twist ties) and labels for forms of energy. Students learn to construct accounts that trace matter and energy separately for each process through a combination of empirical investigations and direct instruction using molecular models and other representations of systems at multiple scales (e.g., atomic/molecular, cellular, organismal, and large-

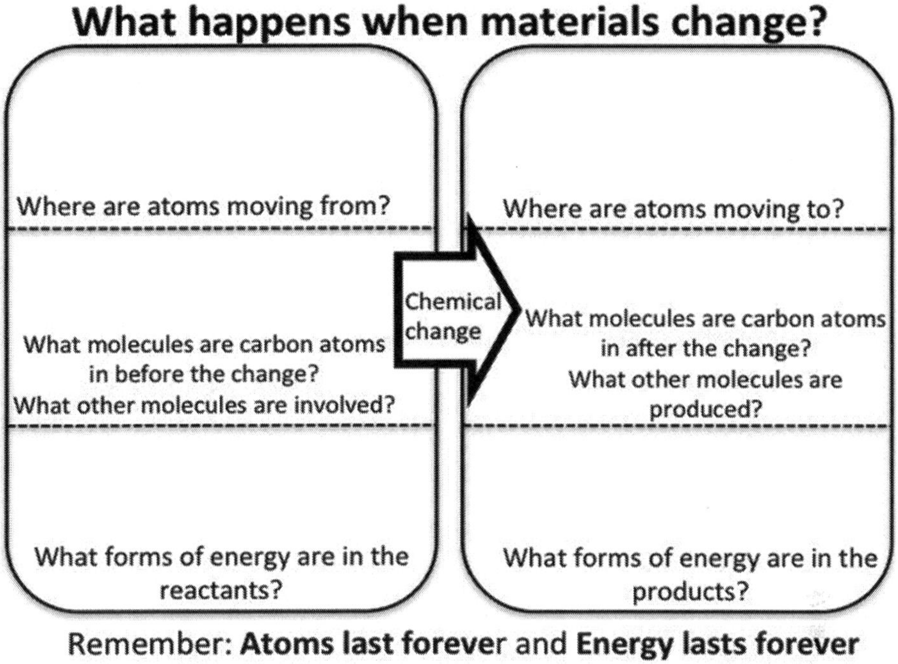

Fig. 4.3 Matter and energy process tool

scale systems). Investigations are necessarily at macroscopic scale and involve observations and measurements of carbon-transforming processes (ethanol burning, mealworms growing and moving, plants growing in the light or respiring in the dark, bread molding). During investigations students can develop partial answers to the "Movement Question" by tracing mass changes in systems; to the "Carbon Question" by detecting CO_2 using probes, soda lime, or bromothymol blue as indicators; and to the "Energy Question" by observing energy indicators such as movement, light, and temperature change. The investigations do not lead to complete answers to the Energy Question because the observed indicators are insufficient to trace energy through the full process.

Following the investigations, molecular modeling exercises help students address some of their unanswered questions that arose in the investigations (for example in combustion: where the carbon in the CO_2 came from, and where the energy was before it was transformed to heat and light). To address these, the students use molecular models to trace carbon atoms in CO_2 back to carbon atoms in fuel molecules to account for carbon atoms before and after the burning. Similarly, students use twist ties to trace heat and light energy back to chemical energy associated with C–C and C–H bonds in the fuel molecules to account for energy before and after the burning. One final instructional scaffold is the "Matter and Energy Process Tool" (Fig. 4.3), which students use to construct accounts that answer all "Three Questions."

The process tool is continually revisited by the students. The students write their answers to the "Three Questions" on the process tool after the investigation, then again after molecular modeling exercises. By the end of the instruction the students have developed an account of a carbon-transforming process that traces matter and energy separately through the chemical change.

4.4 Conclusion

In our work we have developed and applied a learning progression on energy in complex biological systems. We have focused, in particular, on systems where energy transformations involve creation, transformation, and oxidation of organic compounds; this includes all living systems and human technological systems relying on biomass or fossil fuels—more than 90 % of current human energy use worldwide. We have identified three central goals for student learning around which this paper is organized: Students should (1) understand that a primary purpose of energy-based explanations is to identify constraints on systems, (2) identify forms of energy in biological systems, and (3) trace energy separately from matter. Our learning progression research reveals that students can gain access to the power of energy as an analytical tool in these systems only if we deal directly with their associations of energy with cause, need, and vitality, and if we help them to develop a sense of necessity—a commitment to the principle that all systems are inevitably constrained by the laws of thermodynamics. We argue that studying energy in simplified physical systems such as pendulums and roller coasters is neither necessary nor sufficient to accomplish these goals.

In our work we have developed a set of scaffolds and simplifications to make these core insights and practices accessible to middle school and high school students. At the core of these supports are the "Three Questions" (Table 4.2) focusing on movement and transformation of matter and energy in biological systems. We teach students didactically that a good explanation of matter and energy transformation must answer each question in ways that satisfy "Rules to Follow" (Column 2 of Table 4.2, essentially the conservation laws) and "Connect Atoms with Evidence" (Column 3 of Table 4.2, identifying key indicators for forms of energy and chemical changes). This didactic framework is accompanied by specific simplified models focusing on forms of energy and by quasi-quantitative ways of accounting for "energy units."

We argue that these scaffolds and simplifications in instruction will make it easier for students to develop a sense of necessity about conservation of matter and energy and will also facilitate an understanding of carbon-transforming processes that is both practical (i.e., students can use their understanding for meaningful inquiry and application) and productive (i.e., it prepares students to learn more sophisticated models in the future). When students succeed in using these strategies to analyze familiar systems and events, then we feel they will have made substantial progress toward our overall goal: to uncover the chemical basis of biological and socio-ecological systems. Developing this productive and practical scientific discourse of matter and energy in socio-ecological systems is an important piece of learning to act as informed citizens around issues that involve carbon cycling and its role in climate change.

References

Achieve, Inc. (2013). *Next generation science standards*. Washington, DC: Achieve, Inc. http://www.nextgenscience.org/next-generation-science-standards

Cooper, M., & Klymkowsky, M. (2013). The trouble with chemical energy: Why understanding bond energies requires an interdisciplinary systems approach. *CBE Life Sciences Education, 12*(2), 306–312.

Feynman, R., Leighton, R., & Sands, M. (1963). *The Feynman lectures on physics* (Vol. I, chpt 4.1). California Institute of Technology. www.feynmanlectures.caltech.edu

Jin, H., & Anderson, C. W. (2012). Learning progressions for energy in socio-ecological systems. *Journal of Research in Science Teaching, 49*(9), 1149–1180.

Lee, H., & Liu, O. (2010). Assessing learning progression of energy concepts across middle school grades: The knowledge integration perspective. *Science Education, 94*(4), 665–688.

Mohan, L., Chen, J., & Anderson, C. W. (2009). Developing a multi-year learning progression for carbon cycling in socio-ecological systems. *Journal of Research in Science Teaching, 46*(6), 675–698.

National Research Council. (2012). *A framework for K-12 science education: Practices, crosscutting concepts, and core ideas*. Washington, DC: National Academy Press.

Nordine, J., Krajcik, J., & Fortus, D. (2011). Transforming energy instruction in middle school to support integrated understanding and future learning. *Science Education, 95*(4), 670–699.

Rice, J., Doherty, J. H., & Anderson, C. W. (2014). Principles, first and foremost: A tool for understanding biological processes. *Journal of College Science Teaching, 43*(3), 74–82.

Trumper, R. (1990). Being constructive: An alternative approach to the teaching of the energy concept – Part one. *International Journal of Science Education, 12*(4), 343–354.

Trumper, R. (1993). Children's energy concepts: A cross-age study. *International Journal of Science Education, 15*(2), 139–148.

Watts, M. (1983). Some alternative views of energy. *Physics Education, 18*(5), 213–217.

Part II
What Does the Research Say About the Teaching and Learning About Energy?

Energy is a critical concept in every branch of science, thus, much research has been carried out in the past about (i) how students conceptualize energy, (ii) how students' conception of energy changes as a result of instruction, and (iii) how students' conception of energy can be changed through instruction. This research provides a wealth of information about the teaching and learning of energy K through 12.

Research on students' ideas about energy has identified many non-normative ways in which students conceptualize energy. These alternative conceptions include anthropocentric ideas (i.e., energy being only associated with living things), energy as a sort of fluid or substance, energy as a cause of changes, and many more. With the increasing relevance of energy for driving societies and technical progress, new conceptions of energy as a resource emerged – including conceptions about the value of energy for human use. To further complicate the vast array of alternative energy conceptions, research has shown that students commonly hold many (often-contradictory) ideas simultaneously and that the activation of these ideas during their reasoning through energy-related scenarios is strongly context-dependent.

Naturally, researchers have become interested in how students' conceptions of energy can be developed into more scientific conceptions, and there has been particular interest in how students build understandings of energy transfer and transformation as well as energy conservation and degradation. In studying this, researchers have found that students' everyday conceptions of energy are both building blocks and stumbling blocks in their progress toward developing more normative scientific understandings. For example, students who hold an anthropocentric view of energy may be well-positioned to understand the energy requirements of a living organism but have trouble conceptualizing the role that energy plays in non-living systems that do not exhibit locomotion. Conceptualizing energy as a fluid may help students to understand the transfer of energy in an electric circuit, but may also cause them difficulties in differentiating between thermal energy and heat. Because energy is a ubiquitous idea in both scientific and non-scientific contexts, even very young students will come to school with their own intuitive ideas about energy, and these ideas will influence their subsequent learning.

While there exists an extensive base of research into students' non-normative ideas about energy, little is known about how students' alternative conceptions can be successfully developed into the kind of scientific understanding that is outlined in the first part of this book. Key questions remain unanswered. Should early energy instruction incorporate terms such as "forms" and "types" or avoid such terms to stress the unitary nature of energy? How can students most effectively be taught to use the ideas of energy transfer and transformation to analyze systems? How and when should students be introduced to the most important aspect of energy (and one of the most important ideas in science) – its conservation? How should students who might accept energy conservation come to understand that not all energy is readily usable by humans? These questions and many others relating to how students develop an understanding of the nuanced role of energy in everyday systems have yet to be definitively resolved.

This part contains four chapters representing the manifold approaches to investigating the teaching and learning of energy in the past: a review of existing research on the teaching and learning of energy in grades K through 12 by Reinders Duit, a qualitative analysis of standards documents from various countries in order to develop a model of student learning about energy in the context of chemical reactions by Lei Wang, Weizhen Wang and Rui Wei, a quantitative large-scale study on students' understanding of energy in middle and high school by Cari Herrmann-Abell and George deBoer, and a concept mapping approach to investigating how energy is represented in the Boston Public School (BPS) curriculum by Robert Chen, Allison Scheff, Erica Fields, Pam Pelletier, and Russ Faux.

The first chapter in this part, by Reinders Duit, presents a comprehensive review of previous research on the teaching and learning of energy in grades K through 12. Using different ways of conceptualizing energy and the aspects of a scientific energy concept as a point of departure, Duit summarizes the findings from the research into the teaching and learning of energy from the early 1990s as well as the recent research into the development of a learning progression of energy. He concludes with a discussion of the importance of students' everyday conceptions in influencing their learning trajectories as they develop a more sophisticated understanding of energy.

In the second chapter, Lei Wang, Weizhen Wang and Rui Wei describe their efforts in developing a research-based model of students' understanding of energy in the context of chemical reactions. Based on a brief review of scientific perspectives on chemical reactions and the role of energy in chemical reactions, Wang, Wang and Wei analyze the standards documents from nine countries. In their analysis, they identify energy-related topics that students are expected to use as a basis for understanding the role of energy in chemical reactions. Next, Wang et al. develop a category set of performance expectations aligned with these topics. From these findings, they develop a cognitive reasoning model of energy in chemical reactions.

The third chapter describes Cari Herrmann-Abell's and George deBoer's efforts in developing distractor-driven multiple-choice items to assess students' understanding of four key ideas about energy: forms, transfer, transformation and conservation. In a multi-step and process that includes a review of the literature, definition of learning goals, and considerations of validity, Hermann-Abell and deBoer developed items for assessing students understanding of the four key ideas. Following a multi-matrix design (in which items are distributed across multiple test booklets and a set of booklets share common items), they administered their energy items to a sample of almost 24,000 students. A Rasch analysis of the obtained data provides insight into how students progress in their understanding of energy in course of middle and high school.

In the fourth chapter, Robert Chen, Allison Scheff, Erica Fields, Pam Pelletier, and Russ Faux use a concept mapping approach to identify opportunities to teach energy as a cross-cutting concept in the Boston Public School (BPS) curriculum. Chen, et al., gathered 12 teacher leaders with BPS curriculum experience to identify the aspects of energy covered in the BPS curriculum and the relations amongst these aspects across grades and science disciplines. Based upon among 162 connections among 105 curriculum units from grades 1–12, the authors identify key opportunities for teaching energy as a cross-disciplinary scientific idea.

While not exhaustive, the four chapters in this part provide key insights into what is known about how students learn the energy concept over time, and they shed light on potential directions for future research. The corpus of knowledge regarding the teaching and learning of energy is both substantial and enlightening, and future work should build on what is known. A major outstanding question is "How do students' progress in their understanding of key energy ideas over time?". More large-scale studies are needed in order to better understand how students tend to progress toward these key ideas and to identify factors that may affect this progression. However, large-scale studies often neglect the influence of particular curricular interventions. Thus, large-scale studies must be complemented by smaller-scale investigations of how coherent curricula can impact students' progression in understanding key energy ideas.

Chapter 5
Teaching and Learning the Physics Energy Concept

Reinders Duit

5.1 Introduction

It is hardly possible to overestimate the significance of the energy concept in science. It provides a powerful frame for integrating the various science disciplines as it plays key roles in biology, chemistry, physics, and in earth sciences as well. But the energy concept also plays a key role in teaching and learning science. In fact, energy provides a very powerful way of thinking about and modeling processes in nature and technology. In addition, understanding problems of energy supply in modern societies is only possible with a sound insight into key basic ideas of the energy concept. Accordingly, the energy concept is among the "big ideas" forming general frameworks in international monitoring studies like TIMSS and PISA as well as in Science Education Standards around the world. Research on teaching and learning science, however, reveals that students' understanding of key basic ideas of energy is somewhat limited (see e.g., Duit and Häußler 1994; Nordine et al. 2010). It seems that a major reason for this is how energy is taught in schools and at the tertiary level. Interestingly, there appear to be different "rituals" of teaching energy. Already Lehrmann (1973) argued that the traditional approach to introduce the energy concept via force and work narrows down the range of the energy concept taught (see also Papadouris and Constantinou 2011, p. 965; Duit 1986a, p. 71f). Nordine et al. (2010, p. 670) claim that the "*traditional energy instruction often focuses on simple calculations of energy in idealized systems*" (drawing on Bryce and MacMillan 2009).

R. Duit (✉)
Department of Physics Education, IPN – Leibniz Institute for Science and Mathematics Education, Kiel, Germany
e-mail: duit@ipn.uni-kiel.de

R.F. Chen et al. (eds.), *Teaching and Learning of Energy in K-12 Education*,
DOI 10.1007/978-3-319-05017-1_5,

5.2 Energy – A Core Physics Concept

In the following first a brief outline of the scientific energy concept and its links to other key concepts is presented. Subsequently educational concerns that arise when teaching and learning the energy concept are taken into account.

5.2.1 *On the Energy Concept in Physics*

As discussed at greater depth in other chapters of this volume (especially in Chap. 2), in physics energy is usually seen as an abstract accounting principle. It is a quantity denoting a debit and credit balance in nature (see Fig. 5.3 below). Energy may be seen as a fundamental property of matter – the other fundamental properties being inertial and gravitational mass. *Special relativity* claims the equivalence of inertial mass and energy. Energy may be seen as an "active" and inertial mass as a "passive" property of matter. *General relativity* theory draws on the equivalence of energy and gravitational mass. Interestingly, there is also an intimate relation between energy and time. The *Noether Theorem* holds that energy is a quantity denoting the symmetry of natural laws concerning translations of time. In other words, energy is based on the homogeneity of time. If this homogeneity is not guaranteed – which may be the case in the space-time-world of general relativity – the principle of energy conservation does not hold as outlined above (Trautman 1962; Weiss and Baez 2012).

5.2.2 *Four Basic Ideas of the Energy Concept*

For the purpose of designing efficient teaching and learning approaches that may guide students to understand the energy concept, it is essential to analyze which basic (elementary) ideas compose the physics energy concept. The ideas provided in Fig. 5.1 resulted from this analysis. The arrows denote that the four ideas are closely interrelated.

The set of the four basic ideas of the energy concept presented in Fig. 5.1 draws on analyses including the following issues (Duit 1986a): (a) the role of the energy

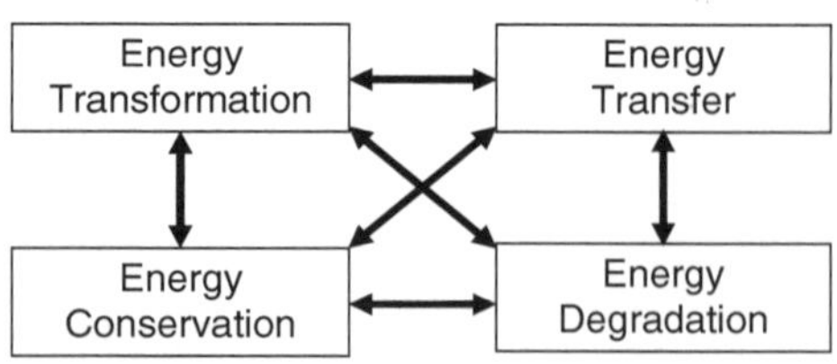

Fig. 5.1 Four basic ideas of the energy concept

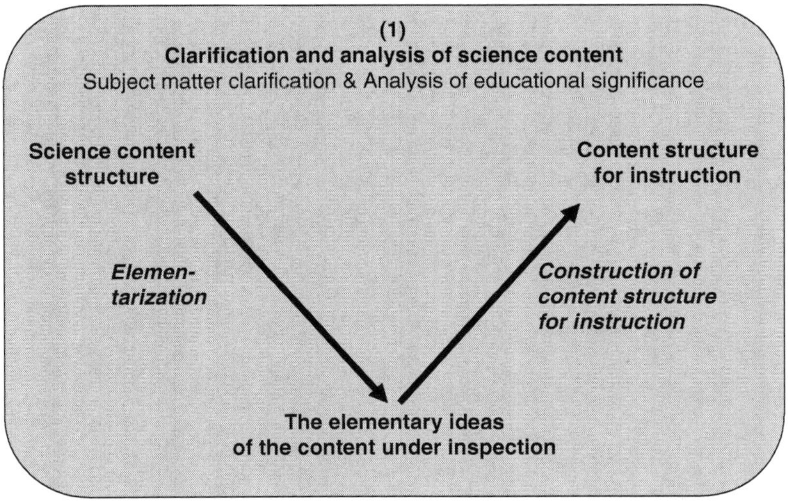

Fig. 5.2 Steps towards a content structure for instruction (according to the MER)

concept in science, (b) the historical development of the energy concept, (c) an analysis of competencies and insights students need to achieve, in order to be able to understand energy issues in their life-world and in society (as part of their scientific literacy), (d) an analysis of students' pre-instructional conceptions of energy and their learning processes towards the energy concept (empirical studies and analysis of the referring literature).

These analyses were carried out in the "spirit" of the Model of Educational Reconstruction (MER) (Duit et al. 2012). Briefly summarized, a key feature of the *Educational Reconstruction* approach is that in planning instruction – by teachers or curriculum developers – the science content to be learned *and* students' cognitive and affective variables, including their learning processes, have to be given equal attention. In addition the "educational significance" (i.e., the contribution of the particular content to student scientific literacy) has to be taken into account.[1]

The science content is not viewed as "given" but has to undergo certain *reconstruction* processes (Fig. 5.2). The *science content structure* (e.g., for the energy concept) has to be transformed into a *content structure for instruction*. The two structures are fundamentally different. In the first step the elementary ideas with regard to the aims of instruction have to be detected by taking into account student perspectives (e.g. their pre-instructional conceptions). After that the content structure for instruction needs to be constructed. Finally, efficient teaching and learning settings have to be designed.

[1]The Educational Reconstruction approach shares key features of current attempts in developing learning progressions for core ideas of science through integrating domain analyses with findings from research on students understanding of those core ideas (e.g. Berland and McNeill 2010).

The four basic ideas of the energy concept in Fig. 5.1 refer to all processes where energy is involved. They are the result of the above outlined analyses of "elementary" ideas of the (in particular, physics) energy concept. It turns out that they are also essential for understanding key energy related issues in nature, technology and society (in particular issues of energy supply). However, if quantum features come into play, additional ideas need to be taken into account. On the one hand the basic ideas stand for constancy amidst change, on the other hand for a loss of energy value (i.e., its potential to be used for various processes) whenever a process takes place.

5.2.3 On the Nature of the Four Basic Ideas

The idea of constancy amidst change was the most important factor leading to the energy concept (Hiebert 1962; Elkana 1974). In the early nineteenth century, the science research program was very much influenced by the romanticist idea of intimate interrelations of the "forces" of nature. Researchers investigated the change of "electricity" into "magnetism" (Oerstedt in 1820), "magnetism" into "electricity" (Faraday in 1831) and various additional changes of phenomena. The energy concept may be viewed as the scientific way of expressing the idea of a unifying, overarching romanticist "force of nature". It is noteworthy, that this idea is still at the core of the contemporary science energy concept. When we speak of energy transformation today, we have to be aware that the changes are occurring at the *phenomenological* level. The very idea of energy is that its amount does not change despite all changes at the phenomenological level. At the conceptual level, there is only a change of energy manifestations that are usually called energy forms. It is essential to point out, that a qualitative conception of energy conservation is outlined here. Clearly, in physics the first law of thermodynamics draws on this qualitative idea but includes the quantitative (mathematical) side.

Interestingly, the energy concept was not "invented" by the established scientists but by "outsiders" such as the physician Julius Robert Mayer (1814–1878) and the son of a brewer James Prescott Joule (1818–1889). These ideas initially were not taken serious by the established scientists – partly for good reasons, as, for instance, the physics knowledge of Mayer actually was quite limited. However, when the idea of energy conservation was more clearly stated the principle was quickly and widely adopted.

If interactions between certain systems occur, the amount of energy in some systems may increase if in others it decreases. In total decrease and increase balance. These changes may be seen as exchange of energy between the systems involved. In terms of the energy flow metaphor it makes good sense to speak of *energy transfer* from one system to the other or from one part of a system to another part. In a nutshell, due to transformation and conservation, energy usually changes locations, i.e. energy becomes manifest in different places while processes are running.

Whatever processes take place in closed systems the amount of energy does not change – but the "usefulness" (or "value") of energy inevitably declines. This means that at the end of every process that occurs in a closed system, the number of further processes that can occur, decrease. In science the degradation idea is usually addressed by the 2nd law of thermodynamics and the entropy concept. The reasons to include *degradation* into the basic ideas of energy are the following. First, energy and entropy are closely related to one another in science. But still in science instruction the major emphasis is given to energy and the conservation idea. Second, the conservation idea will become understandable for students only if the degradation idea is also given attention. In all processes that take place in reality, there is the above interplay of conservation and degradation. This means that real processes are only understandable to students if both ideas are used to explain them. Thirdly, the degradation idea is a key issue in approaches to make students familiar with problems of energy supply and consumption in technology and society.

In order to avoid misunderstandings it should be taken into account that the four basic ideas address different *qualities* of energy. Hence, energy conservation should not be seen primarily in terms of mathematical formulae used to calculate the amount of energy involved in certain processes. When, for instance, a lifted body falls down, the energy in the system composed of the body lifted and the earth attracting one another is first transformed into energy of the moving body. Hitting the ground the energy is transformed into "heat", i.e. the surrounding is heated (a bit). Due to the second law of thermodynamics (the qualitative version being the above degradation idea) all temperature differences even out in the long run – however, the energy provided to the surface of the earth by the impact is not lost.

5.2.4 On the Relation of the Four Basic Ideas to Standards and Instruction

Energy is a key concept of standards around the world, e.g., in a recent document of the National Science Council (NRC 2012) in the USA. The basic ideas presented above denote elementary features of the science energy concept. However, if it is possible to successfully teach them to students, they do not only gain an elaborate understanding of a central science concept and of key views of the nature of science linked to this concept, they also gain mental tools to understand the major problems of energy supply which are given attention in actual scientific literacy approaches (Osborne 2007; Choi et al. 2011). They learn, for instance, that energy may not just be gained from nothing but only from forms of energy stored in certain sources stemming from various processes. The intimate interplay of transformation and conservation points out that energy after use does not just disappear and hence does not need any further attention. On the contrary, energy may not be destroyed but only transformed into other forms of energy or transferred to other places. In other words, one cannot get rid of energy. Energy degradation plays a significant role here. Problems of energy supply are due to the fact that energy is degraded in every process running in the real world. Hence, the value (in the above sense) of energy declines. The reason for this is that in every real process "heat" occurs – i.e. the temperature of certain parts of the system rises. All temperature differences even out in the long run – due to the second law of thermodynamics. There is an additional reason why insight into energy degradation is important. Processes running in the real world may be only adequately modeled by taking into account degradation. The very idea of energy conservation is therefore only understandable for students and may be in accordance with daily life experiences if degradation is also considered: Energy is conserved however its utility value unavoidably decreases.

As mentioned earlier, the above four basic ideas of the energy concept are closely linked. Understanding a single idea is possible only in a process taking the other ideas into account. If this argument holds, it is not possible to start the path to a full understanding of the energy concept primarily with one of the ideas (e.g. energy transformation) and include the other ideas step by step. *All* ideas have to be regarded in some way from the very start. In the final section of the present paper a preliminary approach on how this may be set into practice is discussed.

5.3 Conceptualizations of Energy

Various approaches in introducing energy to students through respective conceptualizations of this concept can be found in the literature (Duit 1986b; Doménech et al. 2007). In the following major features of these conceptualizations are discussed.

5.3.1 Energy Is an Abstract Accounting Quantity

Energy is a key quantity of a system. It may be seen as an accounting quantity in the following sense. In System I (see Fig. 5.3) certain quantities are defined, one

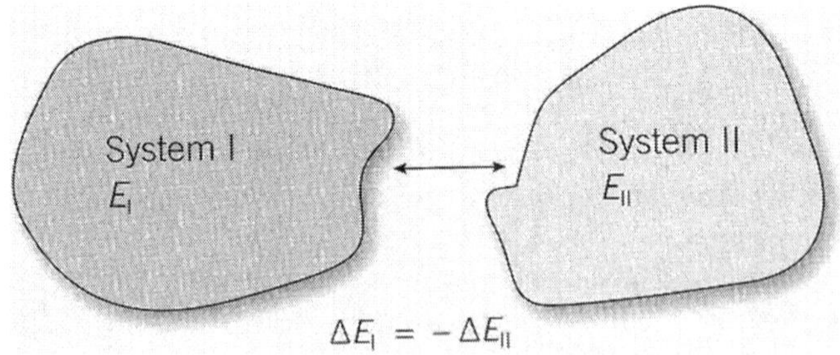

Fig. 5.3 Energy a key quantity of a system

of them being energy. Whatever happens in the system, the energy of the system does not change, i.e., putting the results of certain measurements into the formulas defining the quantity of energy always results into the same amount. If it happens some time that this is not the case anymore, one must search for transformations so far unidentified or for interactions with a System II. If this is the case, the amount of energy loss or gain is balanced by the gain or loss of energy in the other system (Fig. 5.3).

Auerbach (1913) provided the nice metaphor that in the big firm of nature energy holds the position of the bookkeeper who is responsible for keeping the balance of credit and debit. The director, being responsible for the direction the firm should develop is the entropy. It is essential to keep in mind that in science, energy is a quantity defined in a system. It seems that quite frequently this is disregarded, for instance if potential energy is attached to a stone lifted to a certain height. However, the energy may not be attached to the stone (provoking the idea that energy "sits" in the stone). The energy "sits" in the system composed of the stone and the earth (attracting one another). There are various other examples in the literature of neglecting the "system nature" of the concept of energy (e.g., Doménech et al. 2007).

5.3.2 Energy Is the Ability to Do Work

Major problems of the traditional approach drawing on the concept of work are briefly mentioned above in the Introduction. It is argued that this approach narrows down the energy concept just to mechanics and neglects the large family of other energy manifestations (such as energy forms linked to electrical or optical phenomena). It seems that this predominance is due to the introduction of the energy concept in the nineteenth century in which the transformation of mechanical work into heat has played a significant role. Still Max Planck (1913, p. 104), for instance, defined energy in the following way: "*... as the sum of all effects, measured in mechanical units of work ...*".

5.3.3 Energy Is the Ability to Cause Changes

This conceptualization has been used since the energy concept was developed. Rankine applied it in 1853 in the following way: "*Every affection of substances which constitutes or is commensurable with a power of producing change in opposition to resistance*" (quoted from Planck 1913, p. 77). Cassirer (1910, p. 258) described energy explicitly as the ability to cause changes. However, it is not clear what kinds of changes are meant. In addition there are also problems concerning the 2nd Law of Thermodynamics. The ability of a system to cause changes is dependent on the amount of energy, but at the same time it is also determined by the type of energy.

5.3.4 Energy Is the Ability to Produce Heat

Clearly, the idea that energy is the ability to heat something makes some sense (Kedves et al. 1983). In the very end all real processes driven by certain forms of energy come to a rest due to frictions of various kinds. Hence, the final result is that something is heated. However, this conceptualization shares a problem of the classical "energy is the ability to do work" discussed above, namely that it applies to just one class of phenomena.

5.3.5 Energy Is a General Kind of Fuel

In a university textbook for non-physicists Rogers (1965, p. 370 ff) draws on the everyday use of the word "energy". Fuels are needed to perform useful jobs for people (e.g., lifting a weight, towing a car, accelerating a body, heating water). "*For the present you should think of energy as something supplied by fuels, something which can be interchanged between several forms, and something whose interchanged amount is measured by force multiplied by distance* (p. 379)". Clearly, the idea that energy is a something supplied by fuels does not restrict energy just to mechanics and hence has the potential to overcome a major limitation of the traditional conceptualization. However, as this something is measured by force times distance the above outlined problems of the traditional approach in principle also hold for Roger's idea.

5.3.6 The Conceptualist and the Materialist Distinction

"*According to the materialists' view, energy is like a substance, something of the nature of a pervasive fluid, which has objective existence in the same way as (say) a horse, or Winsor castle, has objective existence .... According to the conceptualists, energy is an abstract idea invented by scientists to help in the quantitative investigations of phenomena. It is defined as capacity for doing work, and its importance lies in the fact that for all phenomena hitherto studied thoroughly a rigorous law of conservation is found to be applicable* (Warren 1982, p. 295)". Warren emphatically argues that the materialist view is not adequate from the physics point of view. From the above "abstract accounting quantity" perspective Warren in principle is right. However, to view energy *as if* it were a certain kind of substance may help students to understand energy. Research has shown that students have severe difficulties to understand the basic ideas of energy, in particular, energy conservation (e.g., Duit and Häußler 1994). The conception of energy flow (e.g. in energy flow diagrams) has proven a powerful means to aid student understanding.

Energy is conceptualized as a something flowing from one system to another, this flow is analogue to say a water flow. Those who think about energy in terms of an indestructible substance have gained a sound basis for understanding energy conservation.

5.3.7 Energy Is a Substance-Like Quantity

The above materialist idea of energy is not just tolerated but deliberately developed within a German physics course for secondary school (Falk et al. 1983; Herrmann 2000). This course has had substantial influence in Germany. It is still – sometimes hotly – debated. The major ideas of the approach are described as following:

> Energy is a substance-like quantity: It is distributed in and can flow through space. Since the term "energy form" leaves room for misinterpreting different energy forms as different physical quantities, it should be replaced with a more suitable concept. To this end, we rely upon the experience that energy always flows simultaneously with at least one other substance-like quantity. This shows that one must focus upon the substance-like quantities accompanying the flow of energy if one wants to get a suitable description of energy transfer. Instead of speaking of energy forms, it is more appropriate to visualize energy as a kind of "stuff" which can flow from one place to another only when "carried" by another kind of stuff called an energy carrier. In this picture, energy is not transformed (or converted) from one form into another, but rather, it exchanges its carrier. In this way, one arrives at a picture of an energy transport process which is strictly valid, yet simple and easy to present even at an elementary level (Falk et al. 1983, p. 1076).

This approach leads to a substantially different conceptualization of physics as compared to the predominant approaches. It seems that the approach is sound from the point of view of physics. It also appears that understanding energy conservation (Kesidou and Duit 1991) and key ideas of the heat concept (Starauschek 2001) are easier to learn in this approach. However, as Starauschek also revealed, learning mechanics is not easier in this approach. In addition it results in a conceptualization of physics that is significantly different from what students learn so far in schools and universities.

5.3.8 Energy Forms

Various energy forms are usually distinguished in teaching and learning approaches designed to introduce students to key ideas of the energy concept. Research has shown that for many students, a major outcome of energy teaching is only a more elaborate knowledge of names for the energy forms. Millar (this volume, Chap. 11) argues that quite frequently in the literature about the teaching and learning of energy, a rather superficial picture of energy forms and their transformations and transfers is painted. The way of speaking and thinking about energy in many approaches is incorrect from the physics point of view. Millar argues in favor of focusing on energy *stores* instead of *forms*. As pointed out before, it is essential to clarify that energy forms denote the various "manifestations" of energy in order to avoid the deficiencies Millar identified.

5.4 Findings of Studies on Teaching and Learning Energy

The literature on teaching and learning energy (in schools, universities, and out of school settings) is extensive. In the STCSE bibliography (Duit 2009) some 190 studies are listed. In the following a brief overview of major findings will be presented.

5.4.1 On the State of Research in the Early 1990s

Duit and Häußler (1994) provided a comprehensive review of the state of teaching and learning energy including cognitive and affective issues. The analyses are based on the above outlined framework of the four basic ideas of the energy concept.

5.4.1.1 Students' Conceptions of Energy

Research on students' conceptions flourishing in the 1980s and early 1990s revealed that students' ideas about energy before and also after instruction mainly reflect the use of energy in their life-world domain (Duit and Häußler 1994, p. 188). Interestingly this is also true for many teachers (Baird et al. 1987). Students often conceive energy as a universal kind of fuel. Watts (1983) identified the following set of alternative frameworks: (a) Human centered energy, (b) A depository model of energy, (c) Energy as an ingredient or as a product, (d) Energy as an obvious activity, (e) Energy is functional, and (f) Energy as a kind of fluid.

Interestingly, in various countries quite similar ideas occur – however, there are also differences that seem to be due to the varying colloquial meanings of the energy concept in different countries. In Germany, for instance, the idea that energy has something to do with food occurs significantly less often than in the UK (Duit 1981; Solomon 1983). Trumper (1990) found that among Israeli high school students the percentage holding the fuel idea of energy is much lower than in Lijnse's (1990) study carried out in the Netherlands. Lijnse showed that the energy ideas prevalent in pamphlets, reports and newspaper articles were surprisingly similar to students' conceptions. The central idea was often that we need energy for technical facilities and that energy is consumed when it is harnessed. These findings support the above assumption that students' ideas reflect the use of energy in daily life language. Viewed from the perspective of the above four basic ideas of the energy concept, it is noteworthy in particular that almost no conservation of energy ideas occur in daily life context. However, facets of energy degradation are identifiable – as, for instance, in the statement that energy is used up in processes (Duit and Häußler 1994, p. 190).[2]

5.4.1.2 Results of Learning the Energy Concept During School Science Instruction

It is necessary to point out from the outset that limited success of traditional science teaching is true for whatever field of science and whatever country. In Lijnse's (1990) study, just 17 % of his students, after being instructed about energy, fall into a category the researchers accepted as being correct. There is however evidence that constructivist methods of teaching and learning energy are more successful than "traditional" methods (for studies on energy see Trumper 1990, 1991). In general, research has shown that conceptual change approaches usually are more efficient than traditional ones (Duit and Treagust 2012). Nevertheless, it seems that the following results of energy instruction are quite typical.

1. *Students usually do not learn the basic ideas of the energy concept*
   Duit (1986a) compared energy conceptions of 150 German Grammar School students in grade 6 and in grade 10 after 4 years of science instruction in which the energy concept played a significant role. The main changes that occurred were in the students' energy vocabulary. Students in grade 10 were able to give significantly more examples of energy forms than the students in grade 6. However many students did not learn the idea of energy transformation. An interview study carried out with 35 grade 10 students (Kesidou and Duit 1991) revealed that students failed to learn the basic idea of energy conservation. This principle is quite frequently mentioned in the interviews as an important idea of energy. However it is not interpreted within the physics framework but within the students' *cause-effect* schema: Energy is not lost as an effect (for instance, something was heated or a falling stone left an imprint on the ground) resulted.
2. *Students do not use the energy language as taught in school when explaining processes*
   It is a well known general finding of research on student conceptions that students rarely use the scientific terms (here the science energy language) when they are asked to explain processes (here processes in which energy plays a significant role).

[2]Jin and Wei (this volume, Chap. 9) investigate the meaning of energy as provided in leading English language dictionaries. They reveal that the term energy according to these documents is used in significantly different ways. It is, for instance, associated with a person's physical or mental strength, with life energy of living beings, with vital power of certain places, but also with the ability to do work.

3. *Students do not use the energy concept learned in science instruction in real life situations*
   Arzi (1988) carried out interviews on energy in food. Students had severe difficulties using what they had learned about energy. In a study by Duit and von Zelewski (1979), the German teacher asked a class to write an essay on the significance of energy. Only a few students mentioned ideas they learned about energy in school – and even fewer did this appropriately.

5.4.2 Learning Progressions Towards the Energy Concept

Learning progressions have gained significant attention the past years, first within the US science education community, later internationally. *"Learning progression in science are empirically grounded and testable hypothesis about how students' understanding of, and ability to use, core scientific concepts and explanations and related scientific practices grow and become more sophisticated over time"* (NRC 2007). Development and empirical validation of learning progressions has become a major research field in the past years. The learning progression movement goes far beyond the mere focus on identifying and empirically testing efficient learning pathways. It also includes general attempts towards more efficient science instruction and science teacher education (Duncan and Hmelo-Silver 2009; Corcoran et al. 2009; Duschl et al. 2011). A number of studies deal with the development of the energy concept. Major studies of this kind are briefly discussed in the following. If energy is introduced in early grades of secondary school, energy forms are usually the starting point, energy transformations (from one form into another) and transfers (from one place to another) follow, providing the ground for introducing energy conservation which eventually is supplemented by energy degradation. In the following primary findings of studies investigating the development of energy ideas over a certain time (such as a school level) are discussed.

5.4.2.1 Results of a Study on Deliberately Developing Energy Ideas

Nordine et al. (2010) draw on the extensive research on teaching and learning energy briefly summarized above. They deliberately avoid the problems affiliated with an approach still predominant in middle schools of the USA, namely to focus on simple calculations of energy in idealized systems. They put major emphases on qualitatively conceptualizing energy transformations in everyday, non-idealized, systems. First, *types of energy* are identified and investigated. A set of "factors" and "indicators" for the different energy types are developed. For sound energy, for instance, the "factor" is *loudness*, the "indicator" *emission of sound*. Second, the idea of *energy transformation* is introduced. The observation that an increase in one type of energy is always linked to a decrease of at least one other type, leads to the idea that the types of energy investigated may be transformed into each other. This idea leads to a qualitative understanding of energy conservation, that namely a decrease in one type of energy is always linked to the increase in another type of energy and vice versa. Energy transformation diagrams (flow diagrams) are also introduced to further strengthen the idea of long chains of energy transformations. Finally, earth's energy resources are investigated in order to allow using the energy ideas introduced and to show their relevance in understanding the role of energy in explaining key features of energy supply and consumption. In a nutshell, there is the following teaching and learning sequence: (1) Energy Forms, (2) Energy Transformations and Transfer, (3) Energy Conservation, (4) The role of energy in understanding nature and key issues of energy supply. The evaluation of enactments following this sequence included pre-post and also long term (a school year) measures. It turned out that there was a significant increase of student understanding of the interplay of transformation and conservation after the course but there were also long term effects showing that even one school year later the students that participated in the enactments were superior to other students that did not participate.

5.4.2.2 A Survey on Student Learning Progression Drawing on TIMMS Data

Liu and McKeough (2005) used the US TIMSS database to carry out a survey on the progression of student energy conceptions. The following items were used (Liu and McKeough 2005, p. 501):

> Two items involve an understanding of energy phenomena or causes for activities (Activity/Work); six items involve understanding of energy sources or different energy forms (Source/Form); nine items involve understanding or analyzing various situations of energy transfer (Transfer); four items involve an understanding of energy loss when doing work (Degradation); and finally two items involve an understanding of total energy and the constancy of the total energy (Conservation).

The following sequence of the categories according to the Median Rasch Difficulty occurred (Liu and McKeough 2005, p. 505): Activity/Work (−2.48), Source/Form (−0.91), Transfer (−0.18), Degradation (0.56), and Conservation (1.21). Accordingly, to solve items falling into the "activity/work" category is comparably easy, while items in the conservation category are rather difficult. Further analyses reveal that there is an increase in proficiency due to rising student grade level (Liu and McKeough 2005, p. 509, fig. 4). Interestingly the items addressing degradation are solved only by a certain number of students in grades 7 and 8 as well as by high school students – but not by students in grades 3 and 4. The items on energy conservation are solely solved by a few high school students. In general terms, these findings are in accordance with the analysis of Driver et al. (1994) based on research of students' conceptions. By far the easiest items are those allowing students to draw on personal experiences. The items on energy conservation turned out to be by far the most difficult ones.

5.4.2.3 Development of Energy Ideas During a Series of Units in Chemistry

Liu and Park (2012) carried out a follow up study to investigate whether a similar learning progression as in the previous study (Liu and McKeough 2005) also occurs during a short term intervention. Ten sets of computer model-based assessments of energy were developed. The same students worked on these assessments after each of their ten instructional units on chemistry. As in the previous study, all students were able to deal with the levels of source, form and transfer. However energy degradation did not occur as a distinct level of understanding and could be subsumed under energy conservation. Clearly, the results of both studies point to the particular difficulties of understanding energy conservation and degradation.

5.4.2.4 Development of Energy Ideas from Grade 6 to 10

Neumann et al. (2013) provide the results of a study to investigate the learning progression concerning key basic ideas of the energy concept during grades 6 and 10 of German students (see also Viering 2012). They distinguish between the categories *Forms, Transformation, Dissipation* and *Conservation*. A multiple choice test was developed (The *Energy Concept Assessment*). In total some 1,600 students of grades 6, 8 and 10 participated. Again, basically the same progression as in the above study by Liu and McKeough (2005) occurred. Grade 6 students were predominantly able to answer the items investigating understanding of energy forms and sources. In addition, grade 8 student successfully dealt with items on energy transfer and transformation. Only a few grade 10 students were able to provide the right answers for energy conservation items. Neumann et al. (2013, p. 181) provide the following summary of their findings:

Students of grade 6 tend to typically solve items regarding energy forms and sources. The average grade 6 students solved about one third of the energy forms and sources items, and also some of the least difficult items regarding energy transformation and degradation. Some of the more advanced grade 6 students solved more difficult items of higher conceptions. Grade 8 students typically solve items regarding energy forms and sources of higher difficulty than those energy and sources items solved by most grade 6 students. Grade 8 students also seem to be able to solve the less difficult items of energy transformation and degradation. However, these students typically remain mostly at the form and sources level. Items of energy conservation are solved by some of the grade 8 students, but only the most able. From grade 8 to grade 10 students again move toward more advanced levels of the continuum. Almost all students of grade 10 master the energy forms and sources level, and above average students also master the transformation and degradation levels. However, items regarding the principle of energy conservation appear to be mastered by only the most able 10th grade students. Finally, it needs to be noted that the range of student ability widens from grade 6 to grade 10.

An interesting additional finding is that students seem to develop an understanding of *energy degradation* along with understanding *energy transfer* and *transformation.*

5.4.3 Learning Progressions on Energy – A Summarizing View

On the grounds of research findings on teaching and learning energy, Driver et al. (1994) proposed to start with students' feeling energetic, further develop this idea to various examples (from the living and non-living world), get an idea of stored energy and finally achieve ideas of energy conservation and degradation.

Prima facie, it seems that the sequence of "stepping stones" to the energy concept is quite similar in the above studies. The start always comprise issues of providing experiences that allow students to become familiar with phenomena that are essential as a start to the energy concept, called *feeling energetic* (Driver et al. 1994), *activity/work* (Liu and McKeough 2005) or just "*facts*" (Neumann et al. 2013). Usually, after that *energy forms* follow pointing out that energy may occur in various manifestations. Viewed from the above perspective of the basic ideas of the energy concept, *transformation and transfer* denoting the manifold changes of energy forms come first. Interestingly, *degradation* follows and *energy conservation* seems to be the most difficult basic idea. There are, however, some findings being not (fully) in accordance with this progression. In the study by Liu and Park (2012), for instance, energy degradation and conservation could not be clearly differentiated in the analyses. Neumann et al. (2013) deliberately call degradation *dissipation* as their test items on degradation primarily address the dissipation idea denoting the even distribution of heat. Energy conservation has proven the most difficult idea in all studies. In the study by Liu and McKeough (2005) it seems that the two items proving energy conservation require quantitative considerations resting on certain energy formula. In contrast, for energy degradation just qualitative considerations are needed. It seems that in the study by Neumann et al. (2013) the energy conservation items also require quantitative considerations although no mathematical calculations are needed. In the study by Herrmann-Abell and DeBoer (this volume, Chap. 7) items investigating energy conservation ideas have also proven much more difficult than items investigating energy transfer and transformation – but they used items of a more qualitative type. In a series of drawings, a ball is released and further moves in different curved paths. Students are asked to predict the height the balls reach in the various cases.

Hence, further analyses are needed to analyze the reasons for the particular difficulty of energy conservation.

5.5 Towards Unfolding and Differentiating Students Pre-instructional Ideas

As pointed out from the outset the basic ideas of the energy concept – transformation, transfer, conservation and degradation – are closely interrelated. Each idea, hence, may be understood only if all other ideas are also understood – at least to a certain extent. The philosopher of science Hanson (1965, p. 54 ff) argued, that physics terms and concepts may be fully understood only in the framework of the theory these concepts include. The science educator Jung (1975) used the term "*theory loaded*" concepts. This position, on the one hand, leads to the conclusion that (slightly) different energy concepts are used – depending of the framework of concepts the particular energy concept is embedded. This would mean that different energy concepts exist in different domains, e.g. in physics and biology, but also in different physics sub-domains such as in classical mechanics and in quantum-mechanics. Regarding learning the energy concept there is a certain learning paradox. The usually highly complex set of interrelations between the concepts "defining" a particular energy concept has to be transferred into a (more or less) linear progression of facets being understandable to the learners.

The development of the heat concept may serve as an example. Figure 5.4 outlines the present state. What is called heat in daily live concerns includes facets of the science concepts of temperature, entropy and energy. A brief outline of the history of science is the following (s. for more details: Wiser and Carey 1983; Kesidou et al. 1995). In the seventeenth century scientists in the Academia del Cimento in Florence designed instruments to measure "heat". Wiser and Carey (1983) argue that these early scientists did not differentiate intensive and extensive issues of heat. Hence, they had difficulties to explain the results of their experiments. In particular the ideas of temperature equalization and thermal interaction were missing. Only in the middle of the eighteenth century Joseph Black clearly differentiated intensive and extensive aspects. About another 80 years later Carnot's theory of the steam engine and the "invention" of the energy concept resulted in

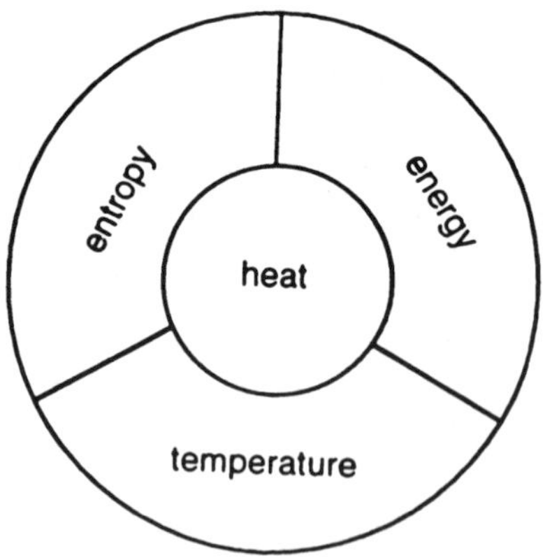

Fig. 5.4 Key concepts of heat

the actual state of Fig. 5.4. The daily life phenomenon of heat is conceptualized by the *intensive* variable temperature and the *extensive* variables entropy and energy. In summary, the development of the science view of heat may be seen as a process of unfolding and differentiating previously undifferentiated ideas.

Teaching and learning the energy concept may be seen in an analog manner as a process of unfolding and differentiating the basic ideas of the energy concept. The studies on the development of energy ideas discussed above, e.g., in research on learning progressions include developments that may be described also in terms of unfolding and differentiating. As mentioned, Neumann et al. (2013) observed that ideas of energy degradation develop alongside the ideas of energy transformation and transport. All approaches towards the energy concept begin with making students familiar with a wide range of phenomena illustrating key facets of the energy concept. Maybe it is time to revitalize the idea of the "*energy circus*" (Brook and Wells 1988). The basic idea of this approach is that students carry out a substantial number of experiments that make them familiar with various energy transformations and transfers. These experiences allow developing preliminary ideas of the four basic ideas of the energy concept that in later years may be step by step further unfolded and differentiated. The learning progressions discussed above provide linear progression from one step to another. It seems that the unfolding and differentiating position allows further developing the learning progression idea from preliminary linear towards recursive progressions.

References

Arzi, H. J. (1988). *On energy in chocolate and yoghurt, or: On the applicability of school science concepts to real life*. Paper presented at the annual meeting of the American Educational Research Association, New Orleans.

Auerbach, F. (1913). *Die Weltherrin und ihr Schatten* [The ruler of the world and her shadow]. Jena: G. Fischer.

Baird, J. R., Fensham, P. J., Gunstone, R. F., & White, R. T. (1987). Individual development during teacher training. *Research in Science Education, 17*, 182–191.

Berland, L. M., & McNeill, K. L. (2010). A learning progression for scientific argumentation: Understanding work and designing supportive instructional contexts. *Science Education, 94*(5), 765–793.

Brook, A., & Wells, P. (1988). Conserving the circus? – An alternative approach to teaching and learning about energy. *Physics Education, 23*, 80–85.

Bryce, T., & MacMillan, K. (2009). Momentum and kinetic energy: Confusable concepts in secondary school physics. *Journal of Research in Science Teaching, 46*(7), 739–761.

Cassirer, E. (1910). *Substanzbegriff und Funktionsbegriff* [The concepts of substance and function]. Berlin: B. Cassierer.

Choi, K., Lee, H., Shin, N., Kim, S.-W., & Krajcik, J. (2011). Re-conceptualization of scientific literacy in South Korea for the 21st century. *Journal of Research in Science Teaching, 48*(6), 670–697.

Corcoran, T., Mosher, F., & Rogat, A. (2009). *Learning progressions in science*. New York: Centre for Continuous Instructional Improvements, Teacher College Press, Columbia University.

Doménech, J. L., Gil-Pérez, D., Gras-Martí, A., Guisasola, J., Martínez-Torregrosa, J., Salinas, J., Trumper, G., Valdes, P., & Vilches, A. (2007). Teaching of energy issues: A debate proposal for a global reorientation. *Science and Education, 16*, 43–64.

Driver, R., Squires, A., Rushworth, P., & Wood-Robinson, V. (1994). *Making sense of secondary science. Research into children's ideas*. New York: Routledge.

Duit, R. (1981). Understanding energy as a conserved quantity. *European Journal of Science Education, 3*, 291–301.

Duit, R. (1986a). *Der Energiebegriff im Physikunterricht* [The energy concept in physics instruction]. Kiel: IPN – Institute for Science Education Series.

Duit, R. (1986b). In search of an energy concept. In R. Driver & R. Millar (Eds.), *Energy matters. Proceedings of an invited conference. Teaching about energy within the secondary science curriculum* (pp. 67–101). Leeds: Centre for Studies in Science and Mathematics Education, University of Leeds.

Duit, R. (2009). *STCSE – Bibliography students and teachers conceptions and science education*. Kiel: IPN – Leibniz Institute for Science Education.

Duit, R., & Häußler, P. (1994). Learning and teaching energy. In P. Fensham, R. Gunstone, & R. White (Eds.), *The content of science* (pp. 185–200). London: The Falmer Press.

Duit, R., & Treagust, D. (2012). How can conceptual change contribute to theory and practice in science education. In B. J. Fraser, K. G. Tobin, & C. J. McRobbie (Eds.), *Second international handbook of science education* (pp. 107–118). Dordrecht: Springer.

Duit, R., & von Zelewski, H. D. (1979). Ohne Energie ist es düster in unserem Leben [Without energy it is dark in our life]. *Naturwissenschaften im Unterricht Physik/Chemie, 27*, 161–164.

Duit, R., Gropengießer, H., Kattmann, U., Komorek, M., & Parchmann, I. (2012). The model of educational reconstruction – A framework for improving teaching and learning science. In J. Dillon & D. Jorde (Eds.), *The world handbook of science education – Handbook of research in Europe* (pp. 13–37). Taipei: Sense Publishers.

Duncan, R. G., & Hmelo-Silver, C. (2009). Learning progressions: Aligning curriculum, instruction, and assessment. *Journal of Research in Science Teaching, 46*(6), 606–609.

Duschl, R., Maeng, S., & Sezen, A. (2011). Learning progressions and teaching sequences: A review and analysis. *Studies in Science Education, 47*, 123–182.

Elkana, Y. (1974). *Discovery of the conservation of energy*. London: Hutchinson Educational.

Falk, G., Herrmann, F., & Schmid, G. B. (1983). Energy forms or energy carriers? *American Journal of Physics, 51*(12), 1074–1076.

Hanson, N. R. (1965). *Patterns of discovery*. Cambridge: University Press.

Herrmann, F. (2000). The Karlsruhe physics course. *European Journal of Physics, 21*, 49–58.

Hiebert, E. N. (1962). *Historical roots of the principle of conservation of energy*. Madison: Department of History, University of Wisconsin.

Jung, W. (1975). Was heißt Physiklernen? Didaktik der Physik zwischen Physik und Wissenschaftstheorie [What does learning physics mean? Physics Didaktik between physics and philosophy of science]. In M. Ewers (Ed.), *Naturwissenschaftliche Didaktik zwischen Kritik und Rekonstruktion* (pp. 133–158). Weinheim/Basel: Beltz.

Kedves, F. J., Kovacs, L., & Kovesdi, P. (1983). Teaching energy in Hungarian schools. Developing the concept over seven years. In E. J. Wenham (Ed.), *New trends in physics teaching* (Vol. 4). Paris: UNESCO.

Kesidou, S., & Duit, R. (1991). Wärme, Energie, Irreversibilität – Schülervorstellungen im herkömmlichen Unterricht und im Karlsruher Ansatz [Heat, energy, irreversibility – Student conceptions in traditional physics instruction and in the Karlsruhe approach]. *Physica Didactica, 18*(2/3), 57–75.

Kesidou, S., Duit, R., & Glynn, S. (1995). Conceptual development in physics: Students' understanding of heat. In S. Glynn & R. Duit (Eds.), *Learning science in the schools* (pp. 179–198). Mahwah: Lawrence Erlbaum.

Lehrmann, R. L. (1973). Energy is not the ability to do work. *The Physics Teacher, 11*, 189–202.

Lijnse, P. L. (1990). Energy between the life-world of pupils and the world of physics. *Science Education, 74*(5), 571–583.

Liu, X., & McKeough, A. (2005). Developmental growth in students' concept of energy: An analysis of selected items from the TIMSS database. *Journal of Research in Science Teaching, 42*(5), 493–517.

Liu, X., & Park, M. (2012). *Progression of students' understanding of the energy concept*. Paper presented at the annual meeting of the National Association for Research in Science Teaching, Indianapolis.

Neumann, K., Viering, T., Boone, W. J., & Fischer, H. E. (2013). Towards a learning progression of energy. *Journal of Research in Science Teaching, 50*(2), 162–188.

Nordine, J., Krajcik, J., & Fortus, D. (2010). Transforming energy instruction in middle school to support integrated understanding and future learning. *Science Education, 95*, 670–699.

NRC – National Research Council. (2007). *Taking science to school: Learning and teaching science in grades K-8. Committee on Science Learning, Kindergarten through eighth grade* (R. A. Duschl, H. A. Schweingruber, & A. W. Shouse, Eds.). Washington DC: The National Academies Press. (cited from Corcoran et al, 2009, 15).

NRC – National Research Council. (2012). *A framework for K – 12 science education*. Washington, DC: The National Academies Press.

Osborne, J. (2007). Science education for the twenty first century. *Eurasia Journal of Mathematics, Science and Technology Education, 3*(3), 173–184.

Papadouris, N., & Constantinou, C. P. (2011). A philosophically informed teaching proposal on the topic of energy for students aged 11–14. *Science and Education, 20*, 961–979.

Planck, M. (1913). *Das Prinzip von der Erhaltung der Energie* [The principle of energy conservation]. Leipzig, Berlin: B.G. Teubner.

Rogers, E. M. (1965). *Physics for the inquiring mind*. Princeton: Princeton University Press.

Solomon, J. (1983). Learning about energy: How pupils think in two domains. *European Journal of Science Education, 5*(1), 49–59.

Starauschek, E. (2001). *Physikunterricht nach dem Karlsruher Physikkurs* [Physics instruction according to the Karlsruhe Approach]. Berlin: LOGOS.

Trautman, A. (1962). Conservation laws in general relativity. In L. Witten (Ed.), *Gravitation: An introduction to current research* (pp. 169–198). New York: Wiley.

Trumper, R. (1990). Being constructive: An alternative approach to the teaching of the energy concept – Part one. *International Journal of Science Education, 12*(4), 343–354.

Trumper, R. (1991). Being constructive: An alternative approach to the teaching of the energy concept – Part two. *International Journal of Science Education, 13*(1), 1–10.

Viering, T. A. (2012). *Entwicklung physikalischer Kompetenz in der Sekundarstufe I – Validierung eines Kompetenzentwicklungsmodells für das Energiekonzept im Bereich Fachwissen* [Development of physics competence in secondary level I. Validation of a model of competence development concerning the energy concept]. PhD thesis, University of Kiel, Kiel, Germany.

Warren, J. W. (1982). The nature of energy. *European Journal of Science Education, 4*, 295–297.

Watts, M. (1983). Some alternative views of energy. *Physics Education, 18*, 213–217.

Weiss, M., & Baez, J. (2012). *Is energy conserved in general relativity*? Usenet Physics FAQ. Retrieved September 11, 2013, http://math.ucr.edu/home/baez/physics/Relativity/GR/energy_gr.html

Wiser, M., & Carey, S. (1983). When heat and temperature were one. In D. Gentner & A. Stevens (Eds.), *Mental models* (pp. 267–297). Hillsdale: Lawrence Erlbaum.

Chapter 6
What Knowledge and Ability Should High School Students Have for Understanding Energy in Chemical Reactions? An Analysis of Chemistry Curriculum Standards in Seven Countries and Regions

Lei Wang, Weizhen Wang, and Rui Wei

6.1 Introduction

There is no doubt that energy issues play a significant role in our daily life. Few can argue the importance of energy education. *Energy Literacy: Essential Principles and Fundamental Concepts* (U.S. Department of Energy 2012) is a guide that includes formal and informal energy education. The notion of energy literacy focuses on the interdisciplinary characteristics of energy and its significance in everyday life. An energy-literate person can not only understand energy but can apply an energy lens to solve problems. Together, these principles can lead to recommendations of what is important to teach about energy in the classroom.

Some have suggested that a core set of ideas—instead of disconnected knowledge and isolated facts—is necessary for effective teaching and learning of energy. This core set of ideas should incorporate both knowledge and practice. In the National Research Council (NRC)'s (2011) *Framework for K-12 Science Education,* energy is discussed as a core idea and as a crosscutting concept. Energy is looked at in many ways. It is looked at as a definition at the macroscopic and microscopic levels; energy transfers, transformations, and law of conservation are examined; and the relationship between energy and force is explored. Within chemistry, energy in chemical processes and everyday life are looked at in the NRC's *Framework.* By the end of high school, students are expected to be able to explain how food and fuel provide energy, and—if energy is conserved—why people say it is produced or used. *The Next Generation Science Standards* (2013) provided more specific disciplinary core ideas and performance expectations about energy in chemical processes from

L. Wang (✉) • W. Wang • R. Wei
College of Chemistry, Beijing Normal University, No. 19 Xinjiekouwai Street, Beijing 100875, China
e-mail: wangleibnu@126.com

R.F. Chen et al. (eds.), *Teaching and Learning of Energy in K-12 Education*,
DOI 10.1007/978-3-319-05017-1_6,

elementary through high school. Student understanding is supposed to span from knowing how energy transfer and transport apply to living organisms to how energy efficiency impacts society, the economy, and the environment.

Energy is a word that is often used in schools but has many misconceptions surrounding it. Watts (1983) categorized varies energy misconceptions and identified seven alternative frameworks of energy. Many of these misconceptions come from the word energy in everyday language (Erickson 1979; Duit 1981). Students consider energy as a concept only related to life, health, and food (Black and Solomon 1983; Solomon 1983). Researchers found that when students say energy is conserved, they mean energy is stored in a system and can be released in an initial form (Brook and Driver 1984; Kesidou and Duit 1993; Solomon 1985). Survey research also shows that endothermic reactions are not thought to be spontaneous, and students consider outside conditions as a spontaneous judgment basis, like heating, rather than Gibbs free energy (Johnstone et al. 1977; Boo and Watson 2001).

Energy has been researched with a focus on students' conceptions, including their conceptions of chemical bonds, chemical change, and their attitudes toward energy (Holden and Barrow 1984; Hesse and Anderson 1992). Considering energy as a concept that should be learned throughout one's life (U.S. Department of Energy 2012), researchers need to begin to think about learning progressions for energy. According to research, high-achieving high school students can use energy conservation as a scientific tool for analysis (Lee and Liu 2010; Jin and Anderson 2012). Students between ages of 15 and 19 can conceptualize energy for various systems and begin to quantify the total energy of a closed system (Liu and McKeough 2005). It is at this level of development that we look at how students develop an understanding of energy conservation.

Teachers also lack an overall understanding of energy and how to teach important ideas about energy. Traditional courses and pedagogical approaches for teaching energy are often overly simplistic and not aligned with energy concepts from the perspective of cross-disciplinary. Without the most up-to-date and appropriate scientific models, students will continue to be challenged by difficult concepts like entropy and the second law of thermodynamics. As a result of teachers lacking a deep understanding of energy as a crosscutting concept, students cannot realize the relation between energy in chemical reactions and in physical processes; nor can they realize the value of energy to life of individuals and societies. Researchers have tried to develop an energy course and new approaches to improve energy teaching (Nahum et al. 2007; Amin et al. 2012). Although teachers are familiar with specific concepts such as entropy, few teachers reflect on the question why they are teaching entropy or how entropy is used in a chemical reaction. Besides the reason that life is full of chemical reactions, when a teacher begins to teach with an energy lens, the topic of energy in chemical reactions (ECR) contributes to a more complete understanding of energy as a concept and can lead to a more energy-literate citizenry.

There has been considerable research on students' naïve conceptions about energy, but little research has been conducted at the middle school and high school levels about what students can learn and understand about energy (Rogat 2011). The purpose of this study was to analyze the topic ECR and try to understand what can be learned at the high school level and to suggest a progression for the teaching and learning of ECR. Three specific problems should be solved: (1) What knowledge is most important that should be understood in this topic? (2) From what perspectives can students understand energy? (3) In high school, from lower to higher grades, in what performances can students be expected to show proficiency?

6.2 Scientific Perspective on Energy in Chemical Reactions (ECR)

The topic energy in chemical reactions (ECR) is closely linked with chemical thermodynamics. To effectively teach ECR, ontological concepts should be included. The purpose of the following sections is to provide a foundation for the selection of concepts in ECR for teaching and learning.

Chemical thermodynamics is the study of the interrelation of heat and work with chemical reactions or with physical changes of state within the confines of the laws of thermodynamics (Fu and Shen 2005). Chemical thermodynamics involves three laws. The first law of thermodynamics is the principle of energy conservation, which is used to interpret and calculate the heat or other forms of energy change in chemical and physical processes. In a system of chemical reactions, the energy that a system absorbs/emits is equal to energy that its surroundings emit/absorb. Considering the relation between the system and its surroundings, we regard them as a pair of cognitive perspectives in learning about energy, which are promoting the deep understanding of energy for students.

The second law of thermodynamics is about the direction of spontaneous processes. This law states that heat can never pass from a colder to a warmer body without some other change, connected therewith, occurring at the same time. The concepts of second law of thermodynamics and entropy help students to understand energy efficiency. For example, they will get a better understanding of why waste heat always exists in energy transfer. The third law of thermodynamics states that the entropy of a system approaches a constant value as the temperature approaches zero.

Energy in chemical reactions is not just limited to thermodynamics. Thermodynamics is useful when thinking about macroscopic systems but is not so useful about the microscopic structure of matter and the reaction mechanism. Thermodynamics can only answer the questions about whether or not a chemical reaction will occur and if so, to what degree. It cannot tell people the time needed for the reaction to occur, the reason for the reaction, or how the reaction occurs. However, if teachers fail to establish a relationship between energy and structure of matter (the particles in matter), students, when learning energy in chemical reactions, may not be able to understand where the energy in chemical reactions comes from exactly.

However, if students understand the particulate nature of matter, then they should have a good understanding of energy at the microscopic level. That is to say, both matter (understanding energy at matter level) and particles (recognizing energy at the particle level) are other two cognitive perspectives for energy learning.

6.3 Curriculum Standards Analysis

From the scientific perspective, the main purpose for understanding energy is to use energy more efficiently and to use concepts and laws of energy for judging the direction of reactions in various chemical reaction contexts. For high school students, what level of energy understanding and how it should be achieved require additional research.

Both students and teachers often turn to textbooks and curriculum materials instead of curriculum standards for guidance. However, many curriculum materials provide little support for the attainment of the key ideas about the flow of matter and energy (Stern and Roseman 2004). With high school students being required to pass high-stake tests, teachers are taking more time to solve algorithmic problems that often are seen as assessments instead of problems that will lead to deep conceptual understanding. As a result, students can solve energy problems that are algorithmic in nature without understanding the underlying concepts (Chiu 2001). Teachers also pay more attention to teaching methods of solving specific problems instead of focusing on teaching scientific concepts and student understanding of those concepts. Many countries and regions have developed or revised their national science education standards several times to reflect specific knowledge, concepts, and learning goals (Ontario Ministry of Education [OMOE] 2000, 2008a, b; Ministry of Education of the People's Republic of China [MOEPRC] 2000, 2003a, b; Ministry of Education, Culture, Sports, Science and Technology [MEXT] 1998, 1999, 2009; Ministry of National Education 2010).

Science instruction is always expected to align with content standards. National curriculum standards reflect the ideas of science education in a country. Many cross-national, standards-based studies have been proposed for studying the science curriculum. The book *Many Visions, Many Aims (Volume 2)* (Schmidt et al. 2002) stated why research curricula:

> Any set of science educational experiences will have some things in common (and many times that vary) among countries. Some differences are deep; some are incidental. It is essential for seeking – within given cultural settings . . . curriculum is the most fundamental structure for these experiences. It is a kind of underlying skeleton that gives characteristic shape and direction to science instruction in educational systems around the world. (p. 4)

In this study, we have attempted to look across countries to find common perspectives of knowledge and expectation performances to build a model of energy understanding to answer our research questions.

6.3.1 Country and Region Selection

We analyzed curriculum standards from seven countries/regions to synthesize the expected levels of energy concepts that students should understand by the end of high school. These countries/regions are the United States, Ontario (Canada), France, mainland China, Taiwan, Japan, and South Korea. We selected these countries/regions in this analysis based on several criteria. The first was the country's/regions' importance to mainland China for economic, political, or cultural reasons. The second was whether the country/region has specific content about energy in its curriculum standards. The third was whether a country/region is a high performer on international assessments.

6.3.2 Methodology

We conducted a quantitative analysis of the curriculum standards for all seven countries/regions included in this study. We first listed the knowledge topics that refer to the content standards of mainland China, and then we modified the list according to the standards of other six countries and regions. We coded all content standards to indicate that a topic is present in the country's or region's standard without regard to how many times it appears. Next, we counted the frequency of every knowledge topic to determine the most common knowledge. As for the analysis of performance expectations, we selected the relevant sentences or verbs which describe the learning goals of knowledge, we classified them into several categories, and we further characterized the categories as the performance expectations of learning in the topic energy in chemical reactions (ECR).

6.3.3 Results

6.3.3.1 Analysis of Knowledge

We counted the frequency of the knowledge topics that were addressed in all the seven countries/regions (see Table 6.1), and we found that first law of thermodynamics, energy forms and thermochemical equations were the most common topics that all these seven countries/regions addressed. As shown in Table 6.1, these three topics are followed by exothermic and endothermic reactions, calorimetry, bond energy, Hess law and efficiency. Enthalpy, the relationship between temperature and heat, and the relationship between changes of state and heat only appeared in the curriculum standards of four countries/regions.

In the NRC's (2011) *Framework*, the question of how energy is transferred and conserved is raised. Learning goals should be designed to answer that question.

Table 6.1 Knowledge topics included in country/region curriculum standards

Knowledge topic	Knowledge score	U.S.	Ontario (Canada)	France	Mainland China	Taiwan	Japan	South Korea
System & surroundings	**3**	√		√				√
Energy forms	**7**	√	√	√	√	√	√	√
First law of thermodynamics	**7**	√	√	√	√	√	√	√
Heat capacity & latent heat	**2**			√			√	
Exothermic, endothermic	**5**	√	√		√	√	√	
Enthalpy	**4**	√	√		√			√
Standard enthalpy of formation	**2**	√	√					
Thermochemical equation	**7**	√	√	√	√	√	√	√
Activation energy	**3**	√	√		√			
Calorimetry	**5**	√	√	√	√		√	
Hess law	**5**		√		√	√	√	√
Temperature & heat	**4**	√	√	√			√	
Changes of state & heat	**4**		√	√			√	
Bond energy	**5**	√	√		√		√	√
Entropy & second law of thermodynamics	**3**	√			√			√
Direction of a chemical reaction	**3**	√			√			√
Efficiency	**5**	√	√	√	√		√	

To answer this question in the topic of energy in chemical reactions (ECR), three aspects can be considered: (1) how chemical reactions provide energy, (2) the amount of energy that is provided, and (3) efficiency. Lee and Liu (2010) used a construct-based assessment approach to measure the learning progression of energy concepts across physical science, life science, and earth science contexts in middle school grades. They listed knowledge integration requirements for energy sources, transformations, and conservation items. Energy sources ask students to identify the source of energy; transformations require students to recognize that one form of energy converts to another form, causing a desired or unexpected change; and conservation is an integral idea of energy that is used to explain and predict energy phenomena.

From the above analysis, problems related to energy in chemical reactions can be divided into three cognitive perspectives: energy sources (where the chemical energy comes from), the forms of transformed energy (what energy forms the chemical energy changes to), and the amount of energy changes (what the amount of energy transferred/transformed in chemical reactions is).

In the first perspective, the concepts of state change and bond energy are used to answer the question of where the energy comes from. State change, a kind of physical change, is an approach of getting energy; chemical bonds breaking and forming, which is chemical process, is also a way of gaining energy.

In the second perspective, energy forms, which are shown in Table 6.1, are included with energy transformation in learning ECR. Energy transformation question items in the research of Lee and Liu (2010) indicated that students are required to understand that chemical energy transforms to heat and other energy forms. The concepts about exothermic reactions, endothermic reactions, temperature, heat and enthalpy can be qualitatively used as descriptions of energy change that are used to trace energy flow, while they can also be used quantitatively in the perspective discussed below. All Energy forms in chemical reactions can be seen as kinetic and potential energies.

In the third perspective, first law of thermodynamics, thermochemical equation, Hess's law, energy efficiency and many other concepts can be quantitatively related with the amount of energy changes. Calorimetry is a practice that requires students to apply physical quantities to describe energy change.

"First law of thermodynamics" listed in Table 6.1 and the score "7" do not mean that the topic was introduced in chemistry class for the first time but might have already been learned in other disciplines. As a core concept of energy learning, energy conservation is an expected learning outcome that is usually introduced in other courses like physics and environmental science (OMOE 2008b; MOEPRC 2003b; Ministry of Education and Human Resources Development, Korea 2007). However, very few students actually have a scientific understanding of the concept (Boyes and Stanisstreet 1990), and much less the ability to use this concept as a tool to solve problems in everyday life. The idea of energy conservation has been found by researchers as a high level of energy understanding (e.g., Liu and McKeough 2005; Lee and Liu 2010; Neumann et al. 2013).

Both the thermochemical equation and Hess's law reflect energy flow and conservation. The thermochemical equation shows both matter and energy flow in chemical reactions, and Hess's law is a direct application of energy conservation. Also, the thermochemical equation shows the relationship of energy and categories, states, and amount of matter in a reaction. From the macroscopic perspective, the equation shows that the quantity of energy is determined by categories, amounts, states, and from the microscopic perspective, interaction of particles in matter. It can be seen in Table 6.1 that Taiwan's standards include less knowledge topics than did other countries/regions, but these two topics, Hess's law and the thermochemical equation, were mentioned in by the Ministry of education, Taiwan (2008).

Energy efficiency is related to both energy change (maximum efficiency cannot reach 100 %) and transformation (energy can be transferred out of the system in unwanted ways like heat). Although the second law of thermodynamics is another idea of thermodynamics that can help improve understanding of energy efficiency (Morrisey and Barrow 1984), it doesn't get enough attention in learning ECR at the high school level. Only three of the seven countries'/regions' curriculum standards in this study suggested learning it.

6.3.3.2 Categories of Performance Expectations

The NRC's (2011) Framework claimed that students learn disciplinary core ideas in the context of science and engineering practices. Eight practices of science and engineering are essential for all students to learn: asking questions (for science) and defining problems (for engineering); developing and using models; planning and carrying out investigations; analyzing and interpreting data; using mathematics and computational thinking; constructing explanations (for science) and designing solutions (for engineering); engaging in argument from evidence; and obtaining, evaluating, and communicating information. The Framework suggested that each disciplinary idea focus on some of these eight but not all disciplinary core ideas are associated with a practice.

We believe that what students can do is the reflection of their understanding of concepts. The performance expectation is a way to measure understanding through application. Not all the eight practices but some of them can be reflected in each concept. Different concepts have different performance expectations. For instance, the concept system and surroundings needs a deep understanding in order to better learn other concepts related to energy sources, changes, and transformation. Entropy is not highly important for high school students to master, but they are required to use it to explain or predict the direction of a reaction. We also note that there might be several performance categories where we can see a progression of learning that reflects Bloom's Taxonomy.

In the seven countries or regions, a performance expectation is composed of verbs and concepts (e.g., use chemical bonding to explain the reason of energy change in reactions). We listed all expectations of each knowledge topic, summed them up (see Table 6.2), and then sorted the performance expectations of ECR learning into following four categories:

1. *Use examples to describe.* Students use examples to describe relationships of concepts and explain what the concepts contain, such as burning coals is exothermic, and energy in the reaction system is transformed into heat to its surroundings.
2. *Use mathematical expressions, develop explanatory models to explain, and predict.* Take for example activation energy is a concept which improves students' qualitative understanding of reaction conditions. For instance, this concept enables students to understand why an exothermic reaction needs to be heated. By the help of a graph that shows the energy of reactants and products, students can explain that the heat is produced by a reaction and predict whether the reaction is exothermic or endothermic. Hess's law is a mathematical expression to explain energy conservation. With the help of a particle model (which shows chemical bonds), students can explain where the energy comes from in reactions. If students are able to calculate the heat using the formula $Q = cm\Delta t$, they can explain that the heat depends on more factors than only on temperature.

Table 6.2 A summary of performance expectations of knowledge topics

Knowledge topic	Performance expectation
System & surroundings	Describe/explain
Energy forms	Describe
Heat capacity & latent heat	Explain the relation
Exothermic, endothermic, heat of reaction	Explain the energy change in terms of endothermic and exothermic
Enthalpy	Understand/use
Standard enthalpy of formation	Use to explain the energy change
Thermochemical equation	Write
Activation energy	Use to explain the change
Calorimetry	Conduct/design and conduct/
First law of thermodynamics	Use forms to explain/use mathematical expression to explain/design investigation to claim
Hess law	Use examples to state/use examples to explain
Temperature& heat	Investigate to explain
Changes of state& heat	Explain
Efficiency	Use examples to describe/analysis and evaluate/design to improve
Bond energy	Use bond energy to explain and predict
Entropy, second law of thermodynamics	Use entropy to explain the direction of energy transfer
Direction of a chemical reaction	In terms of dissipation to explain/use mathematical expression to explain

3. *Conduct an investigation to analyze and evaluate.* Students conduct an investigation to analyze the concepts and evaluate energy flows and factors influencing energy efficiency (e.g., calorimetry is expected in all this seven countries' or regions' curriculum standards in the study). By conducting experiments of calorimetry, students have the opportunity to use temperature to show different amount of energy changes in a variety of reactions. Evaluating the process of conducting the experiments also promotes students' understanding of energy flow and efficiency.
4. *Design devices and approaches to improve.* Students design devices to trace energy flow and consider how to raise energy efficiency. For example, they study the activation energy of a certain reaction to choose a catalyst raising efficiency; different pathways of energy flow can be chosen to produce fuels with higher caloric value and more efficiency.

Because of incomplete specific grade information for every country/region in this study, we could hardly describe performance expectations within each grade band. Instead, we referred to a revised Bloom's taxonomy (Krathwohl 2002) on whose basis we suggested that students are required to move from a low category 1 to a higher learning level in category 4.

6.4 A Proposed Cognitive and Reasoning Model for Learning ECR

Based on the above study, we propose a cognitive and reasoning model for learning about energy in chemical reactions (ECR) (see Fig. 6.1). This model consists of cognitive perspectives based on our analysis of knowledge and performance expectations we have outlined above.

In this square model, the left side is energy analysis dimension, including energy source, the forms of transformed energy, and the amount of energy change. In the reaction system, chemical energy stored in the reaction is transformed to heat, electricity, light, and other energy forms to its surroundings. Then using the relationship of the system to its surrounding, a more sophisticated student should be able to explain any exothermic and endothermic reactions that may take place. At this level, students should be developing a qualitative or semi-quantitative understanding of energy. They can understand the energy symbols "+" and "–" depending on the direction of energy flow and how the system is defined. They can also understand that if energy increases somewhere in the system and its surroundings, then it decreases by an equal amount somewhere else. A progression of learning is used to move from understanding energy sources to understanding energy changes. With the particle model, students can make use of bond energy, and kinetic and potential energy of atoms to explain energy changes. Students understand that energy changes associated with chemical reactions are a result of the

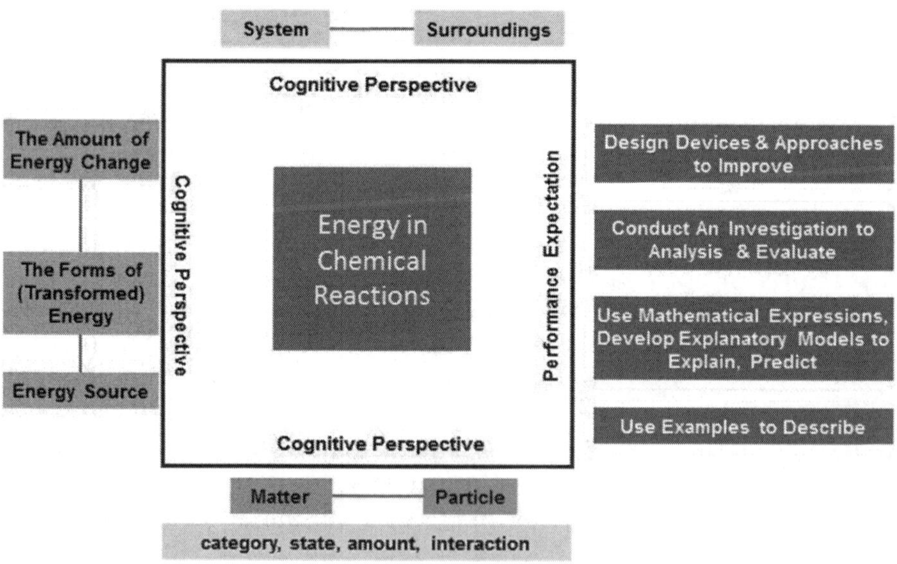

Fig. 6.1 A supposed cognitive and reasoning model for learning energy in chemical reactions (*ECR*)

rearrangement of atoms in a chemical system (The College Board 2009). In grades 9 and 10, students are expected to develop qualitative understanding of energy, and then move toward a more quantitative and complex understanding of reactions by grades 11 and 12. It is this more sophisticated understanding that provides students with the ability to understand the significance of energy in daily life.

The upper side of the square model is the dimension of system and surroundings. An important concept in thermodynamics is the thermodynamic system, a precisely defined region of the universe under study. Everything in the universe except the system is known as the surroundings. In practical teaching, teachers generally represent the energy changes of the system by ΔH. When energy of the system decreases, $\Delta H < 0$. Meanwhile teachers use Q to describe the heat of reactions. When a reaction releases heat to its surroundings, $Q > 0$. Because the decreased energy of the system ($\Delta H < 0$) corresponds to the released heat ($Q > 0$), it usually makes the students confused because of the Q and ΔH being opposite in sign. If the students can effectively distinguish between system and surroundings, they would be able to understand that ΔH describes the change of the system's energy and Q describes the change of the surroundings' energy so that they are opposite in sign. Understanding energy only from the perspective of system means students could consider one reaction or a group of reactions as a system, which is a naïve understanding of energy. While if students could understand the energy of reactions from both the perspectives of system and surroundings, they will show a more sophisticated understanding of energy. For instance, they are able to explain energy transfers and transformations between two or more systems/between the system and its surroundings.

The lower side of this square model is the dimension of matter and particle. These two perspectives reflect different levels of understanding about where energy comes from and what energy is in science. If students understand energy only from a perspective of matter, then they cannot answer why matter has energy. Also they cannot comprehend that all energy forms are either kinetic or potential energies. The amount of energy depends on the category, state, amount of matter, and the interaction of particles.

6.5 Hypothesized Cognitive Levels for Learning ECR

We can hypothesize detailed cognitive levels for learning ECR with the help of this model. Synthesizing the seven cognitive perspectives and four performance categories, we suppose that four cognitive levels exist. The four cognitive levels in detail will support teachers in designing assessment approaches, instruction, and curriculum.

- Level 1: *Describing the energy transfer and transformation between the system and its surrounding.* This is a qualitative and macroscopic understanding of energy. The performance expectations consist of using exothermic, endothermic, heat and temperature changes to describe the energy change of reactions.

- Level 2: *Developing a quantitative understanding of energy.* Students are expected to use the value of enthalpy changes to interpret the reaction as exothermic or endothermic. They can calculate the enthalpy change of reactions by using Hess's law. On this level which is still a macroscopic understanding of energy, students can analysis the energy change only from the perspective of matter.
- Level 3: *Understanding the energy source from a microscopic perspective.* Students' performance might be to recognize the category, amount of matter and particles in reactions, and to use chemical bonds to explain energy source of chemical reactions.
- Level 4: *Developing a systematic understanding of energy.* Students are expected not only to understand energy flow in complex contexts, but also to be able to choose, evaluate, or even design the path of energy flow with the idea of energy efficiency.

The cognitive levels are determined by two factors. The first factor is the number of cognitive perspectives students have; the second factor is students' understanding through the relationship between perspectives. For example, when explaining why a kind of fuel is selected, students who understand reactions from the perspective of matter could answer that the fuel is safe or the production of reactions is environment-friendly. But from the perspective of the amount of energy change, students could understand a kind of fuel is selected because of high heat value. If students get the relationship between perspectives of energy source and the amount of energy change, they are able to learn about how some special kinds of fuel, such as the fuel of rocket, are designed with considering the structure of molecules, and why some traditional kinds of fuel are changed into new substances to use in our daily life, such as coal is processed into gaseous fuel. It can be predicted that the first factor has more influence on cognitive levels. The second factor influences the cognitive level less than does the first one because the number of students' perspectives inevitably influences their understanding through the relationship between these perspectives. Students' understanding through the relationship between perspectives helps students to pick up information more quickly when solving problems.

6.6 Discussion and Implications

We studied the topic energy in chemical reactions by analyzing specific concepts and curriculum standards documents in seven countries/regions to answer three questions: "What knowledge is important to know?" "What are cognitive perspectives for understanding energy?" and "What can students be expected to do?" The two questions are related to the understanding of energy in high school. As a result, based on the answer of these two questions, we propose a cognitive model to describe the understanding of energy in chemical reactions.

In our analysis of knowledge topics in seven countries/regions, we found that first law of thermodynamics is a more prevalent topic than the second law and that energy conservation is the core idea in high school. This result was consistent with those from other research in energy learning progressions. In the topic of ECR—the most common knowledge topic in seven curriculum standards documents—energy forms is included in energy transformations. The knowledge of state changes and bond energy are used to answer the question of where the energy comes from. The concepts exothermic reactions, endothermic reactions, temperature, heat and enthalpy are descriptions of energy changes that can be used to trace and quantify energy flow. Calorimetry is a practice that requires students to use physical quantities to describe energy change. The thermochemical equation, Hess's law and energy efficiency, are used to fully understand energy flow and conservation.

In the process of analyzing common energy concepts, our goal was to determine the cognitive perspectives that influence levels of understanding energy. We propose seven cognitive perspectives, in which, energy sources (where chemical energy comes from), the amount of energy change (what the amount of energy transferred/transformed in chemical reactions is), and the forms of energy (what energy forms chemical energy transforms to) belong to energy analysis dimension. These three perspectives are different from those discussed in Lee and Liu's research (2010), which used a construct-based assessment approach to measure learning progression of energy concepts in middle school grades. They listed knowledge integration requirements for energy sources, transformations, and conservation. These three areas show three significant different levels of understanding. In our study, we propose other four basic perspectives, systems, surroundings, matter and particle, and we think all the seven cognitive perspectives reflect different learning levels and are more likely to be categories of understanding energy. We believe that the level of learning energy depends not only on the understanding of energy from each cognitive perspective, but also on the understanding through the relationship of the perspectives. Lastly, students' performance in learning with complex contexts or problems is a reflection of their level of understanding.

Performance expectations in this study are developed into four categories: (1) use examples to describe; (2) use mathematical expressions and develop explanatory models to explain and predict; (3) conduct an investigation to analyze and evaluate; and (4) design devices and approaches to improve. Performance expectations are used to determine what students can do. In our square model, performance expectations are put on the right side of the square (see Fig. 6.1) that can be used to analyze the expected cognitive levels in high school. In the lower grades of high school, students are expected to develop qualitative understanding of energy. By grades 11 and 12, a quantitative understanding should be developed, and the complexity of the chemical reaction systems that students will think about will change from having one reaction to a group of reactions, which can definitively benefit students' energy understanding. First, the chemical reactions they face in daily life are more complicated than those they discuss in class because in daily life there are usually not only one single reaction but a group of reactions. Second, the understanding of concepts, such as enthalpy and entropy; and the understanding of laws, for example, Hess Law enables students to choose better fuel and analyze industrial processes, such as liquefaction of coal.

The cognitive levels are determined by two factors—the number of cognitive perspectives students have; students' understanding through a relationship between the perspectives—of which, we predict that the first factor has more influence on cognitive level. Without the perspectives of particle, energy source and amount of energy change, students may only solve the problems of Level 1. When the number of students' perspectives increases, they can solve problems of higher cognitive levels. Level 2 problems require students to have a perspective of amount of energy changes, whereas Level 3 problems require a particle perspective. It is necessary to learn energy from all the perspectives if students have reached Level 4.

The significance of our study for chemistry teachers is to show them an overall understanding of energy in high school. Cognitive perspectives support student understanding of the core concepts and knowledge of energy in learning. Teachers are expected to focus on the core concepts and cognitive perspectives, like system and surroundings. The second law of thermodynamics is an idea of thermodynamics that can help improve understanding of energy efficiency, but it doesn't get enough attention in learning ECR at the high school level according to the results of our study. Although the calculation of enthalpy change enable students to develop quantitative understanding of energy, it is over trained because of high-stakes tests. Additionally, our model can be used as an assessment framework. Teachers can assess students' understanding levels with the help of this cognitive model.

We have to admit that this is only a proposed cognitive model. It is necessary to design a questionnaire to clarify and improve the model, and the important future study is to describe the specific cognitive levels of understanding in the topic energy in chemical reactions. We can design specific types of problems or contexts according to specific performance expectations.

Acknowledgments We would like to thank Joseph Krajcik and Xiufeng Liu for their detailed reviews of the drafts of this chapter and their brilliant suggestions. Thanks to Chi-Yan Tsui for his careful edit on the paper. We are also grateful to many other friends for kindly offering us their help.

References

Amin, T. G., Jeppsson, F., Haglund, J., & Stromdahl, H. (2012). Arrow of time: Metaphorical construals of entropy and the second law of thermodynamics. *Science Education, 96*(5), 818–848. doi:10.1002/sce.21015.

Black, P., & Solomon, J. (1983). Life world and science world: Pupils' ideas about energy. In G. Marx (Ed.), *Entropy in the school. Proceedings of the 6th Danube Seminar on Physics Education* (pp. 43–455). Budapest: Roland Eoetvoes Physical Society.

Boo, H. K., & Watson, J. R. (2001). Progression in high school students' (aged 16–18) conceptualizations about chemical reactions in solution. *Science Education, 85*, 568–585.

Boyes, E., & Stanisstreet, M. (1990). Misunderstandings of "law" and "conservation": A study of pupils' meanings for these terms. *School Science Review, 72*, 51–57.

Brook, A., & Driver, R. (1984). *Aspects of secondary student understanding of energy: Full report*. Leeds: University of Leeds, Centre for Studies in Science and Mathematics Education.

Chiu, M. H. (2001). Algorithmic problem solving and conceptual understanding of chemistry by students at a local high school in Taiwan. *Proceedings National Science Council Republic of China D, 11*(1), 20–38.

Duit, R. (1981). Understanding energy as a conserved quantity—remarks on the article by R.U. Sexl. *International Journal of Science Education, 3*(3), 291–301. doi: 10.1080/0140528810030306.

Erickson, G. L. (1979). Children's conceptions of heat and temperature. *Science Education, 63*(2), 221–230.

Fu, X. C., & Shen, W. X. (2005). *Physical chemistry* (5th ed.). Beijing: Higher Education Press. (in Chinese).

Hesse, J. J., & Anderson, C. W. (1992). Students' conceptions of chemical change. *Journal of Research in Science Teaching, 29*(3), 277–299.

Holden, C. C., & Barrow, L. H. (1984). Validation of the test of energy concepts and values for high school. *Journal of Research in Science Teaching, 21*(2), 187–196.

Jin, H., & Anderson, C. W. (2012). A learning progression for energy in socio-ecological systems. *Journal of Research in Science Teaching, 49*(9), 1149–1180. doi:10.1002/tea.21051.

Johnstone, A. H., Macdonald, J. J., & Webb, G. (1977). Misconceptions in school thermodynamics. *Physics Education, 12*, 248–251. Retrieved from http://iopscience.iop.org/0031-9120/12/4/011

Kesidou, S., & Duit, R. (1993). Students' conceptions of the second law of thermodynamics – An interpretive study. *Journal of Research in Science Teaching, 30*(1), 85–106.

Krathwohl, D. R. (2002). A revision of bloom's taxonomy: An overview. *Theory into Practice, 41*(4), Revising Bloom's Taxonomy 212–218. Retrieved from http://www.jstor.org/stable/1477405

Lee, H.-S., & Liu, O. L. (2010). Assessing learning progression of energy concepts across middle school grades: The knowledge integration perspective. *Science Education, 94*(4), 665–688. doi:10.1002/sce.20382.

Liu, X., & McKeough, A. (2005). Developmental growth in students' concept of energy: Analysis of selected items from the TIMSS database. *Journal of Research in Science Teaching, 42*(5), 493–517. doi:10.1002/tea.20060.

Ministry of Education and Human Resources Development, Korea. (2007). *Proclamation of the Ministry of Education and Human Resources Development: Science curriculum*. Retrieved from http://ncic.re.kr/english.kri.org.inventoryList.do;jsessionid=55A0FD691FB6398A223993FAA091005A# (in Korean).

Ministry of Education of the People's Republic of China. (2000). *Chemistry curriculum standard of senior high school*. Beijing: People's Education Press. (in Chinese).

Ministry of Education of the People's Republic of China. (2003a). *Chemistry curriculum standard of senior high school*. Beijing: People's Education Press. (in Chinese).

Ministry of Education of the People's Republic of China. (2003b). *Chemistry curriculum standard of junior high school*. Beijing: People's Education Press. (in Chinese).

Ministry of education, Taiwan. (2008). *Curriculum guidelines of general high school compulsory subjects "Chemistry"*. Retrieved from http://www.edu.tw/userfiles/%E6%99%AE%E9%80%9A%E9%AB%98%E7%B4%9A%E4%B8%AD%E5%AD%B8%E5%BF%85%E4%BF%AE%E7%A7%91%E7%9B%AE%E3%80%8C%E5%9F%BA%E7%A4%8E%E5%8C%96%E5%AD%B8%E3%80%8D%E8%AA%B2%E7%A8%8B%E7%B6%B1%E8%A6%81(1).pdf (in Chinese).

Ministry of Education, Culture, Sports, Science and Technology. (1998). *Course of study for lower secondary school: Science*. Retrieved from http://www.mext.go.jp/a_menu/shotou/cs/1320119.htm (in Japanese).

Ministry of Education, Culture, Sports, Science and Technology. (1999). *Course of study for higher secondary school: Science*. Retrieved from http://www.mext.go.jp/a_menu/shotou/cs/1320231.htm (in Japanese).

Ministry of Education, Culture, Sports, Science and Technology. (2009). *Course of study for higher secondary school: Science*. Retrieved from http://www.mext.go.jp/component/a_menu/education/micro_detail/__icsFiles/afieldfile/2011/03/30/1304427_002.pdf (in Japanese).

Ministry of National Education. (2010) *Programme of physics–chemistry of high school*. Retrieved from http://cache.media.education.gouv.fr/file/special_4/72/9/physique_chimie_143729.pdf (in French).

Morrisey, T., & Barrow, L. (1984). A review of energy education: 1975 to NEED 1981. *Science Education, 68*(4), 365–379.

Nahum, T. L., Naaman, R. M., Hofstein, A., & Krajcik, J. (2007). Developing a new teaching approach for the chemical bonding concept aligned with current scientific and pedagogical knowledge. *Science Education, 91*, 579–603. doi:10.1002/sce.20201.

National Research Council. (2011). *A framework for K-12 science education: Practices, crosscutting concepts, and core ideas*. Washington, DC: The National Academies Press.

Neumann, K., Viering, T., Boone, W. J., & Fischer, H. E. (2013). Towards a learning progression of energy. *Journal of Research in Science Teaching, 50*(2), 162–188. doi:10.1002/tea.21061.

Next Generation Science Standards Lead States (2013). *Next generation science standards (NGSS): Achieve, Inc. on behalf of the twenty-six states and partners that collaborated on the NGSS*. Retrieved from http://www.nextgenscience.org/next-generation-science-standards

Ontario Ministry of Education. (2000). *The Ontario curriculum grades 11 and 12 – science*. Retrieved from http://www.edu.gov.on.ca/eng/curriculum/secondary/science1112curr.pdf

Ontario Ministry of Education. (2008a). *The Ontario curriculum grades 9 and 10 – science*. Retrieved from http://www.edu.gov.on.ca./eng/curriculum/secondary/science.html

Ontario Ministry of Education. (2008b). *The Ontario curriculum grades 11 and 12 – science*. Retrieved from http://www.edu.gov.on.ca./eng/curriculum/secondary/science.html

Rogat, A. (2011). Developing hypothetical learning progressions for other core ideas. CPRE White Paper Series. WP-01.

Schmidt, W. H., Raizen, S., Britton, E. D., Bianchi, L. J., & Wolfe, R. G. (2002). *Many visions, many aims (Vol. 2): A cross-national investigation of curricular intentions in school science*. New York: Kluwer Academic Publisher.

Solomon, J. (1983). Learning about energy: How pupils think in two domains. *European Journal of Science Education, 5*(1), 49–59.

Solomon, J. (1985). Teaching the conservation of energy. *Physics Education, 20*, 165–170.

Stern, L., & Roseman, J. E. (2004). Can middle school science textbooks help students learn important ideas? Findings from project 2061's curriculum evaluation study: Life science. *Journal of Research in Science Teaching, 41*(6), 538–568. doi:10.1002/tea.20019.

The College Board. (2009). *Science college board standards for college success*. Retrieved from http://www.collegeboard.com

U.S. Department of Energy. (2012). *Energy literacy. Essential principles and fundamental concepts for energy education*. Version 1.1. Retrieved from http://www1.eere.energy.gov/education/energy_literacy.html

Watt, D. M. (1983). Some alternative views of energy. *Physics Education, 18*, 213–217.

Chapter 7
Developing and Using Distractor-Driven Multiple-Choice Assessments Aligned to Ideas About Energy Forms, Transformation, Transfer, and Conservation

Cari F. Herrmann-Abell and George E. DeBoer

7.1 Introduction

In today's society, people are asked to use their knowledge about energy to make decisions about what kinds of light bulbs to use, to evaluate arguments related to global climate change, and to think about national energy resource policy. Students learn about energy in their physical science classes and also in life and earth science classes when they are studying topics such as photosynthesis and respiration or weather and climate. But, research has shown that in spite of the instruction they receive, students (and adults) hold a wide range of misconceptions and alternative ideas about energy. Because energy is such an important concept, it must be taught well, and part of teaching it well is understanding what students know and do not know about energy before, during, and after instruction. Having assessments designed specifically to pinpoint students' conceptual problems and their causes is essential for teachers. To respond to this need for high quality diagnostic assessments, in 2004 we began to develop a bank of assessment items that are precisely aligned to middle school science ideas in the life, physical, and earth sciences and that can be used to diagnose common misconceptions and students' difficulties with energy concepts (DeBoer et al. 2008a).

Our work builds on and extends the existing knowledge base about student understanding of energy in a number of ways. First, the assessment results are based on a very large national sample of students in grades 6–12 (N = 23,744). This gives us a more broad-based description than most studies do of how students' understanding of energy changes from grade-to-grade. We can also make comparisons across grades and across concepts because all students in grades 6–12 were tested from a common pool of linked test items. Second, our assessment items are precisely aligned to specific energy concepts including energy forms,

C.F. Herrmann-Abell (✉) • G.E. DeBoer
American Association for the Advancement of Science/Project 2061, Washington, DC, USA
e-mail: cabell@aaas.org

R.F. Chen et al. (eds.), *Teaching and Learning of Energy in K-12 Education*, DOI 10.1007/978-3-319-05017-1_7,

transformation, transfer, and conservation. This enables us to explore problems that students are having understanding particular concepts, not just energy in general. Third, the items are designed not only to test for the correct scientific understanding but also to probe for common misconceptions. This provides the opportunity to make a detailed analysis of the alternative ideas students hold that may be giving them difficulty. Incorporating common misconceptions into answer choices also gives students plausible answers to select from so they are less likely to guess, thus giving us a more valid measure of what they actually know.

In this chapter, we describe the development of assessment items for the energy topic, and then we review the results of national field testing of those items focusing on what students know, what they do not know, and the misconceptions they have.

7.2 Item Development

7.2.1 Selecting a Set of Target Learning Goals

The first step in our item development process is selecting a set of learning goals that will serve as the targets for the assessment items. Energy is a very broad topic with applications in many fields of science, so the topic has to be narrowed down to a coherent set of learning goals before beginning item development. Our goal was to choose a set of ideas about energy that can be connected together and reinforce each other in a coherent storyline, that target particularly problematic ideas about energy that students have, that are appropriate for middle school students and reflect the progression of their understanding of energy based on empirical research reported in the literature, and that are consistent with the recommendations in national standards documents.

7.2.2 Consulting the Research Literature on Students' Understanding of Energy

Before beginning item development, we conducted a thorough review of the research literature to determine which ideas about energy students were most likely to be learning in school and, therefore, appropriate for assessment. One body of research on students' understanding of energy has focused on energy as a single unified concept (Watts 1983; Trumper 1990, 1993; Nicholls and Ogborn 1993), not on the different "forms" or manifestations of energy such as thermal, chemical, and elastic energy. For example, Watts (1983) classified students' ideas about energy into seven general types: anthropocentric, depository, ingredient, activity, product, functional, and flow-transfer. Trumper (1990) later expanded on Watts' work by splitting the depository framework into two, the original passive "depository" framework and an "active" deposit or "cause" framework, and adding the transformation framework, which is the accepted scientific view. Although that work demonstrates the variety of ways that students may think about energy, we chose to focus on the ideas students have about the different ways that energy is manifested in their everyday lives. Thus, forms of energy (e.g., motion, thermal, and gravitational potential energy) and form conversions are key aspects of our assessments.

Research on students' ideas about energy revealed a number of misconceptions. For example, some students associate energy only with obvious activity or movement (Brook and Driver 1984; Finegold and Trumper 1989; Kruger 1990; Kruger et al. 1992; Stead 1980; Summers and Kruger 1993; Trumper 1990, 1998; Trumper and Gorsky 1993; Watts 1983). To them, objects at rest have no energy at all. Regarding thermal energy, students often think that living things have thermal energy but inanimate objects do not (Stead 1980; Solomon 1983; Watts 1983; Finegold and Trumper 1989; Kruger 1990; Trumper 1990, 1993; Kruger et al. 1992; Trumper and Gorsky 1993; Leggett 2003). Ideas about gravitational potential energy frequently include the idea that "potential" energy is the potential to have energy in the future (Stead 1980; Summers and Kruger 1993). Students may also confuse force and energy, especially thinking that objects in motion have a force within them (Fischbein et al. 1989; McCloskey 1983). Students also report that it is "coldness'" that is transferred between two objects at different temperatures (Brook et al. 1984; Clough and Driver 1985; Newell and Ross 1996), not thermal energy. Finally, it is common for students to think that energy can be created or destroyed (Brook and Driver 1984; Kesidou and Duit 1993; Kruger 1990; Loverude 2004; Papadouris et al. 2008; Stead 1980; Trumper 1998). These alternative ideas were built into the assessment items as distractors (Sadler 1998).

Regarding the progression of students' understanding of energy over time, a number of recent empirical studies have shown that students develop ideas about forms of energy first, followed by transformations and transfers, and finally conservation. Liu and Mckeough (2005) used data from selected items from the Third International Mathematics and Science Study (TIMSS) database. Using the partial credit Rasch model, they demonstrated support for their hypothesized sequence of development of the energy concept. In their proposed progression of understanding, students first perceive energy as activity, or the ability to do work. As students' understanding progresses, they begin to distinguish different energy sources and forms of energy. Next comes an understanding of energy transfer, followed by an appreciation of energy degradation, and finally an acceptance of the concept of conservation of energy. Liu and Collard (2005) validated those results in a follow-up study on students in grades 4, 8, 10, 11, and 12 using performance assessments and Many-Facet Rasch measurement. Lee and Liu (2010) found further support for the conclusion that energy conservation was the most difficult concept for students using 10 two-tiered items based on released TIMSS items addressing energy sources, energy transformation, and conservation of energy. The items were administered to a large sample of middle school students, and the results showed that the conservation of energy items required the highest level of knowledge integration compared to the other two concepts. This same trend was also found in a study of German students' understanding of energy (Neumann et al. 2012). Nuemann and his colleagues found that 6th grade students had an understanding of forms and sources of energy, 8th grade students had an understanding of energy transformations and transfer, and only some 10th grade students had reached an understanding of conservation of energy.

7.2.3 Clarification of the Target Learning Goals

To choose the set of ideas for testing, we relied primarily on AAAS Project 2061's previous work developing the strand maps published in the *Atlas of Science Literacy* (Association for the Advancement of Science [AAAS] 2001, 2007). These maps show how student understanding might progress from grade to grade and across concepts to create a complex mental network of interconnected knowledge about the world. The Energy Transformations map in *Atlas of Science Literacy, Volume 2* (AAAS 2007) and Chapter 4 (The Physical Setting), Section E (Energy Transformations) of *Benchmarks for Science Literacy* (AAAS 1993) were used to guide our choice of energy ideas as targets for assessment:

- *Motion Energy (kinetic energy)* is associated with the speed and the mass of an object.
- *Thermal Energy (substance level)* is associated with the temperature and the mass of an object and the material of which the object is made.
- *Thermal Energy (atomic level)* is associated with the disordered motions of an object's atoms or molecules and the number and types of atoms or molecules of which the object is made.
- *Gravitational Potential Energy* is associated with the mass of an object and the distance the object is above a reference point, such as the center of the earth.
- *Elastic Energy* is associated with the stretching or compressing of an elastic object and how easily the object can be stretched or compressed.
- Energy can be *transformed* within a system (e.g. motion to thermal, gravitational potential to motion, etc.).
- Energy can be *transferred* from one object or system to another in different ways: by conduction, mechanically, electrically, or by electromagnetic radiation.
- Energy is *conserved.* Regardless of what happens within a system, the total amount of energy in the system remains the same unless energy is added to or released from the system.

Each of these ideas was further clarified into sub-ideas to state precisely what students would be expected to know and boundary statements to indicate what they would not be expected to know. These clarification statements act as item-writing specifications that ensure a close alignment between the items and the learning goals. For example, the clarification statement for conservation of energy includes the following sub-ideas and boundary statements:

Students should know the following sub-ideas:

1. Regardless of what happens within a system, the total amount of energy in the system remains the same unless energy is added to or released from the system, even though the forms of energy present may change.
2. If the total amount of energy in a system seems to decrease or increase, energy must have gone somewhere or come from somewhere outside the system.
3. If no energy enters or leaves a system, a decrease of one form of energy by a certain amount within the system must be balanced by an increase of another form of energy by that same amount within the system (or a net increase of multiple forms of energy by that same amount). Similarly, an increase of one form of energy by a certain amount within a system must be balanced by a decrease of another form of energy by that same amount within the system (or a net decrease of multiple forms of energy by that same amount).
4. Energy can neither be created nor destroyed, but it can be transferred from one object or system to another and/or be changed from one form to another.
5. If energy is transferred to or from a very large system (or a very complex system), increases or decreases of energy may be difficult to detect and, therefore, it may appear that energy was not conserved.

Boundaries:

1. Students are not expected to quantitatively keep track of changes of energy in a system.
2. Assessment items will avoid using the phrase "energy conservation" or "conservation of energy" because of the misconceptions associated with them.
3. Students are not expected to know about energy-mass conversions such as nuclear reactions or other subatomic interactions.

(Note: Clarifications of the other targeted ideas can be found at http://assessment.aaas.org.)

7.2.4 *Efforts to Ensure Validity*

Each item was developed using a procedure designed to ensure its match to the targeted idea and its overall effectiveness as an accurate measure of what students do and do not know about the idea. We used a set of criteria developed by AAAS Project 2061 to evaluate the content alignment of assessment items and to minimize construct-irrelevant factors that make it difficult to interpret a student's response to an item. The full description of the analysis criteria used during item development can be found at: http://www.project2061.org/research/assessment/assessment_form.htm. Additional details of the analysis procedure can be found in DeBoer et al. (2007, 2008a, b).

Content Alignment Two content alignment criteria were used: necessity and sufficiency (Stern and Ahlgren 2002). The necessity criterion addresses whether the knowledge described in the learning goal is needed to answer correctly. For a multiple choice question, meeting the necessity criterion means that the knowledge in the learning goal is needed to evaluate all of the answer choices, including incorrect answer choices targeting known misconceptions (Sadler 1998). In other words, the knowledge that was targeted by the item had to be necessary to evaluate each answer choice, both correct answers and distractors. For example, an item aligned to conservation of energy would meet the necessity criterion only if it required students to use the knowledge that the total amount of energy in a system remains the same unless energy is added to or released from the system to select the correct answer and eliminate all of the distractors. The second content alignment criterion is sufficiency, which addresses whether the knowledge described in the learning goal is enough by itself to successfully complete the item. The student should not be expected to use additional knowledge not covered by the learning goal to evaluate the answer choices and select a correct answer. In the conservation of energy example above, the item would *not* meet the sufficiency criterion if it required students to use knowledge of chemical reactions to analyze the answer choices. That would be going beyond the knowledge targeted by the item and, therefore, would not a fair measure of students' understanding of the learning goal. In a few cases, we decided to target more than one learning goal in an item, but that was done intentionally, and the target learning goals were specified prior to item development. In general, however, targeting more than one learning goal in the same item makes it more difficult to pinpoint exactly where students have gaps in their knowledge, so we avoided as much as possible alignment to multiple learning goals in what were intended to be highly diagnostic items.

Construct Validity The set of criteria that we used to ensure construct validity and minimize construct-irrelevant features included comprehensibility, appropriateness of task context, and test-wiseness. (1) To meet the comprehensibility criterion, the item had to make it clear what question is being asked, avoid unfamiliar general vocabulary or unnecessarily complex sentence structure, use words or phrases that did not have ambiguous meanings, and present diagrams, pictures, graphs, and tables that could be easily understood. (2) To ensure that the task context was appropriate and fair, it had to be familiar to most students, so that one group of students was not advantaged or disadvantaged because of their familiarity with the context, be clear and easy to understand, use information and quantities that are reasonable and believable, and accurately represent scientific or mathematical realities or make clear when idealizations are involved. (3) When analyzing for test-wiseness, the plausibility of the distractors was considered, along with whether one answer choice differed in length or detail, whether one answer choice was qualified differently, whether one answer choice contained vocabulary at a different level of difficulty, whether a pair of answer choices contained logical opposites that may lead students to eliminate answer choices, and whether the language in one answer choice mirrored the language in the stem.

7.2.5 *Pilot Testing*

After items were drafted, they were pilot tested with students in middle and high school to obtain feedback from them about the items. The pilot test included follow-up questions for each item to give us insight into how well the item was meeting the assessment criteria described above. Questions asked students to describe anything they found confusing about an item, to circle words they were unfamiliar with, and to comment on the helpfulness of diagrams and tables. Student responses to these questions provided information about how well the item was meeting the criteria related to construct irrelevant factors. Students were also asked to write explanations for why they selected or rejected each answer choice, to indicate if they guessed, and to indicate where they had learned about the topic. This allowed us to determine what knowledge students were using to answer the items, which helped us to evaluate the content alignment of the items. After the pilot test, a panel of scientists and science education and assessment experts was convened to review the items using the same criteria that were used in item development. After revisions were made based on the pilot testing and expert reviews, the items were field tested on a large national sample to determine the psychometric properties of the items. A subset of the items used in the field test is available at http://assessment.aaas.org/topics/EG#/.

7.2.6 *Field Testing and Data Collection*

Field testing of the items took place in the spring of 2009 and the spring of 2010. Because we were testing more items than students could finish in a typical class period, we created multiple test forms that contained subsets of the available items. Linking items were included so that we could use Rasch modeling to compare item characteristics across forms and between years. During the field test, students were asked to choose the single correct answer for each item. Items for which students chose more than one answer choice were marked incorrect.

We sent invitations to participate in the field testing directly to teachers or to school and district administrators who then recruited teachers to participate. Teachers were selected to participate on a first-come first-served basis but, when necessary, we adjusted our selections to achieve representation from urban, rural, and suburban schools from different parts of the US. The field tests included 14,484 middle school students and 9,260 high school students from 48 states across the country. (See Table 7.1 for demographic information.) The teachers received the testing materials by mail and administered the field tests to whatever science classes they were teaching (either life, physical, or earth science). Item sampling was used such that each student received 30–44 assessment items, and each item was answered by an average of 2,530 middle school students and 1,665 high school students.

Table 7.1 Demographic information for the students who participated in the field test[a]

Grade	Total % (N)	Female %	Male %	English %	Non-English %
6th Grade	18 % (4,330)	49.6 %	48.7 %	87.1 %	10.3 %
7th Grade	22 % (5,177)	49.1 %	48.6 %	88.0 %	9.0 %
8th Grade	21 % (4,977)	48.9 %	49.3 %	88.2 %	8.7 %
9th Grade	11 % (2,673)	49.5 %	48.0 %	88.7 %	8.9 %
10th Grade	12 % (2,822)	49.3 %	48.6 %	88.8 %	8.1 %
11th Grade	10 % (2,452)	51.6 %	46.7 %	90.6 %	6.9 %
12th Grade	6 % (1,313)	53.5 %	45.1 %	88.7 %	8.6 %
Total	100 % (23,817)	49.6 %	48.3 %	88.2 %	8.9 %

[a]Gender and language columns do not total to 100 % because not all students specified their gender or primary language

7.3 Rasch Modeling

We used Rasch modeling to analyze the field test data. In the dichotomous Rasch model, the probability that a student will respond to an item correctly is determined by the difference in the student's ability and the difficulty of the item, according to Eq. 7.1:

$$\ln\left(\frac{P_{ni}}{1 - P_{ni}}\right) = B_n - D_i \tag{7.1}$$

where P_{ni} is the probability that student n of ability B_n will respond correctly to item i with a difficulty of D_i (Liu and Boone 2006; Bond and Fox 2007). Student ability and item difficulty are expressed in logits, which can range from $-\infty$ to ∞. It is important to note that the ability and difficulty measures are expressed on the same interval scale and that the Rasch model assumes that these two measures are mutually independent, which is not the case for classical percent correct measures. (Note: Rasch modeling uses the term "ability" to refer to the students' understanding of the ideas being targeted by the items. It should not be interpreted as an underlying, innate quality of the student, but more narrowly as the students' understanding of the topic at the time of the field test.)

WINSTEPS (Linacre 2012) was used to estimate student abilities and item difficulties for all of the students and all of the items from both field tests. From these parameters, we were able to determine how well the range of item difficulties matched the range of student abilities and the extent to which each of the items correlated with the entire set of items (point-measure correlation). We also looked to see if the pattern of student responses followed expectations such that the most able students were most likely to answer the most difficult questions correctly. According Rasch (1960):

> a person having a greater ability than another should have the greater probability of solving any item of the type in question, and similarly, one item being more difficult than another one means that for any person the probability of solving the second item correctly is the greater one (cited in Wright and Stone 1999).

Table 7.2 Summary of Rasch fit statistics

	Min.	Max.	Median
Standard error	0.02	0.08	0.04
Infit mean-square	0.85	1.19	0.99
Point-measure correlation coefficients	0.13	0.51	0.38
Item separation index (reliability)		14.29 (1.00)	
Person separation index (reliability)		1.80 (0.76)	

7.3.1 Model Fit

The field test data had a good fit to the Rasch model, suggesting that the items were measuring a unidimensional energy construct. A summary of the fit statistics is shown in Table 7.2. The separation index (14.29) and corresponding reliability (1.0) were high for the item data. The separation index indicates the approximate number of different levels of item difficulty or person ability that can be discriminated. A separation index greater than 2 is considered acceptable according to Wright and Stone (2004). The lower person separation index (1.80) for our data is due to the relatively small number of items available to test students at the extreme ends of the scale, especially the lower end (see Figs. 7.1 and 7.2). This means that there is less information available to measure the ability of very low or very high ability students. In contrast, because there were so many students responding to each item, differences in difficulty level of the items is easier to determine, which can be seen in the very high item separation index and reliability estimate. Additionally, the standard errors for the items were small. The infit mean-square values for all of the items fell within the acceptable range of 0.7–1.3 for multiple-choice tests (Bond and Fox 2007). Infit statistics are reported here because they give more weight to the responses of students with abilities closer to the item difficulty, whereas outfit statistics are unweighted and, therefore, are more sensitive to outlying scores.

7.3.2 Wright Maps

Figure 7.1 shows the Wright map for 91 field test items aligned to ideas about five different forms of energy. Figure 7.2 shows the Wright map for 95 items aligned to ideas about energy transformation, transfer, and conservation. The maps show the range of student abilities on the left side of a vertical line and the range of item difficulties on the right side of the line. The scale runs from low ability/difficulty at the bottom of the maps to high ability/difficulty at the top. The mean of the item difficulties is set at zero. When a student's ability matches an item's difficulty, the student has a 50 % chance of answering the item correctly. The maps show a good match between mean item difficulty and mean student ability, with mean item difficulty being just slightly higher than mean student ability. The maps also

```
MEASURE     PERSON - MAP - ITEM
              <more>|<rare>
  4              .  +
                 .  |
                 .  |
                 .  |
                 .  |
                    |
                 .  |
  3              .  +
                 .  |
                 .  |
                 .  |
                 .  |
                 .  |                          |                        |             |              |
                 .  |                          |                        |             |              |
  2              .  +    Motion Energy         |      Thermal Energy    |   Thermal   |Gravitational| Elastic
                .#  |                          |    (substance level)   |   Energy    |  Potential   | Energy
                 .  |                          |                        |  (atomic)   |   Energy     |
                 .  |                          |                        |             |              |
                .# T|                          |                        |             |              |721
                .#  |                          |                        |             |              |
                .#  |T 065 251                 |                        |             |              |
  1            .##  +  801                     |                        |             |692           |
               .##  |  241                     |                        |             |              |
              .###  |  791                     |                        |             |              |
             .#### S|S                         |                        |             |143           |
            .#####  |                          |273 323 601 622 831 841 |663          |213 931       |
           .######  |  781                     |343                     |452 506      |672 871 901   |
          .#######  |  022 093                 |581 613                 |483 574      |223           |
  0       .#######  +M 921                     |591 292                 |554 742      |134 851       |423
         .########  |  033                     |641                     |492 513 564  |184 195 203   |861
       .########## M|                          |383                     |463 473 524  |153 163 911   |364 403
      .###########  |  103 232                 |284                     |534 542 652  |881 891       |354 442
     .############  |S 053 072 811 812 813     |333 632                 |             |712           |
        .#########  |  032 083 771             |313                     |             |174           |434
         .########  |  016 112 122 761         |303                     |             |              |413
 -1      .########  +  015 044                 |263                     |             |              |373
         .######## S|T                         |                        |             |              |
            .#####  |                          |                        |             |              |
             .####  |
               .##  |
                .#  |
                .#  |
 -2              . T+
                 .  |
                 .  |
                 .  |
                 .  |
                 .  |
                 .  |
 -3              .  +
                 .  |
                    |
                 .  |
                    |         M = mean ability/difficulty
                    |         S = 1 standard deviation away from mean
                    |         T = 2 standard deviations away from mean
 -4              .  +         Each "#" is 163 students.
              <less>|<frequ>  Each "." is 1 to 162 students.
```

Fig. 7.1 Wright map for 91 items on forms of energy included in the field test

show that many of the items are clustered around the mean with only a few items at the higher and lower ends of the student distribution. The map in Fig. 7.2 shows that items aligned to the idea of conservation of energy are more difficult than items aligned to the ideas of energy transformation and energy transfer. This finding is consistent with the results of previous research mentioned earlier that an understanding of conservation comes later than an understanding of forms, transformation, and transfer.

```
MEASURE      PERSON - MAP - ITEM
                  <more>|<rare>
  4                  .  +
                     .  |
                     .  |
                     .  |
                     .  |
                        |
                     .  |
  3                  .  +
                     .  |
                     .  |
                     .  |
                     .  |
                     .  |                                    |                                  |
                     .  |                                    |                                  |
  2                  .  +        Transformations             |            Transfer              | Conservation
                    .#  |                                    |                                  |
                     .  |                                    |                                  |694 902 1041
                     .  |                                    |                                  |
                    .# T|                                    |                                  |
                    .#  |                                    |                                  |672
                    .#  |T                                   |                                  |684 862 922 942
  1                .##  +                                    |                                  |654 823 873 932
                   .##  |                                    |503                               |782 952
                  .###  |  023 142 792                       |483                               |685 962
                 .#### S|S 132                               |                                  |803 813
                .#####  |  173                               |223 232                           |743 883 893
               .######  |  192                               |303 412 582 1011 1031             |702 753
              .#######  |                                    |272 313 463 522                   |
  0           .#######  +M 013 1001                          |242 282 352 474 493 572 612       |852
             .########  |  043 053 104 153 183 202 212 991   |263 333 562                       |
           .########## M|  033 064 083 113                   |262 322 473 542 642               |763 842
          .###########  |  095 163 323                       |                                  |712
         .############  |S 073                               |252 253 392 512 563 564           |833
             .#########  |  122                              |532 622                           |
             .########  |                                    |552 602 1021                      |
 -1          .########  +                                    |                                  |
             .######## S|T                                   |592                               |
                .#####  |                                    |343                               |
                 .####  |                                    |                                  |
                   .##  |                                    |                                  |
                    .#  |                                    |                                  |
                    .#  |                                    |                                  |
 -2                  . T+
                     .  |
                     .  |
                     .  |
                     .  |
                     .  |
 -3                  .  +
                     .  |
                        |
                     .  |
                        |                     M = mean ability/difficulty
                        |                     S = 1 standard deviation away from mean
                        |                     T = 2 standard deviations away from mean
 -4                  .  +                     Each "#" is 163 students.
                  <less>|<frequ>              Each "." is 1 to 162 students.
```

Fig. 7.2 Wright map for 95 items on energy transformations, transfer, and conservation included in the field test

7.4 Grade-to-Grade Differences

We conducted an analysis of covariance (ANCOVA) to investigate if the differences in students' performance from grade to grade were significant. Because our sampling procedures did not ensure that the students in each grade could be considered equivalent in terms of relevant background variables (see Table 7.1), an analysis of

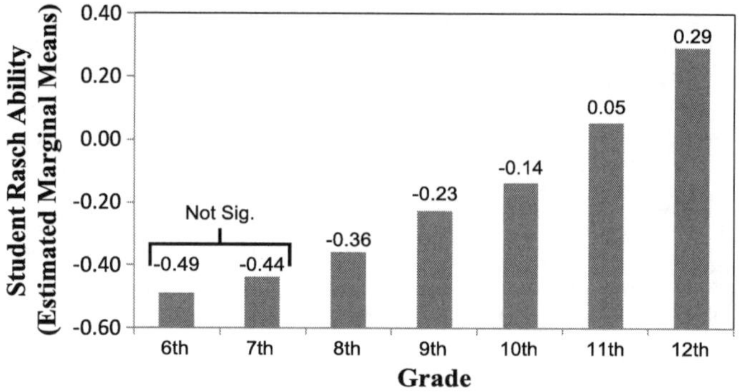

Fig. 7.3 Average student ability by grade for the field test

covariance was performed controlling for gender and whether the students identified English as their primary language. Both gender [F(1, 22698) = 51.1] and English as the primary language [F(1, 22698) = 302] were significantly correlated with student ability measures ($p < .001$). ANCOVA showed that differences in average ability by grade were significant at the .001 level of significance [F(6, 22698) = 247]. The estimated marginal means for the student ability measures are reported in Fig. 7.3. A Bonferroni post hoc test showed that the differences in mean ability for all grades were statistically significant on the .001 level, except for the difference between sixth and seventh grades. Knowledge of the energy ideas increased steadily from seventh to twelfth grades (see Fig. 7.3). This trend of increasing ability can be attributed to more students having the opportunity to learn these energy ideas as they progress through the grades and the greater maturity of students in the higher grades that made them more likely to understand the energy ideas covered by the items, which are often abstract and counterintuitive.

7.5 Students' Knowledge and Misconceptions

In this section, we report on the results of our field test that provided insight into what students know and do not know about energy and what misconceptions they hold. Results for each key idea are presented below.

7.5.1 Motion Energy

Students were tested for their understanding of the idea that motion energy (kinetic energy) is associated with the speed and mass of an object. Although all the motion energy items focused on some aspect of that basic idea, the items also varied in the

Table 7.3 Item difficulties for items aligned to the motion energy sub-ideas from easiest to most difficult

		Difficulty			
Item context	No. of items	Min.	Max.	Median	Mean
Items in which motion energy and speed vary and weight is held constant	10	−1.00	−0.36	−0.82	−0.78
Items in which motion energy and weight vary and speed is held constant	4	0.08	1.09	0.22	0.40
Items in which the stem does not provide enough information to be able to compare the motion energy of two objects	4	0.69	1.12	0.94	0.92

mental processing required of the students. In other words, students were asked to use their knowledge in a variety of ways. For example, in some items, students were asked to determine the motion energy from information about the speed and mass of the objects, and in other items the students were asked to determine the speed from information about the motion energy and mass. In other items, students were asked to compare the motion energy of two objects given information about the speed of the objects but were expected to recognize that they could not make such a comparison because they were not given information about the mass of the objects. Items were more or less difficult for students depending on how they had to reason with the knowledge.

To quantify the effect that different ways of asking about motion energy had on student performance, we divided the items into three categories: (1) items in which the motion energy and the speed of the objects varied and the weight/mass was held constant, (2) items in which the motion energy and the weight/mass varied and the speed was held constant, and (3) items in which the stem did not provide enough information to be able to compare the motion energy of two objects. Table 7.3 summarizes the difficulty of the items in each of these categories. One-way ANOVA revealed statistically significant differences in the means for the three categories [$F(2, 15) = 65.55$, $p < .001$]. Using a Bonferroni post hoc test, the items in which the weight/mass was held constant were significantly easier than the items in the other categories ($p < .001$). Because the weight/mass was held constant, these items only required students to use their knowledge of the relationship between motion energy and speed. Items in the other two categories, which required students to know that motion energy depends on mass as well as speed, were more difficult.

Table 7.4 shows the percentage of correct responses for each of the categories by grade level. As expected, the high school students performed better than the middle school students on the items in all three categories. The high school students were more aware of the dependence of motion energy on both speed and mass and better able to handle the more sophisticated mental operations required for answering correctly.

These results from items aligned to the idea of motion energy indicate that knowing that motion energy depends on speed comes before knowing that motion

Table 7.4 Percentage of correct responses and grade level differences for items aligned to the motion energy sub-ideas

Item context	Middle school	High school	χ^2	Sig.
Items in which motion energy and speed vary and weight is held constant	57.8 %	61.4 %	59	<.001
Items in which motion energy and weight vary and speed is held constant	25.0 %	36.3 %	357	<.001
Items in which the stem does not provide enough information to be able to compare the motion energy of two objects	22.7 %	29.7 %	115	<.001

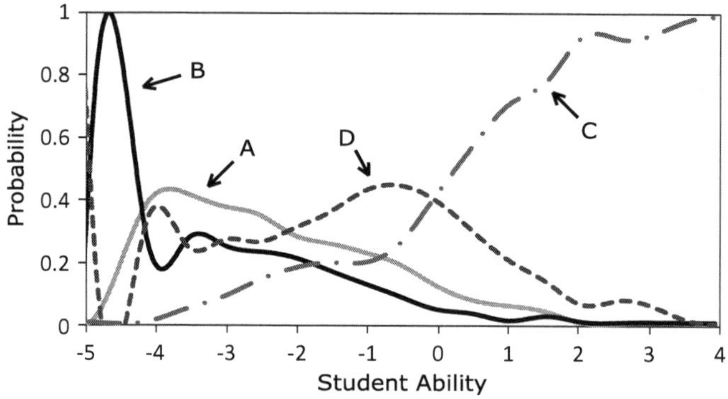

Fig. 7.4 Option probability curves for an item aligned to motion energy

energy depends on mass, in other words, knowing that energy is an extensive property of objects. This is supported by an analysis of the option probability curves for an item shown in Fig. 7.4. Option probability curves show the probability of selecting each answer choice as a function of student ability. For this item, answer choice A corresponds to the misconception that an object has energy because a person gives energy to the object. Answer choice B corresponds to the misconception that inanimate objects do not have energy but people do. Answer choice C is the correct answer, corresponding to the idea that motion energy depends on mass. Answer choice D corresponds to the misconception that motion energy depends only on speed, not on mass.

The misconceptions in answer choices A and B were popular among students at the lower end of the ability spectrum. It is reasonable to assume that as students receive formal instruction on the topic they let go of these human-centered views of energy. Students with abilities between −2 and 0 were more likely to select answer choice D (motion energy depends only on speed) possibly because when motion energy is first introduced, instruction focuses mainly on the speed of objects, with mass held constant. Finally, in the progression of understanding, students with ability level above zero (above the mean) were increasingly more likely to select the correct answer.

We also found from our analysis of items in this set that students held additional misconceptions about motion energy. One item revealed that about a quarter of both the middle and high school students thought that a ball that is thrown has motion energy while it is moving, but a ball that is dropped does not have motion energy while it is moving. Results from another item indicated that 35 % of the students thought that the motion energy of an object depends on both the speed and the direction of motion. These misconceptions may be related to the belief that energy is associated with how difficult it is to move an object or how hard a person has to pull or push it (Brook and Driver 1984). Students who have this misconception would think, for example, that a person walking uphill would have a different amount of motion energy than when walking downhill even if the person was walking at the same speed in both cases. For the item that compared the thrown ball and the dropped ball, students' responses may reflect their belief that the dropped ball has no motion energy because no effort was put into dropping the ball.

7.5.2 *Thermal Energy (Substance Level)*

Similar to our findings related to students' understanding of motion energy, we found that students were less likely to associate thermal energy with the mass of the object than with its temperature. To quantify the difference, we divided the items into two categories: (1) items in which the thermal energy and the temperature of the object or objects varied and the weight/mass and the type of material were held constant, and (2) items in which the thermal energy and the weight/mass varied and the temperature and the type of material were held constant. Table 7.5 presents the average difficulty of the items in each of these categories. A t-test confirmed that the items in which the weight/mass varied were more difficult than items in which the temperature varied [$t(8) = -9.27$, $p < .001$].

Table 7.6 shows the percentage of correct responses for each of the categories by grade level. These results are similar to those found for the motion energy items, showing that a number of students are unaware that the amount of energy an object has depends on how much of the object there is. As with the motion energy items, the high school students performed better than the middle school students on the items in both categories.

We also identified a number of misconceptions that students held about thermal energy. About 43 % of the middle school students and 31 % of the high school students thought that living things, including humans and plants, have thermal energy but dead things, such as dead plants, and inanimate objects, such as pennies, do not. It has been well reported in the literature that students often associate energy with living things, not inanimate objects (Stead 1980; Solomon 1983; Watts 1983; Finegold and Trumper 1989; Kruger 1990; Trumper 1990, 1993; Kruger et al. 1992; Trumper and Gorsky 1993; Leggett 2003). Also, some students associate thermal energy with warmth. For example, on an item involving two pieces of metal, about

Table 7.5 Item difficulties by for items aligned to the thermal energy sub-ideas

Item context	No. of items	Difficulty Min.	Max.	Median	Mean
Items in which thermal energy and temperature vary and weight and type of material are held constant	5	−1.02	−0.42	−0.67	−0.71
Items in which thermal energy and weight vary and temperature and type of material are held constant	5	0.18	0.46	0.40	0.36

Table 7.6 Percentage of correct responses and grade level differences for items aligned to the thermal energy sub-ideas

Item context	Middle school	High school	χ^2	Sig.
Items in which thermal energy and temperature vary and weight and type of material are held constant	56.6 %	59.1 %	18.5	<.001
Items in which thermal energy and weight vary and temperature and type of material are held constant	35.4 %	36.8 %	5.7	<.01

31 % of the middle school students and 26 % of the high school students thought that the piece of metal that feels warm has thermal energy but the piece of metal that feels cold does not.

7.5.3 *Thermal Energy (Atomic Level)*

From the items aligned to the idea that thermal energy is associated with the disordered motions of the atoms or molecules of an object, we learned that most of the students knew that the thermal energy of an object depends on the speed of the molecules the object is made of (67 % of middle school students and 71 % of high school students). Far fewer students knew that the thermal energy is also dependent on the number and type of molecules the object is made of (34 % of middle school students and 35 % of high school students). These results are analogous to the results described above from testing students' understanding of thermal energy at the substance level. Other misconceptions that were revealed in student answer selections included the idea that thermal energy is due to atoms rubbing together, that only living things have thermal energy, and, more generally, that only warm things have thermal energy.

On one item, students were asked if all things have thermal energy and then asked why or why not. The option probability curves in Fig. 7.5 show the probability of selecting each answer choice as a function of student ability. In this item, answer choice A corresponds to the misconception that all things have thermal energy because all things are made up of atoms that are rubbing together (Wiser 1986; Kesidou and Duit 1993). Answer choice B is the correct answer corresponding to the idea that thermal energy is the result of atoms in constant motion. Answer choice C

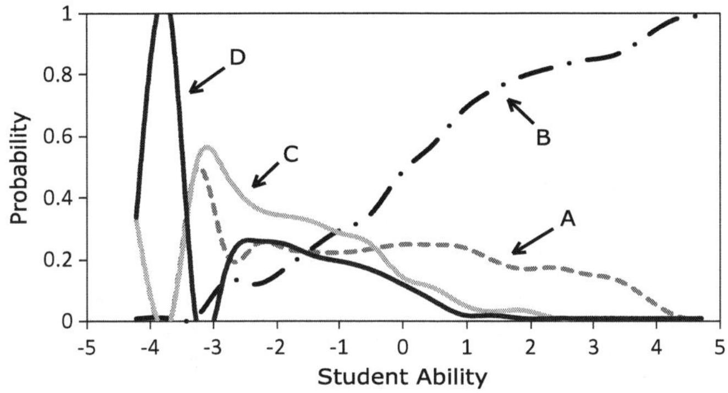

Fig. 7.5 Option probability curves for an item aligned to atomic-level thermal energy

corresponds to the misconception that only warm or hot things have thermal energy, and answer choice D corresponds to the misconception that only living things have thermal energy.

The misconceptions in C and D were popular at the lower end of the ability spectrum, and the probability of selecting these choices decreased to zero at a student ability of approximately 2. The probability of selecting answer choice A was rather constant from ability −3 to 2 and then decreased slowly. The shape of this curve indicates that the misconception that thermal energy is a result of atoms rubbing together is present over a very wide range of ability levels. This misconception most likely results from students thinking about their experiences using friction to warm things, like rubbing their hands together to warm them.

7.5.4 Gravitational Potential Energy

Results of pilot testing the gravitational potential energy items revealed that middle school students confused the phrase "gravitational energy" with the force of gravity (Herrmann-Abell and DeBoer 2009). For an item asking how the gravitational energy of a rocket changes as it gets higher in the sky, about one third of the students chose the answer stating that the gravitational energy decreases as the rocket gets higher. Their written responses confirmed that they were thinking about the force of gravity. For example, students wrote "the farther away you get from the earth, the less gravity" and "its going into space and space has no gravity." In response to a follow-up question asking what "gravitational energy" meant to them, students responded "to me it means gravity" and "it helps us stay on the ground." Students' answer choices and written comments indicated that assessment items using "gravitational energy" may not be a fair judge of middle school students' understanding of gravitational potential energy because the items do not do enough to cue the students away from "gravity" toward an energy context.

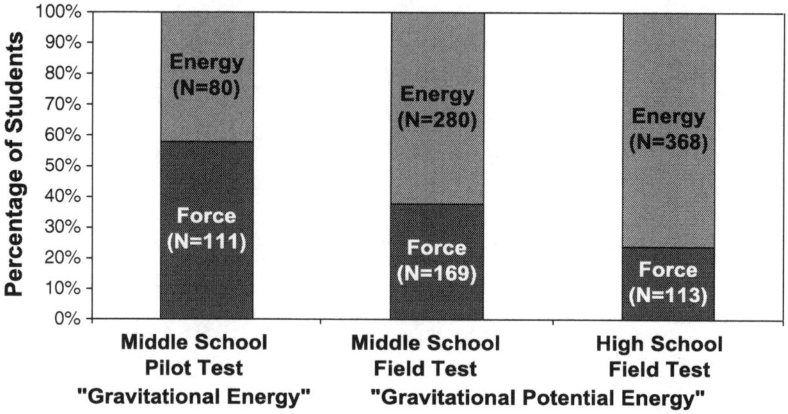

Fig. 7.6 The percentage of students coded "Force" or "Energy" for the pilot and field tests. ($\chi^2 = 74.2$, $p < .001$)

To help reduce this confusion and improve validity, the items were revised after pilot testing and the phrase "gravitational potential energy" was used in place of "gravitational energy." To test the effect of this change, we examined the pattern of responses to estimate how many students were still responding to the items in terms of the force of gravity. We expected that students who were thinking about the force of gravity would select the answer choice that said the gravitational potential energy decreases as the distance increases (thinking that the gravitational force between two objects decreases as the distance between the objects increases) or the answer choice that said that gravitational potential energy remains the same as the distance increases (thinking that the force of gravity is constant near the surface of the earth). Additionally, we expected that students who were thinking about the force of gravity would respond correctly to items that required the knowledge that gravitational potential energy increases as the object's mass increases because the force of gravity does increase as mass increases.

To compare response patterns from the pilot and field tests, we coded students by their answer choice selections on the set of items. Those students who chose both (1) answer choices that said the gravitational potential energy decreases or stays the same as the distance increases and (2) answer choices that said the gravitational potential energy increases as mass increases were coded "Force." Students who responded to the majority of the questions correctly (i.e., those students who were answering the items using a correct understanding of gravitational potential energy) were coded "Energy." Students who did not follow either of these patterns were not coded.

The bar graph in Fig. 7.6 shows the percentage of students who were coded either "Force" or "Energy" for both the pilot test and field test. The field test data were separated by grade level. In the pilot test, where the phrase "gravitational energy" was used in the items, more students' answers were coded "Force" than were coded

Table 7.7 Item difficulties for items aligned to the gravitational potential energy sub-ideas

Item context	No. of items	Difficulty			
		Min.	Max.	Median	Mean
Items in which gravitational potential energy and height vary and weight is held constant	10	−0.74	0.52	0.14	−0.04
Items in which gravitational potential energy and weight vary and height is held constant	5	−0.32	−0.15	−0.19	−0.23

"Energy." In the field test, where the phrase "gravitational potential energy" was used in the items, more middle school students' answers were coded "Energy" than were coded "Force." This suggests that the phrase "gravitational potential energy" did indeed help more students correctly think in terms of energy. With regard to the high school students who participated in the field test, we expected that they would be less likely to confuse gravitational potential energy with the force of gravity because they would have had more instruction on gravitational potential energy than the middle school students. The results shown in Fig. 7.6 support this, as indicated by the fact that more high school students' answers were coded "Energy" than middle school students. All of these results give us confidence that the field test items provide a better measure of students' knowledge of gravitational potential energy than the pilot test items do.

As we noted earlier, for the motion energy and thermal energy ideas, students tended to be less familiar with the effect of mass on energy than the effect of motion or temperature on energy. This was not the case for items related to gravitational potential energy (see Table 7.7). Items in which the gravitational potential energy and weight/mass varied and the height above the ground was held constant were not more difficult than items in which the gravitational potential energy and height above the ground varied and the weight/mass was held constant [$t(13) = 0.875$, $p > .05$]. This is probably because it is just as reasonable to think that the energy of an object held above the ground is a function of its weight as it is to think that it is a function of its height, especially if the student is visualizing the object's energy in terms of the damage it can do when it is dropped. Apparently, it is less obvious to students that thermal energy and motion energy are also associated with mass.

As with thermal energy and motion energy, the most difficult item in the set of items aligned to gravitational potential energy was one in which the gravitational potential energy was held constant and both the height and weight/mass varied (difficulty = 0.99) but, as before, this is as much a function of the sophistication of the mental processing required as it is about the science content being assessed. It is difficult for both middle and high school students to make the mental manipulation that requires them to determine that an object that is higher above the ground must weigh less than an object closer to the ground if the two objects have the same gravitational potential energy. Perhaps it is not surprising that reasoning about an inverse relationship, such as the relationship between height and weight when gravitational potential energy is constant, is more difficult than reasoning about a

Table 7.8 Percentage of correct responses and grade level differences for items aligned to the gravitational potential energy sub-ideas

Item context	Middle school	High school	χ^2	Sig.
Items in which gravitational potential energy and height vary and weight is held constant	39.9 %	49.2 %	487	<.001
Items in which gravitational potential energy and weight vary and height is held constant	46.9 %	50.7 %	31	<.001
Item in which height and weight vary and gravitational potential energy is held constant	23.2 %	29.5 %	38	<.001

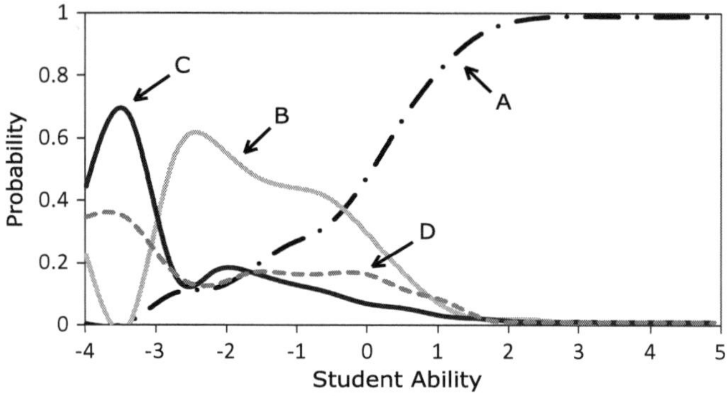

Fig. 7.7 Option probability curves for an item aligned to gravitational potential energy

direct relationship, such as the relationship between gravitational potential energy and height when weight is constant. This also points to the fact that in addition to the science knowledge that is being targeted, each item asks students to reason with that knowledge, which also affects the likelihood of a correct answer.

Table 7.8 shows the percentage of correct responses for each of the categories by grade level. As with motion energy and thermal energy, the high school students outperformed the middle school students in all of the categories.

The field test revealed a number of misconceptions that students held about gravitational potential energy. About 32 % of the middle school students and 26 % of the high school students thought that gravitational potential energy depends on the speed of a moving object. The trend this misconception follows across ability levels is shown in the option probability curves in Fig. 7.7. In the item analyzed there, the students were shown a diagram of a coconut falling from a palm tree and landing on the ground. The students were asked when the coconut has the most gravitational potential energy and why. Answer choice A is the correct answer and explains that the coconut has the most gravitational potential energy before it falls off the tree because that is when the coconut is at the highest point. Answer choice B corresponds to the misconception that the coconut has the most gravitational potential energy while it is falling because gravitational potential energy depends on the speed. Answer choice C says that the gravitational potential energy increases as the coconut gets closer to the ground and answer choice D says that the gravitational potential energy of the coconut is the same all the time because gravitational potential energy depends only on the mass.

Students of low ability (−4 to −3) were most likely to choose answer choice C (gravitational potential energy increases as distance decreases) most likely because they were thinking about the force of gravity as discussed above. Students with abilities ranging from −3 to a little less than 0 had a high probability of selecting the misconception that gravitational potential energy depends on the speed (answer choice B). The probability of this misconception decreased steadily to 0 at an ability level of 2. The correct answer A became the most probable answer choice selected at an ability of 0. The misconception that gravitational potential energy depends only on mass was the second most probable answer choice for students of low ability (less than −3) but it was not the most probable at any ability level.

Another misconception the students had was that gravitational potential energy depends on how likely an object is to fall (16 % of middle school students, 12 % of high school students). These students thought that the gravitational potential energy of a rock resting on a flat surface at the top of a cliff depends on how close it is to the edge of the cliff. This misconception could be related to the misconception that potential energy is the potential to start moving, which was present in 17 % of the middle school students and 16 % of the high school students. Similar misconceptions have been previously cited in the literature. For example, one study found that some students believe that an object on a table has much less gravitational potential energy than an identical object at the same height but not supported by the table because the object not supported by the table has the potential to fall (Loverude 2004). Other studies have shown that students believe that potential energy is the potential to have energy in the future, not now (Stead 1980; Summers and Kruger 1993). For example, a student in Stead's study said "you could use all that water as for hydro-electricity, you know, make hydroelectricity, so it could be used for energy. Potential energy is not energy at all but it could be converted to energy – you could get energy out of it."

7.5.5 *Elastic Energy*

Students were also tested on the idea that elastic potential energy is associated with how much an elastic object is stretched or compressed and with how easily it can be stretched or compressed. Items that tested the idea that when comparing two identical stretched objects the one that is stretched more has more elastic energy (difficulty = −0.87), and the idea that when comparing two identical compressed objects the one that is compressed more has more elastic energy (difficulty = −0.65) were easier than an item that addressed the knowledge that when stretching two elastic objects the one that is harder to stretch has more elastic energy (difficulty = −0.25). When asked directly what elastic energy depends on, 38 % of the middle school students and 32 % of the high school students thought that elastic energy does not depend on how difficult it is to stretch or compress an object.

The most difficult item in this set was one that compared the elastic energy of two springs that were not being stretched or compressed at all (difficulty = 1.43). About 42 % of the middle school students and 38 % of the high school students chose the answer that stated that the longer spring had more elastic energy. These students may be thinking that elastic potential energy is a property of an un-stretched object rather than of an object that has been stretched. Written comments from an earlier pilot test indicated that this was the case (Herrmann-Abell and DeBoer 2009).

7.5.6 *Energy Transformation*

Items aligned to the idea of energy transformation revealed that a little under half of the students knew that energy can be transformed within a system. The percentage of correct responses to the items aligned to this idea was 42 % for the middle school students and 52 % for the high school students, and the mean Rasch difficulty for the 23 items targeting this idea was −0.06. Only a few students selected answer choices that explicitly stated that energy cannot be transformed (14 % middle school and 12 % high school).

Student feedback obtained during the pilot test stage of item development provided evidence that some of the difficulties students had with the energy transformation items can be attributed to a lack of knowledge about the individual forms of energy (Herrmann-Abell and DeBoer 2010). In order to be successful on items testing the transformation of one form of energy into another, students must have an understanding of the individual forms of energy and be able to detect changes in those forms of energy. For example, in order to describe the energy transformations involved when a ball falls to the floor, a student has to know that the motion energy of the ball is increasing because motion energy depends on speed and the speed of the ball increases as it falls. The student also has to know that the gravitational potential energy of the ball is decreasing as it falls because gravitational potential energy depends on the height above the earth, and the height of the ball is decreasing as it falls. Misunderstanding motion energy and gravitational potential energy would cause a student to respond incorrectly to this energy transformation item. For example, one student chose the incorrect answer that said the motion energy is transformed into gravitational potential energy while the ball *falls* and wrote:

> It makes sense the motion energy will decrease & gravitational will increase because it's falling. It will change from one type of energy to the next. The gravitational energy wouldn't decrease until point 3 to 4 [after the ball bounces and is traveling back up into the air].

This student thought that gravitational potential energy decreases when the distance between the floor and the ball increases. This incorrect idea about a form of energy led him to incorrectly identify the energy transformation involved when the ball falls.

Table 7.9 Percentage of correct responses and grade level differences for items aligned to ideas about energy transfer

Type of energy transfer	Middle school	High school	χ^2	Sig.
Electrical	52 %	56 %	30.9	<.001
Radiation	43 %	49 %	174.3	<.001
Conduction	39 %	50 %	918.7	<.001
Mechanical	39 %	42 %	18.7	<.001

7.5.7 Energy Transfer

Students' knowledge of four different ways energy can be transferred from one system to another was also tested. The mean Rasch difficulty for the 43 items targeting the idea of energy transfer was −0.17. Table 7.9 compares the percentage of correct responses from the field test for middle and high school students broken down by type of energy transfer. As with the other topics, the high school students performed better than the middle school students. The largest difference between the grade levels was on items testing students' understanding of conduction (11 percentage points). The smallest gain between grade levels was on items testing ideas about mechanical energy transfer (3 percentage points).

Energy Transferred Mechanically A closer look at the items aligned to ideas about mechanical energy transfer revealed two difficulties that many students had. First, many students did not know that in order for energy to be transferred mechanically there must be a change in position (the push or pull that is required for a mechanical transfer must act over a distance). About 32 % of middle school students and 38 % of high school students selected the answer that states that energy is transferred mechanically whenever one object pushes or pulls on another object even if the objects do not move.

Second, the misconception that both energy and a force are transferred during a mechanical interaction was widespread at all of the grade levels tested (32 % middle school and 43 % high school). Support for this is shown in the option probability curves for one of the items targeting this misconception (see Fig. 7.8). In this item, a black marble rolls across the floor and hits a white marble. After the collision, the black marble stops rolling and the white marble starts to roll. The students were asked what is transferred during this interaction. Answer choice A says that a force is transferred. Answer choice B says that motion energy (kinetic energy) is transferred. Answer choice C says that both a force and motion energy are transferred, and answer choice D says that neither are transferred. As shown in Fig. 7.8, students of low ability (less than −1) were most likely to choose that a force is transferred. The probability of selecting that both a force and energy are transferred (answer choice C) is significant over a wide range of abilities (from −2.5 to 3.5), although students with abilities of approximately 1 and higher were more likely to select the correct answer than this misconception.

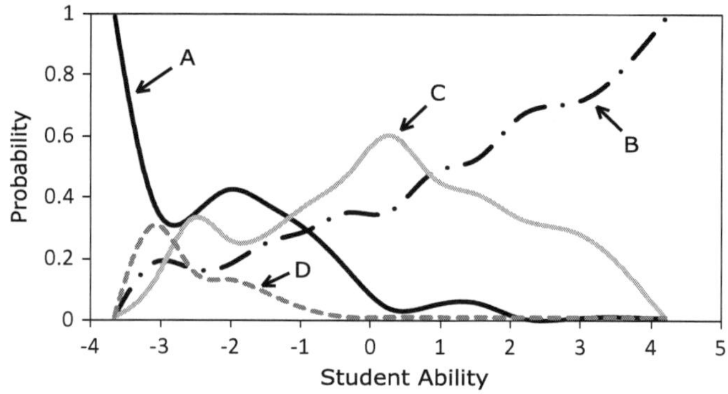

Fig. 7.8 Option probability curves for an item aligned to energy transferred mechanically

The misconception that an object has a force within it, or that a force becomes part of an object when it is thrown or hit, has been documented in previous studies (Fischbein et al. 1989; McCloskey 1983). In our assessment work on the topic of force and motion, we found that middle school students chose this impetus misconception 67 % of the time (AAAS 2013). A similar misconception was revealed by an energy transformation item, where 47 % of the middle school students and 45 % of the high school students thought that the motion energy of a book that has been shoved across a table is transformed into both a force and thermal energy.

Transferring Energy by Radiation Students also had difficulty with the idea that all objects transfer energy by means of electromagnetic radiation, whether the object is in contact with another object or not. Distractors that were aligned to the misconception that only objects that are glowing radiate energy were selected 23 % of the time by the middle school students and 21 % of the time by the high school students. Perhaps this is due to students' linking radiation to visible light and not considering radiation that cannot be detected by the eye. Additionally, some students thought that the objects must be in contact in order to transfer energy by radiation. For one item that asked students to recognize a statement of the general principle of energy transfer by electromagnetic radiation, 29 % of the middle school students and 24 % of the high school students selected the incorrect answer that stated that objects give off energy in the form of electromagnetic radiation at any temperature, but they must be touching each other in order for the energy to be transferred between objects. The idea that energy can be transferred by radiation to objects not in contact was especially difficult for students in an item that asked them to describe what happens to energy in the context of hot food cooling on a counter. The most popular answer choice for this item was that energy was transferred only to the things the food is touching, like the air and the counter, and not to things the food is not touching, like the kitchen walls (40 % of middle school students and 45 % of high school students). In this real-world example of hot food, students seemed to be focused more on conduction than on radiation.

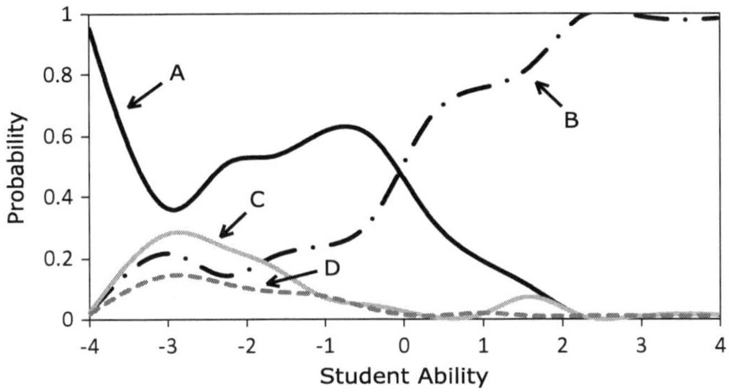

Fig. 7.9 Option probability curves for an item aligned to energy transferred by conduction

Transferring Energy by Conduction With respect to conduction, students performed well and showed the most growth from middle school to high school of all of the energy transfer ideas tested (see Table 7.9). Nevertheless, students at both levels still held several misconceptions. Prior research has shown that one of the most common misconceptions students hold about conduction is that when a cold and a warm object are placed in contact with each other, the warm object gets colder and the cold object gets warmer because "coldness" is transferred from one object to the other (Brook et al. 1984; Clough and Driver 1985; Newell and Ross 1996). We tested the prevalence of this misconception in several items. Overall, the misconception was chosen 31 % of the time by middle school students and 23 % of the time by high school students. The misconception was particularly strong in situations involving frozen objects. For example, Fig. 7.9 shows the option probability curves for an item involving an ice pack and a warm can of juice in a lunch bag. Answer choice A, which corresponds to the "coldness" misconception, was the most popular distractor at all ability levels. Students with an ability level over zero were more likely to choose the correct answer that stated that thermal energy was transferred from the can of juice to the ice pack, but the probability of selecting answer choice A didn't reach zero until an ability level of around 2. Middle school students selected answer choice A 59 % of the time, and high school students selected the misconception 46 % of the time. The other distractors C and D stated that the can of juice got cold because lunch bags are used to keep food cold and that no energy was transferred, respectively.

Transferring Energy Electrically Over half of the students knew that energy is transferred electrically from an electrical source to an electrical device only when the electrical circuit is complete. The most popular distractors were ones that said that energy could still be transferred when the circuit was not complete. For example, 26 % of the middle school students and 27 % of the high school students thought that a power plant would transfer some energy to a lamp in a house even when the lamp is off and the circuit is not complete. Results from the pilot test indicated that these students thought that the lamp, like most modern-day electrical devices, uses some energy at all times. Students wrote "a little bit of energy is used while any electronic is plugged in," "although the lamp uses very little power, as long as it is connected to the socket, power flows," and "that is why you must unplug your electrical units when you don't need them."

7.5.8 Conservation of Energy

Items aligned to the conservation of energy idea had a mean Rasch difficulty of 0.70 and were the most difficult items in the set. The percentage of correct responses to the items aligned to this idea was 28 % for middle school students and 37 % for high school students. These results are consistent with previous research that has indicated that learners do not fully grasp conservation of energy until very late in the developmental progression (Liu and Collard 2005; Liu and McKeough 2005; Neumann et al. 2012). However, on an item that involved identifying a statement of the general principle of conservation of energy, 42 % of middle school and 56 % of high school students answered correctly. These results suggest that the problem lies primarily in not being able to apply the principle of conservation to real-world events. Although many students can recognize a correct statement of the principle of conservation of energy, they are less able to draw upon that basic idea to analyze specific situations where it applies. For example, the most difficult conservation items required students to use the idea of conservation of energy to predict the speed of objects. On one of these items, students were asked to predict (in an idealized environment) the speed of a ball after it goes over a hill on a track in which there is no energy transferred between the track and the ball or between the ball and the air around it. The item required students to recognize that because energy is conserved, and because the heights of the track before and after the hill are equal, the ball must be traveling at the same speed. On three items of this kind, the percent correct was 13 % for middle school students and 19 % for high school students. The additional cognitive load involved in interpreting the scenario, realizing that the conservation principle needs to be used, accepting a situation in which there is no friction, and drawing logical inferences made these items much more difficult than simply recognizing the truth of a general statement about conservation of energy. Similar results have been reported by Chabalengula et al. (2012). Most of the students in their study knew the principle of conservation of energy but were unable to apply it to biological contexts.

Past research has also shown that it is common for students to think that energy can be created or destroyed (Brook and Driver 1984; Kesidou and Duit 1993; Kruger 1990; Loverude 2004; Papadouris et al. 2008; Stead 1980; Trumper 1998). In our sample, distractors involving the creation of energy were chosen 30 % of the time by middle school students and 25 % of the time by high school students. Distractors involving the destruction of energy were chosen 23 % of the time by middle school students and 20 % of the time by high school students.

7.6 Implications for Instruction

The findings from this work do more than simply confirm much of the existing research on students' understanding of energy, or provide a set of assessment items precisely aligned to core ideas about energy. It is our expectation and hope that these findings can be used to inform changes to instruction about energy. To cite just a few examples, our results show how important it is to make clear to students, particularly at the middle school level, the differences between gravitational potential energy and gravity as a force. This can also serve as an opportunity to discuss the differences between forces and energy in general. We also suggest that to help avoid confusion when talking about energy, use of the full phrase "gravitational potential energy" may help to cue students to the idea of the potential energy that a raised object has rather than to the force between two objects. In our study, we found convincing evidence that making this distinction decreased the likelihood that students would think about the force of gravity when asked about energy.

We also found that many students may know general energy principles but not be able to apply the principles to new contexts. This disconnect between teaching scientific principles separate from their application to real-world phenomena is perhaps one of the most significant failings of science education today. To correct this problem, we recommend that energy instruction make the application of general principles to real-world phenomena explicit. Instruction about energy should include a variety of phenomena from different science disciplines so that students have experience applying their energy knowledge in physical, life, and earth science contexts. And students should be asked to explain those specific events of the world in terms of the general principles that apply.

Our work also shows that many students are not aware of the factors that the different forms of energy depend on. This is particularly important in the cases of motion energy and thermal energy, where students are not aware of the mass component. We suggest that instruction include activities that allow the students to examine the effect that changing different variables has on the amounts of the different forms of energy. This is a particularly good opportunity to be explicit about the fact that the energy of an object depends on how much mass the object has, whether it is thermal energy, motion energy, or gravitational potential energy. Some example curriculum units that focus students on discovering the factors or indicators of the different forms of energy are *How can I use trash to power my stereo?* and *Why do some things stop and others continue going?* (Fortus et al. 2005, 2012).

7.7 Moving Forward

Although our work has focused on learning goals from the AAAS *Benchmarks for Science Literacy* (AAAS 1993) and to some extent from the *National Science Education Standards* (National Research Council [NRC] 1996), there is also considerable overlap with the National Research Council's *A Framework for K-12*

Science Education (2012) and the *Energy Literacy Framework* (U.S. Department of Energy 2012). For example, both frameworks include ideas about energy transfer, conservation of energy, and energy degradation. (We have recently started developing assessments aligned to ideas about energy dissipation and degradation.) Places where the overlap is not as strong include ideas about force fields, power, and engineering. As we move forward with our own work on the development of an instrument to test students' understanding of energy concepts from elementary grades through high school, we will consider incorporating a number of these additional ideas recommended in the *Framework for K-12 Science Education* and in the *Energy Literacy Framework*.

7.8 Conclusions

Even after years of instruction, many students have a very limited and unsophisticated understanding of the formal conventions for thinking and talking about energy. For example, students may know that motion energy is related to motion and that thermal energy is related to temperature, but they are unaware that both are also related to mass. They often think that potential energy is related to the potential to have energy; they confuse force and energy, as in the case of the force of gravity and gravitational potential energy; and they say that it is "coldness" that is transferred between two objects at different temperatures rather than thermal energy.

Given the widespread significance of these fundamental ideas about energy, it is critical that students understand them and are able to apply that knowledge in a variety of different contexts. It is also critical that we are able to effectively assess students' difficulties with these energy concepts as well as find out what they do understand. This chapter details our use of precisely aligned, distractor-driven assessment items to investigate students' understanding of energy and the misconceptions and alternative ideas they have that impede their learning. Overall, we found that students' understanding of energy ideas improves steadily from 7th grade to 12th grade. This continuous improvement is most likely attributable to a combination of the formal instruction they are receiving about energy in school, the out-of-school experiences they are having, and their increased ability to interpret the scenarios they are presented with based on their overall cognitive maturation. Certainly, it is encouraging to see growth, but the growth is modest, and there is no point where the improvement begins to accelerate, where students finally "get it." In fact, we found that the high school students held many of the same misconceptions that the middle school students held. Option probability curves revealed that some misconceptions quickly decrease with increasing ability levels while others were popular over wide ranges of abilities. This knowledge of which misconceptions are the most prevalent for which groups of students can be valuable for improving energy education. Additionally, because the items developed for this study were designed to be aligned to key energy ideas but not to any single curriculum or instructional strategy, researchers and curriculum developers will be able to use the items, along with the field test results described in this chapter, to inform the development and test the effectiveness of various interventions that they use. The items, clarification statements, misconceptions, and field test results are available on the AAAS Project 2061 Science Assessment website (http://assessment.aaas.org/).

Acknowledgements The research reported here was supported by the National Science Foundation, through Grants # ESI 0227557 and ESI 0352473, and the Institute of Education Sciences, U.S. Department of Education, through Grant R305A120138 to the American Association for the Advancement of Science. The opinions expressed are those of the authors and do not represent views of the Institute, the U.S. Department of Education, or the National Science Foundation.

References

American Association for the Advancement of Science. (1993). *Benchmarks for science literacy*. New York: Oxford University Press.

American Association for the Advancement of Science. (2001). *Atlas of science literacy, volume 1*. Washington, DC: AAAS/NSTA.

American Association for the Advancement of Science. (2007). *Atlas of science literacy, volume 2*. Washington, DC: AAAS/NSTA.

American Association for the Advancement of Science. (2013). AAAS Project 2061 Science Assessment website. Retrieved October 31, 2012. Available from http://assessment.aaas.org/

Bond, T. G., & Fox, C. M. (2007). *Applying the Rasch model: Fundamental measurement in the human sciences* (2nd ed.). Mahwah: Lawrence Erlbaum Associates.

Brook, A., & Driver, R. (1984). *Aspects of secondary students' understanding of energy: Full report*. Leeds: The University of Leeds, Centre for Studies in Science Education and Mathematics Education.

Brook, A., Briggs, H., Bell, B., & Driver, R. (1984). *Aspects of secondary students' understanding of heat: Full report*. Leeds: The University of Leeds, Centre for Studies in Science Education and Mathematics Education.

Chabalengula, V. M., Sanders, M., & Frackson, M. (2012). Diagnosing students' understanding of energy and its related concepts in biological context. *International Journal of Science and Mathematics Education, 10*(2), 241–266.

Clough, E., & Driver, R. (1985). Secondary students' conceptions of the conduction of heat: Bringing together scientific and personal views. *Physics Education, 20*(4), 176–182.

DeBoer, G. E., Herrmann-Abell, C. F., & Gogos, A. (2007, March–April). *Assessment linked to science learning goals: Probing student thinking during item development*. Paper presented at the National Association for Research in Science Teaching annual conference, New Orleans, LA.

DeBoer, G. E., Herrmann-Abell, C. F., Gogos, A., Michiels, A., Regan, T., & Wilson, P. (2008a). Assessment linked to science learning goals: Probing student thinking through assessment. In J. Coffey, R. Douglas, & C. Stearns (Eds.), *Assessing student learning: Perspectives from research and practice* (pp. 231–252). Arlington: NSTA Press.

DeBoer, G. E., Lee, H. S., & Husic, F. (2008b). Assessing integrated understanding of science. In Y. Kali, M. C. Linn, & J. E. Roseman (Eds.), *Coherent science education: Implications for curriculum, instruction, and policy* (pp. 153–182). New York: Columbia University Teachers College Press.

Finegold, M., & Trumper, R. (1989). Categorizing pupils' explanatory frameworks in energy as a means to the development of a teaching approach. *Research in Science Education, 19*(1), 97–110.

Fischbein, E., Stavy, R., & Ma-Naim, H. (1989). The psychological structure of naive impetus conceptions. *International Journal of Science Education, 11*(3), 327–336.

Fortus, D., Krajcik, J. S., Nordine, J. C., Plummer, J., Rogat, A., & Switzer, A. C. (2005). *How can I use trash to power my stereo?* Ann Arbor: University of Michigan Center for Highly Interactive Classrooms, Curricula, & Computing in Education.

Fortus, D., Weizman, A., Nordine, J., Jin, H., & Abdel-Kareem, H. (2012). Why do some things stop and others keep going? In J. Krajcik, B. Reiser, D. Fortus, L. Sutherland, & D. Fortus (Eds.), *Investigating and questioning our world through science and technology (IQWST)*. Norwalk: Sangari Active Science.

Herrmann-Abell, C. F., & DeBoer, G. E. (2009, April). *Using content-aligned assessment to probe middle school students' understanding of ideas about energy*. Paper presented at the National Association for Research in Science Teaching annual conference, Garden Grove, CA.

Herrmann-Abell, C. F., & DeBoer, G. E. (2010, March). *Probing middle and high school students' understanding of energy transformation, energy transfer, and conservation of energy using content-aligned assessment items.* Paper presented at the National Association for Research in Science Teaching annual conference, Philadelphia, PA.

Kesidou, S., & Duit, R. (1993). Students' conceptions of the second law of thermodynamics – An interpretive study. *Journal of Research in Science Teaching, 30*(1), 85–106.

Kruger, C. (1990). Some primary teachers' ideas about energy. *Physics Education, 25*(2), 86–91.

Kruger, C., Palacino, D., & Summers, M. (1992). Surveys of English primary teachers' conceptions of force, energy, and materials. *Science Education, 76*(4), 339–351.

Lee, H. S., & Liu, O. L. (2010). Assessing learning progression of energy concepts across middle school grades: The knowledge integration perspective. *Science Education, 94*(4), 665–688.

Leggett, M. (2003). Lessons that non-scientists can teach us about the concept of energy: A human-centered approach. *Physics Education, 38*(2), 130–134.

Linacre, J. M. (2012). Winsteps® (Version 3.75.0) [Computer Software]. Beaverton: Winsteps.com. Retrieved from http://www.winsteps.com/

Liu, X., & Boone, W. J. (2006). Introduction to Rasch measurement in science education. In X. Liu & W. J. Boone (Eds.), *Applications of Rasch measurement in science education* (pp. 1–22). Maple Grove: JAM Press.

Liu, X., & Collard, S. (2005). Using the Rasch model to validate stages of understanding the energy concept. *Journal of Applied Measurement, 6*(2), 224–241.

Liu, X., & McKeough, A. (2005). Developmental growth in students' concept of energy: Analysis of selected items from the TIMSS database. *Journal of Research in Science Teaching, 42*(5), 493–517.

Loverude, M. E. (2004, August). *Student understanding of gravitational potential energy and the motion of bodies in a gravitational field.* Paper presented at the Physics Education Research conference, California State University, Sacramento, CA.

McCloskey, M. (1983). Intuitive physics. *Scientific American, 248*(4), 122–130.

National Research Council. (1996). *National science education standards.* Washington, DC: National Academy Press.

National Research Council. (2012). *A framework for K-12 science education: Practices, crosscutting concepts, and core ideas*. Washington, DC: National Academies Press.

Neumann, K., Viering, T., Boone, W. J., & Fischer, H. E. (2012). Towards a learning progression of energy. *Journal of Research in Science Teaching, 50*(2), 162–188.

Newell, A., & Ross, K. (1996). Children's conception of thermal conduction – Or the story of a woolen hat. *School Science Review, 78*(282), 33–38.

Nicholls, G., & Ogborn, J. (1993). Dimensions of children's conceptions of energy. *International Journal of Science Education, 15*(1), 73–81.

Papadouris, N., Constantinou, C. P., & Kyratsi, T. (2008). Students' use of the energy model to account for changes in physical systems. *Journal of Research in Science Teaching, 45*(4), 444–469.

Rasch, G. (1960). *Probabilistic models for some intelligence and attainment tests.* Chicago: University of Chicago Press.

Sadler, P. M. (1998). Psychometric models of student conceptions in science: Reconciling qualitative studies and distractor-driven assessment instruments. *Journal of Research in Science Teaching, 35*(3), 265–296.

Solomon, J. (1983). Messy, contradictory and obstinately persistent: A study of children's out-of-school ideas about energy. *School Science Review, 65*(231), 225–229.

Stead, B. (1980). Energy (Working Paper No. 17). In *Learning in science project*. Hamilton: Science Education Research Unit, University of Waikato.

Stern, L., & Ahlgren, A. (2002). Analysis of students' assessments in middle school curriculum materials: Aiming precisely at benchmarks and standards. *Journal of Research in Science Teaching, 39*(9), 889–910.

Summers, M., & Kruger, C. (1993). *Long term impact of a new approach to teacher education for primary science*. Paper presented at the annual meeting of the British Educational Research Association, Liverpool, England.

Trumper, R. (1990). Being constructive: An alternative approach to the teaching of the energy concept – Part one. *International Journal of Science Education, 12*(4), 343–354.

Trumper, R. (1993). Children's energy concepts: A cross-age study. *International Journal of Science Education, 15*(2), 139–148.

Trumper, R. (1998). A longitudinal study of physics students' conceptions of energy in preservice training for high school teachers. *Journal of Science Education and Technology, 7*(4), 311–318.

Trumper, R., & Gorsky, P. (1993). Learning about energy: The influence of alternative frameworks, cognitive levels, and closed-mindedness. *Journal of Research in Science Teaching, 30*(7), 637–648.

U.S. Department of Energy. (2012). *Energy literacy: Essential principles and fundamental concepts for energy education*. Washington, DC: Author.

Watts, D. M. (1983). Some alternative views of energy. *Physics Education, 18*(5), 213–217.

Wiser, M. (1986). *The differentiation of heat and temperature: An evaluation of the effect of microcomputer teaching on students' misconceptions* (Technical Report). Cambridge, MA: Harvard Graduate School of Education.

Wright, B. D., & Stone, M. H. (1999). *Measurement essentials*. Wilmington: Wide Range.

Wright, B. D., & Stone, M. H. (2004). *Making measures*. Chicago: Phaneron Press.

Chapter 8
Mapping Energy in the Boston Public Schools Curriculum

Robert F. Chen, Allison Scheff, Erica Fields, Pamela Pelletier, and Russell Faux

8.1 Introduction

To understand how the natural world works, people require a deep understanding of the fundamental concepts of science, a broad base of scientific observations, and a multi-dimensional and coherent approach to learning. Yet science curricula, course materials, university science departments, and college majors are predominantly organized into single scientific disciplines. Teachers of science, whether at the K-12 or university levels, normally view their curricula from a disciplinary perspective; thus students of science learn the fundamental concepts of science constrained within disciplinary contexts and often cannot transcend these boundaries to gain a deep appreciation of science as a whole. Teachers and students, especially at the intermediate level (grades 3–8), are re-teaching and re-learning the same concepts in isolated, disciplinary silos (Kali et al. 2008). These conceptual disconnects do not allow teachers to take advantage of the innate curiosity and real-world connections

R.F. Chen (✉)
Environmental, Earth and Ocean Sciences, University of Massachusetts Boston, Boston, MA, USA
e-mail: Bob.Chen@umb.edu

A. Scheff
Massachusetts Department of Higher Education, Boston, MA, USA
e-mail: ascheff@bhe.mass.edu

E. Fields
Educational Development Center, Waltham, MA, USA

P. Pelletier
Boston Public Schools, Boston, MA, USA

R. Faux
Davis Square Research Associates, Somerville, MA, USA

R.F. Chen et al. (eds.), *Teaching and Learning of Energy in K-12 Education*,
DOI 10.1007/978-3-319-05017-1_8,

that children at this age experience daily (Smith et al. 2006; Duschl et al. 2007). It is possible that this segregated way of teaching and the resulting disconnect from real-world examples contributes to the well-documented drop-off of student interest in science during the intermediate years (Osborne et al. 2003; Logan and Skamp 2005). New integrated approaches to learning and teaching have been developed to counteract this disciplinary segregation (Bardeen 2000; Nordine et al. 2011; Gentile et al. 2012).

Interestingly, huge gains in science and technology have recently resulted from a new emphasis on "interdisciplinary" or "cross-disciplinary" research (e.g. molecular biophysics, biogeochemistry) by funding agencies and innovative graduate programs (e.g. NSF "Cross-Cutting Programs", 2009). We hypothesize that similar gains in science learning can be made for teachers and their students from a new emphasis on the foundational concepts that straddle all science disciplines in addition to emphasizing concepts within disciplinary silos (Hough 1994; Gentile et al. 2012; NGSS 2013). Various curricula can be taught more efficiently and effectively when having to teach the fundamental concepts only once, with examples being drawn from a variety of contexts (Steele 2011; Klein et al. 2011). As the Next Generation Science Standards roll out seven cross-cutting concepts in addition to the disciplinary core ideas and science and engineering practices, teachers and school districts will need to find ways to address teaching and learning of these cross-cutting concepts.

Energy is a cross-cutting concept that can be used to integrate all scientific disciplines. However, most curricula are not organized, taught, or learned from this integrated energy perspective. As science becomes increasingly interdisciplinary, K-8 science teachers often teach all scientific disciplines, and many scientific concepts are taught repeatedly in different curricula, we hypothesize that using an integrated energy approach allows for more efficient learning of science as a whole. Here, we take the approach of utilizing existing teacher expertise and knowledge of their existing curriculum to identify energy connections across discipline and grade level. By explicitly making energy connections between curriculum units, teachers can bridge across the disciplines using cross-cutting concepts that are mandated by the Next Generation Science Standards. By identifying the most obvious connections, teachers help visualize the connections within the curricula, and help the movement from teaching four distinct scientific disciplines to teaching science as a single discipline.

> Crosscutting concepts have application across all domains of science. As such, they are a way of linking the different domains of science. They include: Patterns, similarity, and diversity; Cause and effect; Scale, proportion and quantity; Systems and system models; **Energy** and matter; Structure and function; Stability and change. The Framework emphasizes that these concepts need to be made explicit for students because they provide an organizational schema for interrelating knowledge from various science fields into a coherent and scientifically-based view of the world. (NGSS 2013; emphasis on energy applied by the authors of this chapter).

8.2 Methods

Through the development and teaching of a graduate level course, Energy I, Integrating the Sciences Through Energy, at UMass Boston by an instructional team comprised of a physicist, a biologist, a chemical oceanographer, and a middle school teacher, four energy themes emerged: Forms and Transformations, Systems, Conservation, and Resources. The Energy I course for in-service teachers was organized around these four themes, and teachers could better categorize their own lesson plans and make connections horizontally and vertically given this framework. These four themes align well with typical energy categorization: activity/work, source/form, transfer/transformation, degradation, and conservation (Liu and McKeough 2005; Nordine et al. 2011; Neumann et al. 2013) with a slight bias towards the use of energy in real-life complex systems (systems) and in society (resources) in place of physical applications (activity/work) to make energy concepts more engaging for teachers. Forms and transformations refers to the idea that there are various forms or sources of energy that can be transformed from one to another, and that energy can be transferred from one object to another. Systems refers to the idea that complex systems can be better understood by examining the energy flow through the system and the relationship this flow has to the function of the system. Conservation refers to the idea that within a given closed system, energy is conserved, but when a system is not closed, energy is often dissipated. Resources refers to the need for energy in society and provides a real-life answer to the question "Why do we care about energy?" These four sub-themes align quite nicely with the developmental sequence of activity/work, to energy source/form, to energy degradation, to energy conservation proposed by Liu and McKeough (2005) in moving from forms to transfers within a system, to behavior in a system (i.e. degradation and conservation), ending with an application of degradation of energy resources. While the Energy I course helped teachers make connections using the energy themes, an overarching map of energy connections throughout the curriculum did not emerge. Both instructors and participants thought that a map of energy connections across grade bands and science disciplines would help them make explicit connections for their students and better teach the cross-cutting concepts promoted by the Next Generation Science Standards. To meet this demand, the Energy Institute was designed to specifically examine energy in the Boston Public School curriculum.

8.2.1 Boston Public Schools (BPS) Curriculum

The Boston Public Schools (BPS) district is an urban district with about 60,000 K-12 students with a significant amount of movement among schools from year-to-year making a uniform curriculum important. BPS has adopted district-wide curricula composed of Full Option Science System (FOSS) or Science and Technology for Children (STC) "kits" in grades 1–8, and Active Physics (9th), Living by Chemistry (10th), Biology-A Human Approach (11th) and Environmental Science (12th grade) in high school. For the following mapping exercise, we have focused on individual "investigations". Each elementary and middle school grade level is comprised of 2–4 curricula (FOSS or STC kits) each composed of 3–23 individual investigations (here called "curriculum units"). For example, the grade 7 curriculum/FOSS kit, Diversity of Life, is comprised of 10 "investigations" (curriculum units): What is life?; Introduction to the microscope; Microscopic life; The cell; Seeds of life; Transpiration; Plant reproduction; Snails; Roaches; and Kingdoms of Life. In high school, yearlong curricula were composed of 3–7 curricula (sometimes chapters or sections of a single text), each comprised of 3–16 curriculum units. Breaking down the curriculum in this way, a total of 306 curriculum units are taught to each student progressing through grades 1–12 in BPS.

8.2.2 *The Energy Institute*

In order to identify the major energy connections in the BPS curriculum, one teacher leader from each grade (1–12) was selected to participate in the Energy Institute. Over the course of 3 days, teachers were tasked to represent the BPS curriculum of a single grade (grades 1–12), were challenged to identify curriculum units within their grade that exhibited energy concepts prominently, and then were instructed to highlight the connections between these units. Creating an energy map for the BPS included three phases: Identifying curriculum units that contained energy concepts, identifying energy connections between curriculum units, and mapping the energy connections using social network analysis. The first two phases occurred during the 3-day Energy Institute; the third phase followed using the generated data for analysis.

8.2.2.1 Identifying Units Containing Energy

On Day 1 of the Energy Institute, each teacher, representing the curriculum in a single grade (while teachers may teach multiple grades, they were assigned a grade and only examined the curriculum units in that single grade), examined the curriculum units that used energy concepts most prominently. Each teacher identified each unit containing energy on a single Energy Unit Form, described the unit, the energy theme (Forms and Transformations, Systems, Conservation, or Resources), the energy concept (teacher described), and a description of the activities in the unit. At the end of the day, 79 units of the possible 306 were identified as having energy in them. Each unit had attributes of grade level, energy theme, activity description, and energy concept.

8.2.2.2 Identifying Energy Connections

On Days 2 and 3, teachers had the opportunity to examine all 79 units identified as having energy in them, and to identify connections to the units at their own grade level. For each connection, they filled out an Energy Connection Form that included the unit that was connected to the original unit and a description of the connection. For example, in Grade 5, students make a web of living and non-living things in an aquarium that they maintain in the *Ecosystems-5: Observing the Completed Aquarium* unit. The 8th grade teacher identified the connection through the idea that *energy flows through an ecosystem* with the *Populations and Ecosystems-5: Finding the Energy* unit in 8th grade where students use colored beads to model the transfer of energy in the Mono Lake ecosystem they are studying. In this case, there is a clear connection between biology units in two different grades and two different curricula that could be enhanced if the 5th grade teacher prepares his/her students for the 8th grade activity, and the 8th grade teacher refers to the learning that occurred in 5th grade.

In another example, the idea of potential energy, that energy can be stored in a rubber band wound around the axle of a vehicle to be used later (*Motion and Design-6: Evaluating Vehicle Design: Looking at Rubber Band Energy*; 4th grade) was found to be connected to the *Human Body Systems-13: Releasing Energy from Food* (6th grade) unit where students burn a marshmallow and a sunflower seed releasing stored chemical potential energy. In this case, the energy concept represents a cross-disciplinary bridge that can be enhanced by both teachers.

In all, 293 connections were identified by the 12 teachers over 2 days. However, due to some incomplete data entry and multiple connections between the same 2 units, a final 162 unique connections between 105 unique units were identified. In order to make connections, teachers, identified an additional 26 units that contained energy concepts that they had not originally identified ($79 + 26 = 105$). While the connection identification exercise was by no means complete, teachers prioritized during the process by selecting the most obvious energy connections that they could find in their curriculum, so that the resulting connections and units most likely contain the strongest energy connections.

8.2.3 Content Network Mapping

After the 3-day Energy Institute, a 105×105 matrix of connections was created among the 105 curriculum units identified by teachers to have energy concepts that connected with other curriculum units. For this initial examination, the connections were binary rather than weighted, and the data were symmetrized to render all links reciprocal. In other words, a declared link from one unit to another would be treated as reciprocal, even in cases where the teachers did not report the links in both directions. The symmetrized data were then analyzed in Ucinet and visualized in NetDraw (Borgatti et al. 2012) for two basic node-level measures, namely, degree centrality and betweenness. Additional statistics were done using R statistics software (Dalgaard 2008).

The network analysis takes the data provided by the teachers and, using graph theoretic techniques, distributes the units over a geodesic space. The degree centrality measure indicates the number of ties each unit was found to have with other units. Units with a great number of ties may command an especially important position in the overall curriculum, particularly if the ties are unique. The removal of a highly central unit (and thus the ties that the unit has with other units) is likely to fragment the network into clusters and damage the overall cohesion of the curriculum. On the other hand, the removal of a unit with few or no ties will likely have little effect on the connections between the other units. There are, of course, counter-examples to both instances. However, as such counter-examples did not appear in the BPS content network, they will not be discussed here.

Turning from the rather static notion of degree centrality, the analysis also looked at "travel" between nodes. Assuming that one can move from one unit to another in the network, what does that path look like and what sorts of conceptual "hops" are necessary to get from one unit to another? This is a rather large consideration, and, as a first step, the analysis examined the distribution of betweenness values for all nodes. The betweenness measure indicates how often a given unit will be determined to be on the shortest path between any two other units. Where there is more than one shortest path between two nodes, the assigned value is divided accordingly. Nodes with a high betweenness value serve an important link between the other units, whereas units with low betweenness values are likely to be somewhat off the main track of the curriculum. The removal of high-betweenness nodes will likely substantially alter the overall structure of the network, while the removal of low-betweenness nodes will have little effect on the structure. A node with high centrality may have a low betweenness value and vice versa, depending on the degree to which ties are redundant.

8.3 Results

Theenergy investigations and connections among them that were identified in the BPS curriculum through the Energy Institute comprise a network of nodes (curriculum units) and links (connections). This energy content network can be visualized as a "sociogram" where each curriculum unit and connection is shown, and spatial distribution of the units is determined by the software based on spreading out the units to best see the connections. In Fig. 8.1, all 105 unique units are shown with all 162 connections. Colors represent grade bands, and the size of the circle represents degree centrality. It is clear that the most central nodes are found in each grade band. Degree centrality can be used to identify the eight most influential curriculum units that connect the BPS curriculum through energy vertically (across grade bands) and horizontally (across scientific disciplines) (Table 8.1).

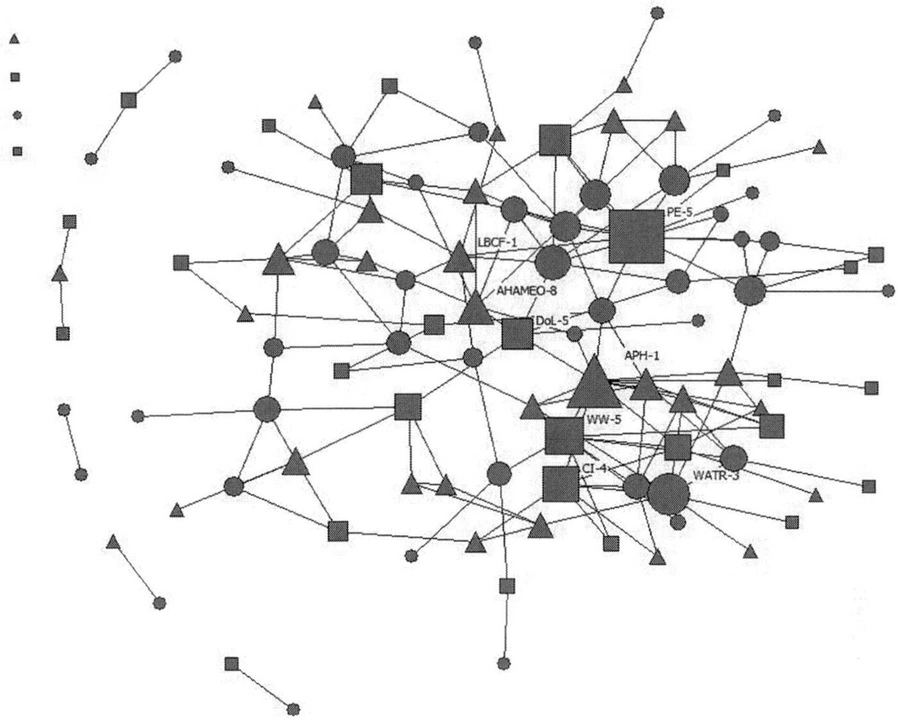

Fig. 8.1 An energy curriculum map showing connections between individual BPS curriculum units. Curriculum units are identified by grade band (*circles* – elementary school, *squares* – middle school, *triangles* – high school) and centrality (*size of shape*). The eight most connected units are labeled. A few relatively unconnected units are shown outside the main network

Of the 162 energy connections determined by teacher leaders over the course of the 3-day Energy Institute, eight units were the most connected (Fig. 8.1) as defined by having the highest degree centrality. As can be seen in Table 8.1, these curriculum units represent all disciplines, represent all four energy themes, and are distributed throughout elementary, middle, and high school grades.

The energy descriptions, themes, and concepts were determined by teachers. Energy experts might categorize curriculum units in different ways, but it is important to note that the energy map generated here (Fig. 8.1) is how teachers perceived the map, and therefore is more likely to be used by teachers. More generally, the energy concepts that were found to be critical to the energy content network in BPS were: the sun is a major source of energy on earth, energy can be stored in food or other matter, organisms need energy to grow, kinetic energy increases as matter is heated, and hot air rises. While these concepts may not be the most important energy concepts theoretically, they are the most important energy concepts that connect disparate curriculum units throughout the entire BPS

Table 8.1 The eight units with the highest degree centrality in decreasing order

Curriculum unit	Grade	Discipline	Activity description	Energy theme	Energy concept
Active physics home-1: designing the universal dwelling	9	Physics	Students build a model home and study heat transfers.	Resources	The sun is a source of solar energy and heat which can be transferred by conduction, convection, and radiation.
Populations and ecosystems-5: finding the energy	8	Biology	Students burn a cheese ball and record the temperature change of a cup of water above the flame.	Conservation	Energy stored in food can be measured. Energy is conserved as it is transformed from one form to another.
Diversity of life-5: seeds of life	7	Biology	Students germinate seeds in light and dark.	Forms and Transformations	Seeds store energy that can be used to germinate.
Living by chemistry fire-1: evidence of change	10	Chemistry	Students mix materials of different temperatures.	Forms and Transformations	Energy is conserved and tends to disperse.
Water-3: water vapor	3	Chemistry	Students observe two cups with melting ice or soaked paper towels and record evidence of phase changes.	Forms and Transformations	Energy is required for phase changes.
Biology-A human approach: matter, energy, and organization-8: the cellular basis of activity	11	Biology	Students examine photosynthesis through an experiment and cellular respiration by designing a marathoner's snack.	Forms and Transformations	Energy is stored in matter.
Weather and water-5: convection	6	Earth Science	Students layer different temperature waters in a vial and create convection by heating.	Systems	Temperature affects the density of air and water. Hot air rises. Temperature differences can transfer energy through convection.
Chemical interactions-4: kinetic energy	8	Chemistry	Students make a simple thermometer with water and observe expansion with heating.	Systems	Matter expands when kinetic energy increases.

Descriptions, themes, and concepts were determined by teachers. The 8 units in this table are among the nine units with the highest betweenness ($r^2 = 0.84$ for a correlation between degree centrality and betweenness; see Table 8.2 and Fig. 8.2)

curriculum. They are quite well-aligned with the main performance expectations of the Next Generation Science Standards, however (NGSS 2013).

The eight curriculum units with the highest number of connections (Fig. 8.1, Table 8.1) are described as examples of what students learn and do that serve as foundations for more peripheral energy learning. Students germinate seeds, burn a cheese ball, and observe water droplets condense to build an understanding of how energy can take many different forms and how energy can be transformed from one form to another. Students observe the behavior of layered water and examine what happens to matter as heat is applied to it, and thus gain an ability to study complex systems by probing their emergent properties, and to use energy to account for system function. Students use their understandings of transformations of energy and energy in systems to think quantitatively as energy is transformed from energy contained in food (e.g. cheese ball) to thermal energy in a calorimeter or from a marathoner's snack to the kinetic energy of running. Students design a model home to study heat transfer and determine which design is most energy efficient. It is interesting to note that although each energy theme occurs in all grade levels, there does appear to be a hierarchy or learning progression that mirrors the organization of the Energy I course from Forms and Transformations to Systems to Conservation to Energy Resources (Neumann et al. 2013). Forms and transformations are more often the focus of curriculum units in elementary grades, and are the key components used to understand complex systems. Given an introduction to systems, the idea of conservation can be used to examine interactions among the system components, something that is seen more often in high school curriculum units. Finally, forms and transformations, systems, and conservation are powerful ideas that can be used to address energy resources, something that is not often addressed until undergraduate courses. While it is clear that each of these themes are visited multiple times throughout a student's K-12 experience, and many if not all the curriculum units can be viewed as addressing more than one energy theme, there does appear to be a bias towards focusing on forms and transformations and systems at lower grades and conservation and resources at higher grades.

Figure 8.2 visualizes the same 162 connections among 105 curriculum units with the size of the circle representing betweenness. The nine units that are most often among the shortest path between other units are labeled in Fig. 8.2. Eight of these units also have the highest degree centrality (Table 8.1), but a new unit emerges as well (Table 8.2). Degree centrality and betweenness are significantly correlated ($r^2 = 0.84$) for this curriculum map.

Again, critical curriculum units and activities as defined by betweenness cross disciplines and grade bands. It appears essential that energy is taught in different grades and different disciplinary curricula to help students make connections across all science. In addition, it is apparent that students' learning of energy is not as hierarchical in the BPS curriculum as we would have expected, but rather appears time and time again in different disciplines and different grade levels. The composite 105 curriculum units that were identified by teachers to focus on energy have multiple connections that can be explicitly stated by teachers to address the cross-cutting theme of energy in the Next Generation Science Standards (NGSS 2013).

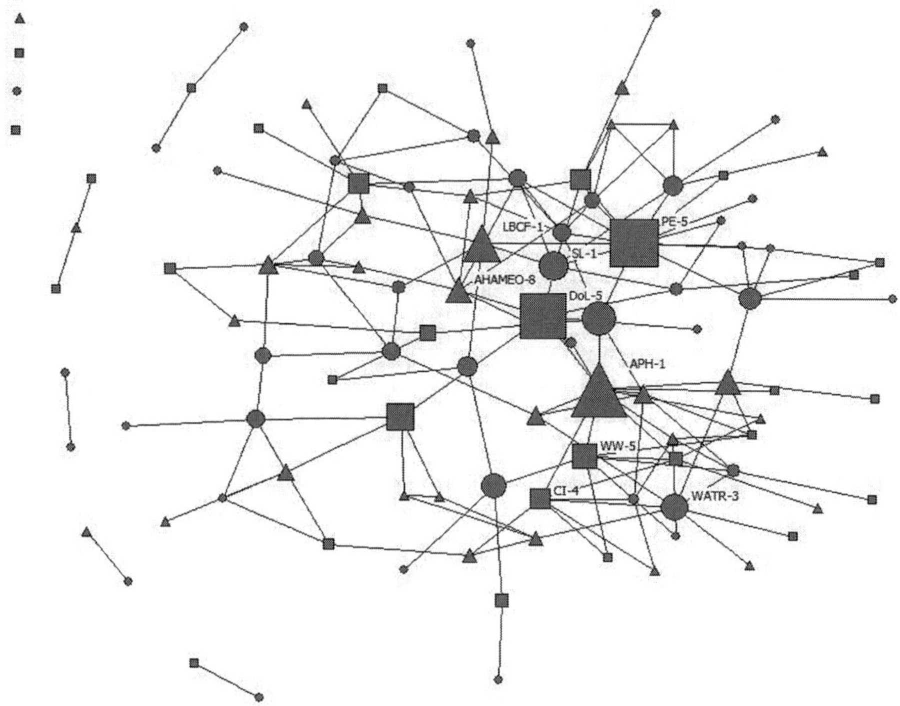

Fig. 8.2 An energy curriculum map showing connections between individual BPS curriculum units. Curriculum units are identified by grade band (*circles* – elementary school, *squares* – middle school, *triangles* – high school) and betweenness (*size of shape*). The nine curriculum units with the highest betweenness are labeled. A few relatively unconnected units are shown outside the main network

Table 8.2 In addition to six units in Table 8.1, this unit is among the seven units with the highest betweenness values

Curriculum unit	Grade	Discipline	Activity description	Energy theme	Energy concept
Structures of life-1: observing and describing two solids	1	Biology	Students observe a seed which has been soaked in water, which then begins to germinate.	Forms and transformations	Organisms need energy to grow.

When examining the energy content network to determine which grade band is most critical to learning about energy, there are no statistically significant differences in degree centrality between grade bands of curriculum unitsidentified

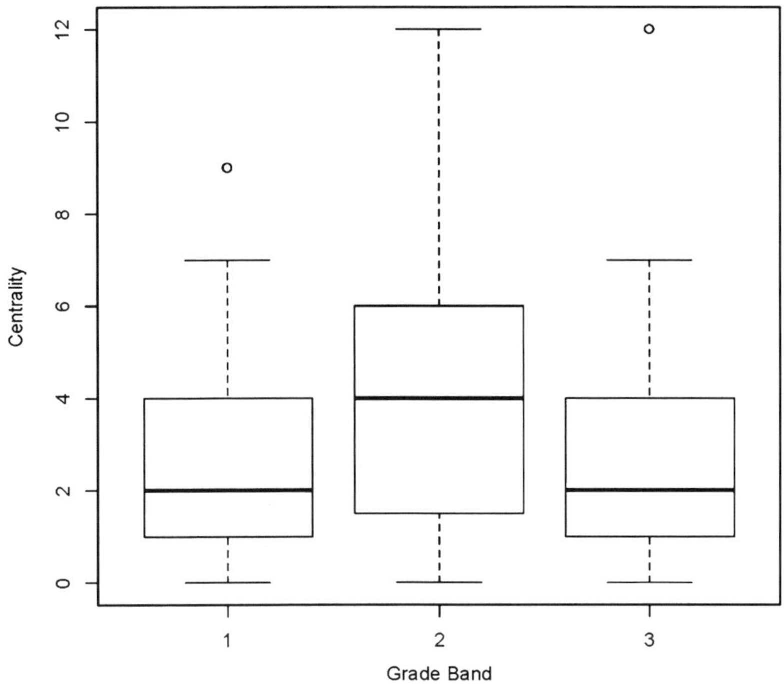

Fig. 8.3 Degree centrality by grade band. (1) Grades 1–5; (2) grades 6–8; (3) grades 9–12. Centrality is shown as a boxplot (median is *thick line*, 25th–75th percentile is shown in the *box*, entire range is bounded by *error bars*, and outliers are small *open circles*

by participating teachers (Fig. 8.3). All grade bands are equally well-connected. While this demonstrates the need for teachers of all grades to make explicit energy connections to support the learning of energy concepts, it appears that the higher mean and variance in centrality in the middle school band suggest that teaching energy in middle school is an essential link between elementary and high school grade bands.

Figure 8.4 shows the betweenness of the individual units by grade band. It is again clear that there is no difference between the importance of any grade band over any other in terms of the connectedness of this curriculum map. This demonstrates the critical nature of teaching energy consistently at all grade bands.

Figure 8.5 shows degree centrality by discipline. All disciplines are equally connected statistically. This demonstrates that energy connections are seen in all disciplines and energy connections need to be taught with equal emphasis in all four science disciplines to support learning the energy cross-cutting concept (NGSS 2013).

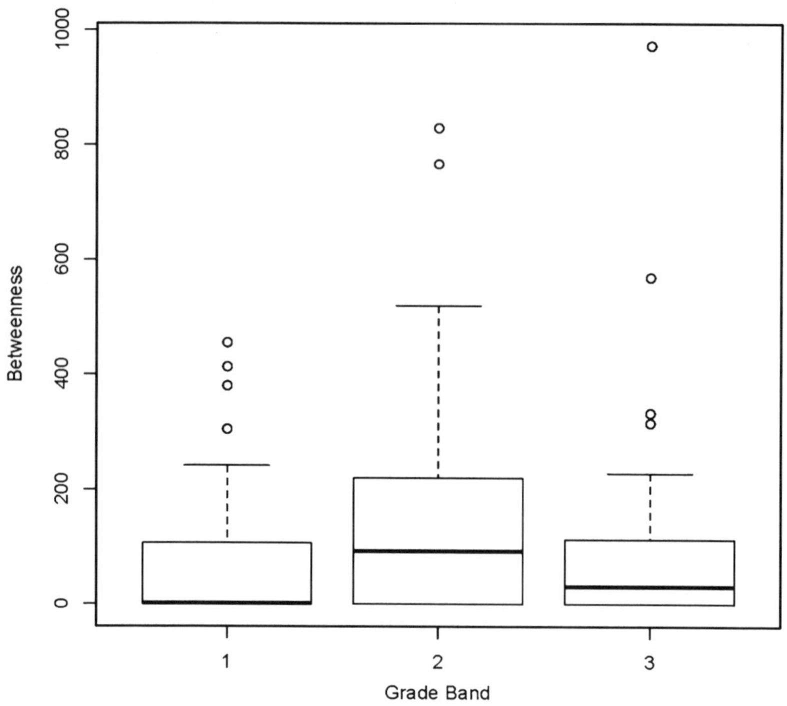

Fig. 8.4 Betweenness by grade band. (1) Grades 1–5; (2) grades 6–8; (3) grades 9–12. Betweenness is shown as a boxplot (median is *thick line*, 25th–75th percentile is shown in the *box*, entire range is bounded by *error bars*, and outliers are small *open circles*

8.4 Discussion

The Energy Institute had multiple outcomes, including the identification of major energy concepts and how they are taught in BPS as well as positive impacts on teachers, an initial energy map for the BPS curriculum, an analysis of what energy curriculum was most significant to learning across the grade bands and scientific disciplines, and identification of challenges for wide-spread application of the mapping techniques.

8.4.1 Major Energy Concepts in BPS

As seen in Fig. 8.1, there are many energy concepts taught throughout the BPS curriculum. However, the energy concepts in the most connected curriculum units have been identified by teachers leaders participating in the Energy Institute to be: (1) The Sun provides heat energy that can be transferred by conduction, convection, and radiation, (2) Energy is required for phase changes, (3) Matter expands when

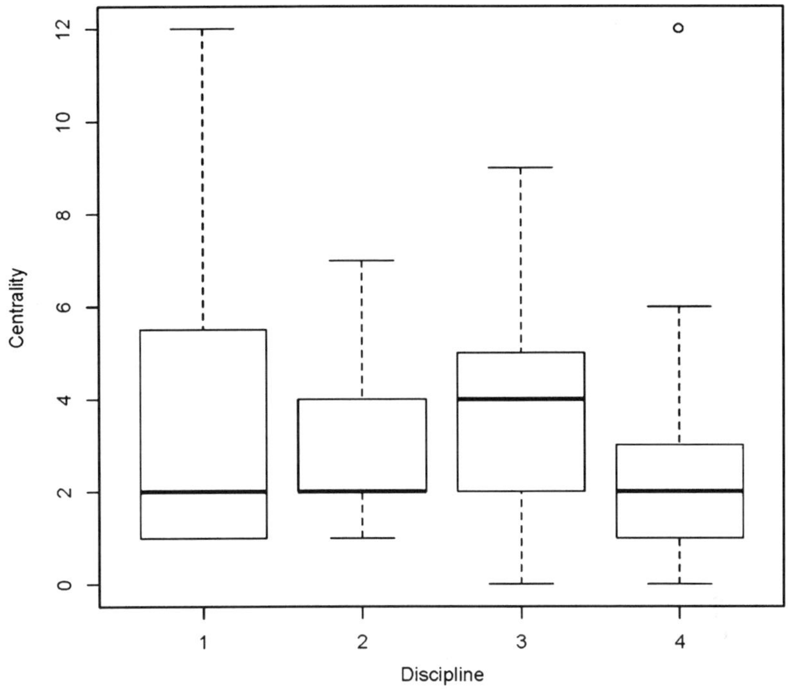

Fig. 8.5 Degree centrality by discipline (*left* to *right*: (1) biology; (2) chemistry; (3) Earth science; (4) physics)

kinetic energy increases, (4) Differential heating can distribute energy through convection, (5) Energy is stored in food, (6) Energy is conserved as it is transformed from one form to another, and (7) Organisms need energy to grow. Taken together, these concepts represent three of the four energy themes: Forms and Transformations, Conservation, and Systems. The curriculum mapping approach identifies specific activities and contextualized examples of these energy themes already within the curriculum, so in a sense allows the overall conceptual understanding about energy (what students need to know about energy) to be taught through identified investigations. Specifically, all BPS students experience activities that address these concepts, and these experiences are well-connected to other learning experiences throughout the BPS curriculum so that taken together the identified curriculum units lead to a graduating senior's understanding of energy.

The BPS energy concept map agrees well with the Next Generation Science Standards. In the NGSS, the first three energy themes are well-distributed through the Life Science, Earth and Space Science, and Physical Science strands. In general, that energy can be transferred from one object to another is introduced in elementary grades; that energy can take different forms, that it flows in and out of systems, that it drives the cycling of matter, and that it can be tracked as it flows through systems are all to be taught at the middle school grades; and that energy flow can be modeled and that energy is conserved is reserved until high school. Thus the progression of energy themes in the BPS curriculum and the NGSS are similar. The only topic explicitly related to energy resources is introduced when nuclear reactions are considered in high school physical science. This fourth energy theme, Energy Resources, is of societal importance and is probably best taught mostly at the undergraduate level.

The Energy Institute generated a concept map that showed that middle school grades are critical to linking energy curriculum units, that teaching energy is distributed among all scientific disciplines, and that there is a progression of the four energy themes from forms and transformations to systems to conservation to resources. These findings are entirely consistent with what is included in the NGSS.

8.4.2 The Energy Institute

While Energy Institute facilitators including a scientist, an educational policy expert, and the BPS Director of Science helped elicit energy connections and concepts among the participating teachers, it should be clear that the Energy Institute was an exercise by teachers for teachers. By using teachers' perceptions, knowledge, and experience to construct the energy map rather than external expertise, the map is more likely to be used by teachers within their district curriculum. It may be that each district creates a slightly different map that is more useful to them than a single expert-derived conceptual map that does not originate from a district's curriculum and teachers. The impact on teachers was evaluated by an external evaluator/observer, the energy map was created by the teachers, and the important energy concepts and future challenges were identified by the Energy Institute facilitators.

Overall, the 3-day Energy Institute was a success. By gathering 12 teachers, focusing on a single cross-cutting theme, and supplying instructions to first identify curriculum units that featured energy concepts and then seek connections between these units, an energy content map could be created for the BPS curriculum. Major outcomes of this professional development event were (1) Teachers' increased appreciation for science teaching at other grades, (2) Teachers' increased appreciation for the interconnectedness of the BPS curriculum through energy, and (3) Teachers' increased content knowledge about energy through interactions with other teachers while focusing on this topic. Teachers at all grade levels became more aware of the content that was being taught throughout the district's curriculum. Teachers in the upper grades were especially surprised by the amount of science to which younger students in the district were exposed. They began to see how the work done in earlier grades could get lost if teachers were not making connections to students' prior and future energy learning. It became clear that by identifying connections to other grade levels and science experiences, teachers would be more able to reinforce what their students had already learned and to prepare them for future science classes, essentially making both teaching and learning more efficient and effective.

Teachers at the Institute also became more aware of the extent of energy in the BPS curriculum. While teachers at all grade levels had certainly taught energy-related concepts before, the exercise of identifying the energy connections throughout the curriculum made it very apparent that this cross-cutting concept did indeed span the entire curriculum. They began to recognize that not only would the connections allow them to teach energy concepts more effectively, but the interconnectedness of the entire curriculum would become illuminated for their students and would unify the curriculum as a whole.

The process of identifying energy concepts within the BPS curriculum, along with identifying the connections throughout, exposed teachers' knowledge not only of the curriculum but of the content as well. Some teachers were easily able to identify the energy concepts and put them into words, while others struggled with the content and the language as well. During the initial exercise, the variations in language used to describe the same phenomena emerged as an obvious challenge. Teachers in high school realized that if they connected scientific terms to language used by teachers in earlier grades, students would not only be able to connect new lessons to previous lessons, but they also might be less intimidated by more advanced scientific terminology. Similarly, if elementary teachers introduced their students to scientific terms, the connections in later grades would feel more seamless. As a result, the language used in the concept map for both the energy concepts and the connection descriptions is basic enough for teachers at all grade levels to understand them.

This mapping approach to determining the most important energy concepts that students learn is a bit different than most theoretical or scientist-driven approaches. While scientists and energy educators will discuss, debate, and finally determine which energy concepts are most critical and what the learning progression may be, the curriculum mapping approach studies what is actually taught in a district, and identifies where the most critical energy learning and connections might take place. The advantage of this approach is that curriculum is already developed, adopted, and taught, so that only connections need to be made to enhance learning about cross-cutting concepts. New energy-centric curricula and adoption of new curricula are difficult, and at the very least will take several years to develop and implement. Of course, the disadvantage is that as energy learning progressions, assessments of energy understandings, and research on how students learn energy are developed, the energy curriculum map may change, and it may be discovered that the existing curriculum is not designed in the most effective way to teach energy.

8.4.3 Challenges

Several major challenges remain for this energy mapping in the curriculum approach. First, scientists, energy education specialists, and energy educational researchers had little influence on this process. There may be better learning experiences and better learning progressions for energy than currently exist in the BPS-adopted science curriculum. It would be interesting to analyze a variety of other curricula, compare these to the current BPS energy map, and to examine the differences between curriculum maps and energy learning progressions. In fact, these seven concepts and many more that are identified in the 105 units containing energy are reasonably well-aligned with the Department of Energy's Essential Principles of Energy Literacy (DOE 2012). Nonetheless, it might be necessary to add to the current list of most important, connected energy curriculum units based on energy education research.

While this process has led to some recommendations about energy in the BPS curriculum, each district's curriculum is different, so it might be necessary to conduct an Energy Institute in every district. Additionally, many school districts do not have a uniform curriculum, so individual teachers or schools could rely on state or national science standards. The questions resulting from this research with the BPS science curriculum are: How replicable is the energy map given a different set of 12 teachers? How do we translate what we have learned here for implementation in other districts? Can the energy map be generalized to national standards and benchmarks or the Next Generation Science Standards, rather than BPS curriculum units? And, if we can create such a generalized map, how would this translate into teacher professional development in different districts?

Finally, it should be noted that this energy map (Fig. 8.1) was created based on individual teacher's work over 3 days. If one teacher was particularly better at making energy connections or clear in their identification of energy in their grade level, a single teacher's influence could be seen in the map and the identification of key energy concepts. The Energy Institute would have to be repeated with different teachers several times to make the resulting map more robust.

8.4.4 Future Steps

Some time at the end of the Energy Institute was used to ask teachers what the most useful products from this activity would be for their teaching in the classroom. Overall, teachers were looking for a clearly-defined, easily-understood set of connections that they could make during critical lessons they taught. So, in addition to the distribution of the overall energy map to understand where each teacher's curriculum units fit into the Grade 1–12 curriculum, the Energy Institute facilitators plan to develop laminated energy connection cards for each curriculum unit that contains energy. So, as teachers prepare to teach the upcoming unit, they will be able to read these cards whose content was developed at the Energy Institute of what to mention while teaching the unit that would refer back to experiences and learning that students had previously and some preparation for what students would learn in the future. These connections in the form of cards included in the FOSS and STC kits would help to enhance implicit connections in the curriculum, and make them explicit.

In conjunction with the development of the connections cards, teacher professional development on energy and using energy connection cards would have to be rolled out throughout the district. Based on the energy content network, professional development for science teachers of all grades bands and all disciplines is necessary for supporting the learning of energy as a cross-cutting concept (Figs. 8.3, 8.4, and 8.5).

Additionally, the energy curriculum map needs to be replicated within BPS, replicated in other districts, and reviewed by energy education experts to be refined and made more generalizable. This refinement, review, and dissemination would increase the reach of this work, validate the most valuable energy connections within curricula, and allow teachers to effectively address part of the Matter and Energy cross-cutting concept that is included in the NGSS.

While the Energy Institute was deemed valuable for teachers, the impact on student understanding of energy, understanding of big ideas in science more broadly, and achievement on standardized tests was not examined by this work. An observation protocol has been developed to examine impacts of professional development around energy on students in the classroom, and this research is underway (Levy, 2013, personal communication).

Finally, there are seven new NGSS cross-cutting themes. To teach these all effectively, instruction must evolve from teaching individual curriculum kits and disciplinary units to increasing teachers' ability to make connections across all science and to supporting these cross-disciplinary concepts with cross-cutting supplements like the proposed energy connection cards. Institutes can be designed for each cross-cutting theme, and teachers can be better prepared to make connections among big ideas in science in an effort to teach science more efficiently and effectively.

8.5 Conclusions

This study relied on the input of teacher leaders with expertise in teaching the prescribed BPS curriculum in each grade. By tapping these teachers' knowledge, an energy map was created that showed nodes (curriculum units including energy concepts) and their connections (how teachers could connect one unit to another across discipline and grade level). Social Network Analysis techniques were applied to this energy content network that identified the most connected units (highest degree centrality) as well as the grade bands, disciplines, and energy themes that were most critical to the overall energy learning network for students. Overall energy connections pervaded all grades and all disciplines equally. Further, this study offers an apparent ordering of the four energy themes (1) Forms and Transformations, (2) Systems, (3) Conservation, and (4) Resources, that is consistent with the Next Generation Science Standards. The Energy Institute and Energy Content Mapping strategies should lead to a capacity for districts to integrate cross-cutting concepts into existing curriculum relatively quickly.

References

Bardeen, K. (2000). *Improving student science literacy through an inquiry-based, integrated science curriculum and review of science media* (81 pp). MA Project, St. Xavier University.

Borgatti, S. P., Everettt, M. G., & Freeman, L. C. (2012). *Ucinet for Windows: Software for social network analysis*. Harvard: Analytic Technologies.

Dalgaard, P. (2008). *Introductory statistics with R* (2nd ed., 364 pp). New York: Springer.

Department of Energy (DOE). (2012). *Energy literacy: Fundamental concepts and essential principals for energy education*. http://eere.energy.gov/education/energy_literacy.html

Duschl, R. A., Schweingruber, H. A., & Shouse, A. W. (Eds.). (2007). *Taking science to schools: Learning and teaching science in grades K-8*. Washington, DC: National Academy Press.

Gentile, L., Caudill, L., Fetea, M., Hill, A., Hake, K., Lawson, B., Lipan, O., Kerckhove, M., Parish, C., Stenger, K., & Szajda, D. (2012). Challenging disciplinary boundaries in the first year: A new introductory integrated science course for STEM majors. *Journal of College Science Teaching, 41*, 44–50.

Hough, D. L. (1994, April 4–8). *PATTERNS: A study of the effects of integrated curricula on young adolescent problem-solving ability*. New Orleans: AERA.

Kali, Y., Linn, M. C., & Roseman, J. E. (2008). *Designing coherent science education: Implications for curriculum, instruction, and policy*. New York: Teachers College Columbia University.

Klein, S., Armillas-Tiseyra, M., & Kuerbis, A. (2011). Listening to students: Building bridges. *The Magazine of Higher Learning, 43*, 57–58.

Liu, X., & McKeough, A. (2005). Developmental growth in students' concept of energy: Analysis of selected items from the TIMSS database. *Journal of Research in Science Teaching, 42*, 493–517.

Logan, M., & Skamp, K. (2005). Students' interest in science across the middle-school years. *Teaching Science, 51*(4), 8–15.

Neumann, K., Viering, T., Boone, W. J., & Fischer, H. E. (2013). Towards a learning progression of energy. *Journal of Research in Science Teaching, 50*, 162–188.

Next Generation Science Standards (NGSS). (2013). For States, by States. http://www.nextgenscience.org/next-generation-science-standards

Nordine, J., Krajcik, J., & Fortus, D. (2011). Transforming energy instruction in middle school to support integrated understanding and future learning. *Science Education, 95*, 670–699.

Osborne, J., Simon, S., & Collins, S. (2003). Attitudes towards science: A review of the literature and its implications. *International Journal of Science Education, 25*(9), 1049–1079.

Smith, C. L., Wiser, M., Anderson, C. W., & Krajcik, J. (2006). Implications of research on children's learning for standards and assessment: A proposed learning progression for matter and the atomic molecular theory. *Measurement: Interdisciplinary Research and Perspectives, 4*(1&2), 1–98.

Steele, B. (2011). *Cross-disciplinary collaboration between two-science disciplines at a community college*. Ed.D. dissertation, Arizona State University.

Part III
Challenges Associated with the Teaching and Learning of Energy

How many times have you heard a child—or an adult—say that sleep gives them energy? This common reference highlights one of the most common challenges—the way we use everyday language—associated with understanding the energy concept. In this part, five chapters highlight some common challenges associated with teaching and learning of energy. The major challenge that faces the energy education community is how to reconcile common, everyday language, experience, and understanding of energy with the scientific concept of energy.

Energy is an abstract and important idea. It has been important since humans first discovered fire through today as countries negotiate for more energy resources. As we have struggled to understand energy, we have tried to develop a language to describe it both in everyday usage and for scientific discourse. While language can be helpful in creating an accepted definition of energy, we have continued to struggle to find acceptable language that spans energy in everyday life and the scientific view of energy. This presents a true challenge for the teaching and learning of energy: is it best to start with everyday phenomena, scientific language, or common usage? These five chapters explore challenges associated with each of these strategies.

Regardless of which strategy is chosen, to lead students to a rich, scientific understanding of energy, we need a learning progression or framework to be articulated so that we are purposeful in what we teach about energy. It needs to be developmentally appropriate and builds upon itself. The challenge here is that for a student to have a coherent understanding of energy requires the learning progression to be cross-disciplinary, meaning that we should have one learning progression for energy and not one for energy in biology and a second one for energy in physics. To make the situation even more complex and challenging, a true understanding of energy as a concept should span not only the four major areas of science (biology, chemistry, physics, earth science), but it should also incorporate the socio-economic and political understandings of energy. Each of these chapters in this part will address some of the challenges that we have outlined here.

In the first chapter in this part, Jin and Wei compare the scientific uses of energy with informal uses of energy to explicitly uncover differences that might act as a barrier to students learning energy concepts. They first examine how energy concepts were historically developed to explain fire and life drawing a parallel between how the scientific community made sense of the world and how students construct their understandings of energy. Then they examine dictionary definitions of energy to construct various views of energy in everyday usage. They then use these two studies to construct an energy learning progression that begins in students' personal and emotional associations with energy and ends in a scientific understanding of the energy concept. The challenges for learning energy uncovered in their chapter focus on language (informal vs. scientific uses of energy), the separation of the concepts of matter and energy, and the increased restriction of uses for energy as students construct a scientific understanding of energy.

Liu and Park also argue that energy should be taught from a larger perspective than just the sciences. These two authors highlight three challenges associated with the teaching and learning of energy: the tendency to define energy before it is developmentally appropriate for a child, the multidisciplinary nature of energy, and the relationship between knowledge and changes in behavior. The central theme of this chapter is that energy should not be taught only within science classes because energy is a concept that has impacted history and culture, as well as technology. The exploration of an energy learning progression identifies that one must have a good foundational understanding of energy before learning a formal definition. Often introducing a definition too early in a child's schooling, such as energy is the ability to do work, has helped to promote misconceptions instead of assist in the mastering of the concept. The challenge is to build a strong progression that promotes understanding and the connection of ideas that make up the energy concept rather than submit to the pressures of teaching students too early. Further, human behavior and choices are often made in search of energy needs and resources. The challenges present for us as educators in educating for an informed citizenry is how to structure learning when energy is an all-encompassing concept to learn. Energy should be taught through many different lenses and subjects, such as biology and history, but it will take a multidisciplinary team take on this charge. Continually, we must ensure that connections are made between the subject areas with regards to energy so that, while energy is being taught, knowledge is not being put into silos. Lastly, Liu and Park remind us that an increase in knowledge, in this case about energy as a resource, does not necessarily correlate to making decisions that are better for society as a whole. As educators, we know how to increase content knowledge, but we are not trained to change the affect of a student.

Millar continues to discuss the tension between every day and scientific understanding in his discussion of some of the challenges of teaching and learning energy. Because energy is an abstract idea that cannot be defined, the language of energy is used loosely in everyday discourse, and the idea of energy is used in different ways in different scientific disciplines, there have been many debates among science educators about how to teach energy ideas. Teachers often try to bridge between the everyday discourse and the scientific ways of thinking about energy, and while teachers are sometimes criticized for not being scientifically accurate, Millar also suggests that teaching must start with students' understanding and colloquial use of energy as a starting point moving them towards a more scientific understanding as students become scientifically literate. Based on his analysis of the predominant literature, Millar suggests a teaching sequence starting at age 9 that moves from traditional discussions of forms and transformation of energy to a more accurate framework of energy stores and pathways of energy transfer. He claims that conservation of energy is really the key, but that energy dissipation must also be taught to maintain consistency with students' prior understanding of energy. Finally, he stresses that the energy education community should agree upon common assessment of energy learning outcomes based on student responses to specific age appropriate questions or tasks.

Papadouris and Constantinou get to the very crux of the matter when they posit that it is the very nature of energy that makes it difficult to teach. Defining energy is difficult because it is not a tangible property or thing. Instead it is a framework that can be used to analyze behavior in systems across all phenomena in all sciences. Instead of providing an operational definition to determine what is energy, why is it needed and how do we use it, the authors suggest pursuing a philosophical approach to science so that as new understandings about energy are learned, they are incorporated into a larger theoretical framework. Lastly, Papadouris and Constantinou call for a learning progression of energy and recognize the challenges associated with this.

In the final chapter in this part, Vigeant et al., expand on the challenges associated with teaching undergraduate engineering students. They find that there are significant misconceptions that are difficult to overcome, especially in four areas: The second Law of Thermodynamics, internal energy vs. enthalpy, temperature vs. energy, and rate vs. amount. Further, they use a heat and energy concept inventory to suggest that even though engineering students can compute traditional test answers and get good grades in their undergraduate courses, fundamental conceptual understanding is more difficult and represents a barrier in their development from novice to expert. They implement two 20 min activities to address each of these misconceptions in ten institutions of higher education and find that in each case, these inquiry-based activities show increased understanding on their concept inventory assessment given pre/post in the course. Challenges remain to convince university instructors to implement these activities in a variety of classroom settings and translating these activities to age appropriate learning at the K-12 levels.

This part of the book elucidates a set of challenges and proposes some strategies to address them. A comparison of what students need to know (first part), and what the research says (second part) with strategies (last part) to address the major challenges (this part) will help the reader get a full sense of the state-of-the-art in the teaching and learning of energy.

Chapter 9
Using Ideas from the History of Science and Linguistics to Develop a Learning Progression for Energy in Socio-ecological Systems

Hui Jin and Xin Wei

9.1 Introduction

Energy is identified as a crosscutting concept in *A Framework for K-12 Science Education* because it can serve as a unifying and organizational framework for students to connect knowledge from the various disciplines into a coherent and scientifically-based view of the world (National Research Council [NRC] 2011). A thorough understanding of energy is fundamental to all science disciplines and therefore essential to scientific literacy. As anthropogenic carbon emission is becoming the major contributor to the global climate change (Intergovernmental Panel on Climate Change [IPCC] 2007), using a scientific view of energy to understand human's impact on climate has also become a critical component of environmental literacy (NGSS Consortium of Lead States 2013).

Energy, as a concept central to scientific literacy and environmental literacy, has been emphasized in science standards across all grade levels for many years (e.g., NRC 1996). However, current teaching in schools tends to focus on quantitative calculation and does not prepare students to apply knowledge of energy in real-life situations (Nordine et al. 2011). Empirical studies have documented many intuitive energy conceptions students hold. For example, both students and their teachers tend to confuse energy with force and power (Trumper 1998; Watts 1983) as well as effort (Driver and Warrington 1985). In addition, they often think that energy only exists in the bodies of living things (Gilbert and Watts 1983; Watts 1983) or when motion is involved (Trumper 1998). As students are learning about different forms of energy, they tend to think that different kinds of energy exist (Gilbert and Watts 1983; Kaper and Goedhart 2002; Schmid 1982). Among the different forms of energy, heat and

H. Jin (✉)
Department of Education and Human Ecology, Ohio State University, Columbus, OH, USA
e-mail: jin.249@osu.edu

X. Wei
Department of Physics, Beijing Normal University, Beijing, China

R.F. Chen et al. (eds.), *Teaching and Learning of Energy in K-12 Education*,
DOI 10.1007/978-3-319-05017-1_9,

chemical energy are the most difficult ones. Students usually see temperature as the measure of heat (Baierlein 1990; Laburu and Niaz 2002; Lewis and Linn 1994; Loverude et al. 2002). They often equate chemical energy with organic molecules (Jin and Anderson 2012a; Mohan et al. 2009). In biological contexts, students often think that energy is the vital power of living organisms (Barak et al. 1997).

Regarding energy principles, researchers found that students seldom use the energy conservation principle to solve problems about mechanical systems (Driver 1994; Duit 1984) or to explain biological and chemical events (Barak et al. 1997; Boo 1998). They usually do not recognize the connections between energy conservation and energy degradation; therefore, students see these two principles as contradictory (Duit 1984; Pinto et al. 2005). They seldom recognize heat dissipation in food chains and often use matter-energy conversion to reason about biological processes (Jin and Anderson 2012a; Lin and Hu 2003; Mohan et al. 2009).

As elaborated above, students encounter tremendous difficulties in learning about energy. Why is the concept of energy so difficult for students? How do students use ideas of energy to understand real-world phenomena? How can instructional approaches help students develop a coherent and sophisticated understanding of energy? This chapter explores these questions as they relate to socio-ecological systems—coupled human and natural systems (Liu et al. 2007; Long Term Ecological Research Network [LTER] 2007). In particular, we focus on environmental events that affect global climate. These events include an oak tree growing, a baby girl growing, people losing weight, a dead tree decaying, a flame burning, and a car running.

In this chapter, we first conduct a historical analysis to better understand how the scientific view of energy differs from the various views of energy that appeared earlier in the history of science. There is a parallel between conceptual change in the history of science and students' development of scientific concepts (Carey 1985). Therefore, an examination of how the concept of energy was constructed in the history of science will provide significant implications for our understanding of students' intuitive ideas of energy. Second, we conduct a linguistic analysis to examine how the scientific view of energy differs from the informal views of energy that are embedded in colloquial English. Energy is not just a scientific term. It is also a common word used in everyday language. An examination of colloquial meanings of energy will enable us to better understand common intuitive ideas that hinder student learning of the scientific view of energy. Finally, we describe how we used the ideas from the historical analysis and linguistic analysis to develop a learning progression for energy in socio-ecological systems. This chapter provides an example of how ideas about the history and nature of science and ideas from linguistics can inform learning progression research. The empirical study on how we developed the learning progression for energy in socio-ecological systems is reported in another paper of the project (Jin and Anderson 2012a).

9.2 Historical Analysis

We study how students use the energy concept to understand environmental events such as an oak tree growing, a baby girl growing, a flame burning, a car running, etc. These environmental events are about fire and life. Therefore, in the historical analysis, we specifically explore how scientists constructed the concept of energy in their inquiries into fire and life. For centuries, scientists have wrestled with essential questions about fire and life. What is it? What is the cause? Why does it happen this way? Is there anything that is conserved? Scientists' inquiry into these questions gradually differentiated the energy concept from matter and life, and eventually established energy as a universally conserved quantity.

9.2.1 *Inquiry into Fire: How Energy Was Differentiated from Matter*

Since ancient time, humans have been asking questions about fire: Why do some materials burn while others do not? What is fire? Why is it hot? Aristotle believed that flammable materials contain a "fire element". Based on this idea, the alchemists of the 1600s developed the phlogiston theory (Cobb and Goldwhite 1995). According to this theory, all flammable materials contain phlogiston, a substance that is given off in burning; the ash of the burnt material always weighs less due to the emission of phlogiston. In retrospect, it is apparent that alchemists conflated matter and energy into an undifferentiated concept of phlogiston.

About a century later, the French chemist Lavoisier challenged the phlogiston theory through experiment (Cobb and Goldwhite 1995). He demonstrated the important role of oxygen in combustion and formulated the law of mass conservation. His work laid the foundation for modern chemistry. However, one question remained unanswered: Why do some materials burn while others do not? To answer this question, Lavoisier proposed the caloric theory: Flammable materials contain caloric, a special form of matter; caloric can pass freely through the pores of dense materials and becomes manifest in explosions; because caloric is imponderable, no change in weight is observed in the reaction (Morris 1972). The caloric theory indicates the beginning of matter-energy differentiation. Unlike phlogiston, caloric has almost no weight. However, matter and energy are not completely differentiated, because caloric is a fluid or semi-matter that flows from one place to another. This view is different from the modern energy view, from which heat is an abstract quantity associated with the kinetic motion of atoms and molecules. After Mayer and Joule discovered the mechanical equivalent of heat, the caloric theory was superseded by the motion theory of heat (Coopersmith 2010). This indicates a complete differentiation between energy and matter.

9.2.2 Inquiry into Life: How Energy Was Differentiated from Life

The study of life began with several essential questions: What is life? How are living things different from non-living things? Why can living things grow and reproduce, but non-living things cannot? Again, the first theory of life is generally credited to Aristotle. According to him, all living things have "soul", and soul is the cause of life. Plants have "vegetative souls" that cause growth and reproduction. Animals have "sensitive souls" that cause not only growth and reproduction, but also motion and sensation. People have "rational souls" that enable us to do all the above and reasoning (Shanks 2001). Soul is a concept that does not differentiate biological entities from psychological ones.

In his book, *The web of life: A New Scientific Understanding of Living Systems*, Capra (1997) traces the establishment of biology as a distinct discipline and details the paradigm shift in biology. The following description is drawn from Capra's book. In the nineteenth century, the study of life bifurcated into two disciplines. Mechanistic biology studied life in terms of mechanical and chemical structures and processes, whereas psychology investigated thinking and reasoning. Modern biology was thus established and separated from psychology but at the expense of reducing life into mechanical processes. Many biologists believed that physics and chemistry were insufficient to understand life phenomena. Among them, vitalists proposed that vital power was the cause of life events such as cell reproduction. Vital power was once considered as the "energy" in biology. Organismic biologists proposed a rival theory that viewed life as an emergent property of autopoietic systems. The atoms and molecules that compose the cells do not have life. However, when they form a living network (cell), life is emerged out of the special organization of atoms, molecules, and organelles. It is also important to note that organic molecules in the cell provide energy, but they do not have "life". As emphasized by Capra, life is a pattern. In this sense, the organismic theory differentiates energy completely from life, and it differentiates both energy and life from psychological entities such as soul.

As discussed above, it took scientists hundreds of years to understand fire and life. Why are these two everyday phenomena so difficult to understand? The reason is because there always seemed to be a quantity that determined what was possible and what was impossible; this quantity was always present but it was never visible. Now we know this quantity is energy. Energy is a powerful concept to understand various environmental events. In particular, our historical analysis indicates two important aspects of the contemporary scientific view of energy. First, energy is an abstract quantity; it should be differentiated from matter, life, and psychological entities. Second, energy is about constraint rather than cause; energy is always conserved and yet degrades. This is a law that constrains our explanations of any environmental events. Similar ideas are also discussed in several other chapters in this book from different angles (See Chaps. 5, 7, and 8).

9.3 Linguistic Analysis

Many scientific terms have been adopted from colloquial languages. The scientific meanings of these terms could be very different from their vernacular senses. Ordinary words with scientific meanings are a major source of students' confusions and learning difficulties (Fang 2006). Energy is one of those tricky words. It has a specialized meaning in science, but it is used in non-scientific ways in everyday language. Therefore, understanding the colloquial meanings of the word energy will enable us to better understand students' naïve conceptions and learning difficulties.

9.3.1 Definitions of Energy in English Dictionaries

We therefore conducted an analysis of the various meanings of energy in English. Reliable sources for colloquial meanings of words are dictionaries because they contain precise, intelligible, and complete definitions of words. When creating definitions for a word, lexicographers use a variety of strategies to capture the essence of the word's meanings as well as the word's unique roles in language (McKeown 1991). These strategies include describing semantic relations among words using synonyms and antonyms, using a strictly controlled vocabulary to define all entries in the dictionary, and using example sentences and collocations (the company in which words customarily appear) to depict word meanings in linguistic contexts. Different dictionaries may use one or more of these strategies. For example, the Longman Dictionary of Contemporary English uses a strictly controlled vocabulary of 2,000 words to define all words, which leads to a greater clarity compared to other dictionaries. The Merriam-Webster Thesaurus uses synonyms and antonyms to depict the semantic boundaries precisely. Therefore, including entries from different dictionaries will allow us to achieve a valid and comprehensive interpretation of the meanings of the word energy. We chose four dictionaries as our data sources: New Oxford Dictionary (3rd edition), Merriam-Webster Thesaurus (online), Dictionary.com, and Longman Dictionary of Contemporary English (online). These dictionaries are widely used today. They also cover the major strategies that lexicographers adopt to depict word meanings.

In the analysis, we first found entries of energy in all four dictionaries. We used the thematic analysis technique (Boyatzis 1998) to analyze the data. The coding units were the definitions of the word energy in the selected dictionaries. We first read and familiarized ourselves with the definitions and generated a set of initial codes to identify important features of the definitions. Then, we used an iterative process to code the data and revise the codes. We found that the dictionaries depict the meanings of energy in terms of three categories: sources of energy, nature of energy, and causal reasoning. The coding scheme was developed based on this finding. It is presented in Table 9.1.

Table 9.1 Coding scheme

Category	Codes			
Sources of energy (SE)	People	Living things including people	Non-living things	Living and non-living things
Nature of energy (NE)	Psychological entity	Physical entity	Abstract quantity	
Causal reasoning (CR)	Cause	Constraint		

The results of the data analysis are presented in Table 9.2. Our linguistic analysis indicates five definitions of energy: (1) a person's physical or mental strength or power, (2) life energy of living things, (3) vital power of places, (4) energy sources utilized by people, and (5) the ability to do work.

9.3.2 Informal Views of Energy

In the paragraphs that follow, we describe the informal views of energy embedded in the above definitions in terms of three categories: sources of energy, nature of energy, and causal reasoning.

9.3.2.1 Sources of Energy

Four definitions of energy explicitly state the sources of energy. Definitions One and Two describe energy as a type of vital power possessed by living things. Definition Three describes energy as a type of vital power existing in certain places. As shown in many of the example sentences (e.g., There was a lot of energy in the room this morning. Did you feel it?), these definitions associate energy with living things or places based on feelings. This is very different from the scientific view, in which energy is associated with its indicators (e.g., light, special chemical structure, movement, etc.) rather than feelings. Definition Four describes multiple sources of energy in ways very close to the scientific view of energy. However, it does not explicitly distinguish between energy and its sources, which could cause confusion especially in situations involving foods and fuels. Foods and fuels are organic matter that provides energy in carbon-transforming processes; they are not energy. Definition Five does not explicitly state where energy comes from, but it defines energy as an "ability", which could lead students to think that only living things possess energy. This is because the word ability in colloquial English is often associated with living things. For example, we often say that living animals have the ability to grow and to move, whereas dead animals do not have this ability.

Table 9.2 Meanings of the word energy in English dictionaries

Definitions of energy	Entries of the word energy in dictionaries			
	New Oxford American Dictionary	Merriam-Webster Thesaurus	Dictionary.com	Longman Dictionary (online)
1. A person's physical or mental strength or power **SE: people** **NE: psychological entity,** **CR: cause**	1. The strength and vitality required for sustained physical or metal activity: *Changes in the levels of vitamins can affect energy and well-being. 2. (energies) A person's physical and mental powers, typically as applied to a particular task or activity.	1. Active strength of body or mind: *for a woman of advanced years, she has remarkable energy; [*Synonyms:* beans, bounce, brio, dash, drive, dynamism, energy, esprit, gas, get-up-and-go, ginger, go, gusto, hardihood, juice, life, moxie, oomph, pep, punch, sap, snap, starch, verve, vim, vinegar, vitality, zing, zip;] [*Antonyms:* lethargy, listlessness, sluggishness, torpidity] 2. The ability to exert effort for the accomplishment of a task: *I'm so tired that I don't think I have the energy to take another step; [*Synonyms:* energy, firepower, force, horsepower, might, muscle, potence, potency, puissance, sinew, strength, vigor] [*Antonyms:* impotence, impotency, powerlessness, weakness]	1. The capacity for vigorous activity; available power: *I eat chocolate to get quick energy. 2. An adequate or abundant amount of such power: *I seem to have no energy these days. 3. Often, energies; a feeling of tension caused or seeming to be caused by an excess of such power: *to work off one's energies at tennis. 4. An exertion of such power: *She plays tennis with great energy. 5. The habit of vigorous activity; vigor as a characteristic: *Foreigners both admire and laugh at American energy.	1. The physical and mental strength that makes you able to do things: *Helping people takes time and energy. *Boy, where do those kids get their energy? *She was full of energy after her vacation. *a waste of energy *have the energy to do something *I don't have the energy to deal with it right now. *Try to put more energy into your defensive game. *nervous energy (=too much energy that you have when you are nervous) 2. Somebody's energies, the effort and interest that you use to do things apply/devote/channel your energies into/to something: *She's devoting all her energies to the wedding plans.

(continued)

Table 9.2 (continued)

Definitions of energy	Entries of the word energy in dictionaries			
	New Oxford American Dictionary	Merriam-Webster Thesaurus	Dictionary.com	Longman Dictionary (online)
2. Life energy of living things **SE: living things, NE: psychological entity** **CR: cause**		3. A spiritual force that is held to emanate from or give animation to living beings: *many Eastern cultures believe in the significance of life energy in the healing processes [*Synonyms*: aura, chi (or ch'i also qi), ki, vibe(s), vibration(s)]		3. A special power that some people believe exists in their bodies and in some buildings: *There was a lot of energy in the room this morning – did you feel it?
3. Vital power of places **SE: non-living things** **NE: psychological entity** **CR: cause**				
4. Energy sources **SE: non-living things** **NE: physical entity** **CR: cause**	3. Power derived from the utilization of physical or chemical resources, esp. to provide light and heat, or to work machines.	4. Something with a usable capacity for doing work: *some of the power needs of the house are provided by solar energy [*Synonyms:* energy, power]		4. Power that is used to provide heat, operate machines etc.: *nuclear/solar etc. energy, *research into renewable energy sources, *the world's energy resources, *energy consumption
5. Capacity to do work **NE: abstract quantity** **CR: cause and constraint**	4. (physics) The property of matter and radiation that is manifest as a capacity to perform work (such as causing motion or the interaction of molecules): *a collision in which no energy is transferred. 5. A degree or level of this capacity possessed by something or required by a process.			5. (technical in physics) The ability that something has to work or move: *kinetic energy

"*" shows the example sentences and collocations

9.3.2.2 Nature of Energy

Definitions One, Two, and Three define energy as a psychological entity; they also associate energy with life, feelings, or emotions (e.g., being energetic and excited). This is very different from the scientific view of energy, in which energy is not differentiated from life, feelings, or emotions. For example, in everyday situations, we may say: I have a lot of "energy" to begin the day, because I had a good night's sleep. However, from the scientific view, the body has less energy due to heat dissipation in cellular respiration over the night. In Definition Four and Definition Five, energy is defined as a physical entity—the power provided by certain sources or the ability to do work. Although these two definitions differentiate energy from psychological entities, they do not explicitly define energy as an abstract quantity. In particular, by defining energy using another abstract term (work), Definition Five does not provide any useful information to students.

9.3.2.3 Causal Reasoning

As elaborated in the historical analysis, the scientific view of energy emphasizes energy as a constraint. However, an informal view embedded in the five definitions is that energy is a cause. Definitions One, Two, Three, and Four all describe energy as the cause of a variety of effects such as life, certain feelings, movement, machines working, etc. In Definition Five, energy is described as both cause and constraint. One example sentence used in Definition Five is about energy being transferred, which indicates a sense of "constraint". That is, energy must go somewhere. However, this definition also describes energy as the cause of motion or interactions of molecules.

In summary, the linguistic analysis indicates two patterns. First, the scientific view differentiates energy from matter, life, and psychological entities, whereas the informal views do not. This is reflected in the different ways of association. In the scientific view, energy is associated with its indicators in specialized ways, whereas in the informal views, energy is often associated with life, feeling, perceptions, or emotions. Second, the scientific view highlights energy as a constraint, whereas the informal views treat energy as a cause. When we treat energy as constraint, we use energy conservation and degradation to constrain our accounts about events. That is, we trace energy in a specialized way, namely separately from matter and including heat dissipation. When we treat energy as cause, we do not trace energy consistently. Instead, we trace a cause-and-effect chain; energy is often treated as the cause in this chain. This relation can be presented as: energy $\rightarrow$ effects such as machines moving, people running, etc.

9.4 The Learning Progression for Energy in Socio-ecological Systems

Our historical and linguistic analyses indicate that the scientific view of energy differs from everyday views of energy in two important ways. First, energy is an abstract quantity from the scientific view, whereas it is often equated with matter, life, and psychological entities in colloquial English. Second, energy is about constraint from the scientific view, whereas it is often treated as cause in everyday situations. Within the scope of the Environmental Literacy Project (Michigan State University), we have conducted several learning progression studies to examine how students use energy to explain environmental events in socio-ecological systems (Jin and Anderson 2012a, b; Jin et al. 2013). Our data indicate that students tend to use the informal views of energy to explain environmental events. Based on the historical and linguistic analyses, we identified two progress variables to assess and measure student performance. The first progress variable is association. Scientists use specialized ways to associate energy with its indicators (see more explanation about energy indicator in Nordine et al. 2011) whereas students often associate energy with perceptions, feelings, and life. The second progress variable is tracing. Scientists trace energy in a specialized way, whereas students often lack the ability to trace energy consistently and successfully. This specialized way of tracing emphasizes that energy must come from somewhere and go somewhere, and that heat is always released, so it is aligned with the idea of "energy as constraint". The scientific ways of association and tracing are listed below:

- *Association.* Energy is associated with a set of indicators: kinetic energy → motion; light energy → light; heat energy → differences in temperature; electrical energy → electricity; chemical energy → organic molecules.
- *Tracing.* There are specialized ways of tracing energy in carbon-transforming processes: tracing energy separately from matter and with the recognition of conservation and heat dissipation.

9.4.1 The Learning Progression for Energy

We have used these two progress variables to develop a learning progression for energy (Table 9.3). In the learning progression, each progress variable contains four achievement levels. Each level is about a specific way of association and a specific way of tracing. Levels one, two, and three are about students' informal views of energy; they describe idiosyncratic ways of association and tracing. Level 4 represents the specialized ways of association and tracing emphasized in the scientific view of energy.

Table 9.3 The learning progression framework

Levels	Level description	Examples from interviews and written assessments
Level 4	Association: Associate energy with its indicators (e.g., sunlight, organic molecules); Differentiate energy from matter.	I: Do you think sand can burn, or limestone? S: No, sand is not made of any of the high-energy bonds. It can be melted from heat converting it to a liquid. It cannot, however, be burned for any energy stored in it because there isn't any energy except a little bit of heat energy from the sun or whatever. And that molecule is slightly shifting around. But there isn't any stored energy in it that can be released through burning.
	Tracing: Trace energy separately from matter and with heat dissipation	... ... I: So, you were talking about the energy of the match, right? So, when the flame is burning, where does that energy go? S: It is released as either the light of the flame or as heat, which is not a lot when it's just the size of a match, but if you put your hands near it, you can feel it is releasing heat.
Level 3	Association: Associate energy with familiar organic molecules, but does not differentiated energy from matter (i.e., the organic molecules).	*Question A: Please describe how one glucose molecule from the grape you eat helps to move your finger.* Response: When you eat the grape you are giving yourself glucose. For cellular respiration, you need glucose. ADP + P and oxygen and this make ATP, which your cells can use for cell work, which you can use to move your finger.
	Tracing: Matter-energy conversion; tracing energy without heat dissipation	*Question B: Does the same glucose molecule also help you to maintain your body temperature?* Response: No, because the glucose is a part of the ATP, but another glucose molecule can be used.
Level 2	Associate energy with physical entities such as power, force, and matter. Do not associate energy with psychological features such as feelings and sleep.	I: So where does the gas [gasoline] go? S: The gas [gasoline] is used up by all the parts. It's also exhausted. It's exhausted through the gas pipe or the exhaust pipe I mean. And it goes back into the air. I: So do you think the car needs energy in order to move? S: Yes. The gasoline is their form of energy ... ...

(continued)

Table 9.3 (continued)

Levels	Level description	Examples from interviews and written assessments
	Trace the energy → process chain: Energy causes hidden processes (e.g., carbon dioxide and oxygen conversion, matter transmutation, etc.); do not trace where energy goes after the event is over.	I: Ok. So when the gasoline is used up or become exhaust, where does the energy go? S: The energy goes with it into the air – back into the air. I: What form of energy is that? S: [silence] I: Oh. That's fine. S: I can't think anymore. I: Yeah. Ok. S: But I guess it – the energy is used in all of the different parts and that's where it goes. But all the energy that's like left out . . . that the car doesn't need goes through the exhaust pipe and back into the air. It pollutes our air.
Level 1	Associate energy with feelings, motions, perceptions, materials, etc.	I: Does the girl's body use food for energy? S: It uses the stuff that is energy that helps you, like for candy, sure the sugar makes you hyper but then it settles you right down. With healthy foods, it keeps you active until the day ends, basically. I: How about water? Does the girl's body use water for energy?
	Trace the cause-effect chain—when the actor has all its needs, it grows or moves.	S: Yeah, because if you've been running and you feel that you need something to drink, you might get dehydrated and you can get sick. I: How about . . . You said somebody cares for the girl or for the child. Do you think somebody gives the child energy? S: They don't basically give it energy, but sleep helps you with the energy, and your parents put you in a bed at a certain time, and that helps you get energy.

I Interviewer, *S* Student

9.4.2 *Trends of Development*

The learning progression framework suggests two trends of development. In the paragraphs that follow, we use examples in Table 9.3 to describe these trends.

9.4.2.1 From a Broad Association to a Restricted Association

Regarding the association progress variable, there is a trend from broad to restricted association. At level one, students tell stories about the actor (i.e., a living organism, flame, or a car) and its needs. Energy usually does not play a role in their stories.

However, when being asked to define what energy was and whether energy was involved in the events, all of our participant students were able to incorporate energy into their explanations. They tended to associate energy with conditions, feelings, and emotions. These characteristics are illustrated in the interview episode at level one. The student associated energy with feelings and conditions such as being healthy not sick, being hyper, and being active.

At level two, energy is treated as a physical necessity, an "essence" required to power hidden processes. Energy is no longer associated with psychological characteristics such as feelings and emotions; it is only associated with physical and mechanical characteristics. However, energy is not differentiated from matter in general. As illustrated in the interview episode at level two, the student stated that gasoline was a form of energy.

At level three, the association of energy is even more restricted. Students are able to associate energy with its common indicators such as light and motion. They are also able to relate energy to organic molecules, but they often state that organic molecules such as glucose and ATP are energy.

At level four, students understand energy as an abstract quantity that is associated with its indicators in specialized ways. They are able to recognize that organic molecules provide chemical energy, but they are not energy. This restricted reasoning is illustrated in the interview episode at level four. The student explained that sand and limestone were not fuels because they did not contain high-energy bonds and therefore did not provide energy.

9.4.2.2 From Tracing the Cause-and-Effect Chain to Tracing Energy Separately from Matter and with Heat Dissipation

Regarding the tracing progress variable, the trend is from tracing the cause-and-effect chain to the specialized way of tracing energy—tracing energy separately from matter and with heat dissipation. At level one, students trace the cause-and-effect chain: When the actor has its needs, it grows and moves. This characteristic is illustrated in the interview episode at level one. The student explained that the causes of the growth of the baby's body are the enablers/needs such as foods, water, and enough sleep.

At level two, students trace a cause-and-effect chain that involves energy. Students begin to develop the idea that a physical necessity such as energy is required to power hidden processes, but they usually do not spontaneously think about where energy goes when the event is over. When being asked to explain where energy goes, they often come up with some plausible explanations. In the interview episode at level two, the student explained that energy/gasoline was burned to power the car movement. When being asked to explain where the energy went after that, he first admitted that he couldn't figure out where energy went, and then he *guessed* that energy could be released from the exhaust pipe as a pollutant.

At level three, students begin to trace matter and energy, but without making necessary distinctions between them. They also do not trace energy with recognition of heat dissipation. In the written example at level three, the student stated that it is impossible for one glucose molecule to provide energy for finger movement and heat at the same time. The student did not recognize that when chemical energy transforms into kinetic energy in cellular respiration, heat is always released as a by-product.

Level four represents the scientific view. Students are able to trace energy in a specialized way—tracing energy separately from matter and with recognition of conservation and heat dissipation. In the interview episode at level four, the student traced matter and energy separately. He was able to explain that the energy of the match was transformed into light energy and heat.

9.5 Implication for Teaching Energy

The learning progression presented above provides rich information about student intuitive ideas about energy in socio-ecological systems. Based on the learning progression, we propose that an effective teaching approach could focus on association and tracing of energy. A detailed report on this teaching approach is described in another chapter of this book (see Chap. 4).

We suggest that an effective instructional approach to energy in socio-ecological systems should contain two components. The first component is using "forms of energy" to teach the specialized ways of association. Several researchers point out that the term "forms of energy" is problematic, because it implies the existence of many different kinds of energy (Gilbert and Watts 1983; Kaper and Goedhart 2002; Schmid 1982). This point is also articulated in the NRC framework (p. 122). We argue, however, in order for students to learn the abstract energy concept, a bridge between the abstract meaning of energy and daily experience is indispensible; this bridge is "forms of energy". We do agree that traditional ways of teaching "forms of energy" are problematic. In science classrooms, many forms of energy are taught, but the overlaps among some forms of energy (e.g., solar energy and light energy, kinetic energy, wind energy, and sound energy, etc.) and the distinction between energy and its indicators/manifestations are seldom explicitly addressed. As the result, students often hold very vague ideas about forms of energy. We suggest teaching "forms of energy" with the focus on the specialized ways of association. That is, students understand that energy is an abstract quantity and that quantity is associated with a limited number of indicators (Nordine et al. 2011) in specialized ways. In the socio-ecological systems, the following forms of energy and specialized ways of association are critical: kinetic energy (associated with motion), light energy (associated with light), heat energy (associated with temperature change), and chemical energy (associated with C–C and C–H bonds of organic molecules). Learning these specialized ways of association is very important, because the lower anchor and intermediate levels of the learning progression indicate that students tend to associate energy with a broad range of phenomena including feelings and perceptions, and that they cannot successfully differentiate foods and fuels

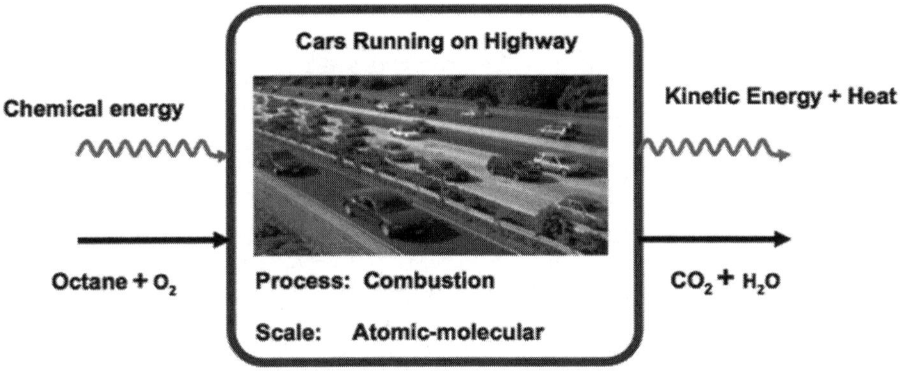

Fig. 9.1 Tracing matter and energy in processes

from materials that do not provide energy. By teaching the specialized ways of association, the teacher will be able to present the concept of "forms of energy" in depth and clarity and therefore help students identify energy in environmental events.

In another chapter of this book, Millar proposes to use forms of energy ONLY for the different ways in which energy can be "stored". According to Millar, light energy should not be used, because it is often about pathways not stores. In another article, Millar (2005) provides more detailed description of the problem with light energy. He discusses two types of situations. In most situations, it is important to know the rate at which energy is being transferred from one place to another by light (i.e., pathways). In other situations, which are rare, it is important to calculate the amount of energy provided by photons. In socio-ecological systems, light energy is an important form of energy involved in photosynthesis. Specifically, the first stage of photosynthesis is light absorption, in which a photon strikes a pigment molecule and passes on part of its energy to the electrons of that pigment molecule. In this sense, light energy in socio-ecological systems is about stores rather than pathways. Therefore, our approach is not contradictory to Millar's approach.

The second component is teaching the specialized ways of tracing matter and energy. We suggest teaching the three fundamental principles of matter and energy—matter conservation, energy conservation and energy degradation at the same time rather than in any particular sequence. As presented in Fig. 9.1, the three principles can be integrated into a framework that emphasizes two specialized ways of tracing: tracing energy separately from matter and with conservation and degradation, tracing matter with conservation. The learning progression indicates that students usually cannot successfully trace matter and energy. They use many informal ways of tracing when explaining environmental events. They often trace the cause-and-effect relations rather than matter and energy. They usually cannot differentiate between matter transformation and energy transformation, and therefore use matter-and-energy conversion to reason about phenomena. When tracing energy, they often do not recognize heat dissipation. By introducing the specialized ways of tracing matter and energy, the teacher will be able to help students better understand the connections among the three fundamental principles and use energy as a conceptual tool to analyze environmental events.

Acknowledgement The authors wish to thank Reinders Duit (Leibniz-Institute for Science and Mathematics Education) for his insightful comments and inspiring discussions. The authors also wish to thank Knut Neumann (Leibniz-Institute for Science and Mathematics Education) and Bob Chen (University of Massachusetts Boston) for their comments, and Kimberly Groshong and Deborah Lan (The Ohio State University) for the language editing of the manuscript.

References

Baierlein, R. (1990). The meaning of temperature. *The Physics Teacher, 28*(2), 94–96.

Barak, J., Gorodetsky, M., & Chipman, D. (1997). Understanding of energy in biology and vitalistic conceptions. *International Journal of Science Education, 19*(1), 21–30.

Boo, H. K. (1998). Students' understanding of chemical bonds and the energetics of chemical reactions. *Journal of Research in Science Teaching, 35*(5), 569–581.

Boyatzis, R. E. (1998). *Transforming qualitative information: Thematic analysis and code development*. Thousand Oaks: Sage Publications.

Capra, F. (1997). *The web of life: A new scientific understanding of living systems*. New York: Anchor Books.

Carey, S. (1985). *Conceptual change in childhood*. Cambridge, MA: MIT Press.

Cobb, C., & Goldwhite, H. (1995). *Creation of fire: Chemistry's lively history from alchemy to the atomic age*. New York: Plenum Press.

Coopersmith, J. (2010). *Energy, the subtle concept: The discovery of Feynman's blocks from Leibniz to Einstein*. Oxford: Oxford University Press.

Driver, R. (1994). *Making sense of secondary science: Research into children's ideas*. London/New York: Routledge.

Driver, R., & Warrington, L. (1985). Students' use of the principle of energy conservation in problem situations. *Physics Education, 20*(4), 171–176.

Duit, R. (1984). Learning the energy concept in school: Empirical results from the Philippines and West Germany. *Physics Education, 19*(1), 59–66.

Fang, Z. (2006). The language demands of science reading in middle school. *International Journal of Science Education, 28*(5), 491–520.

Gilbert, J., & Watts, D. (1983). Concepts, misconceptions and alternative conceptions: Changing perspectives in science education. *Studies in Science Education, 10*, 61–98.

Intergovernmental Panel on Climate Change [IPCC]. (2007). *Contribution of Working Group III to the fourth assessment report of the Intergovernmental Panel on Climate Change*. Cambridge: Cambridge University Press.

Jin, H., & Anderson, C. W. (2012a). A learning progression for energy in socio-ecological systems. *Journal of Research in Science Teaching, 49*(9), 1149–1180.

Jin, H., & Anderson, C. W. (2012b). Development of assessments for a learning progression on carbon cycling in socio-ecological systems. In A. Alonzo & A. W. Gotwals (Eds.), *Learning progressions in science: Current challenges and future directions* (pp. 151–182). Rotterdam: Sense Publishers.

Jin, H., Zhan, L., & Anderson, C. W. (2013). Developing a fine-grained learning progression framework for carbon-transforming processes. *International Journal of Science Education, 35*(10), 1663–1697.

Kaper, W. H., & Goedhart, M. J. (2002). "Forms of energy", an intermediary language on the road to thermodynamics? Part 1. *International Journal of Science Education, 24*(1), 81–95.

Laburu, C. E., & Niaz, M. (2002). A Lakatosian framework to analyze situations of cognitive conflict and controversy in students' understanding of heat energy and temperature. *Journal of Science Education and Technology, 11*(3), 211–219.

Lewis, E. L., & Linn, M. C. (1994). Heat energy and temperature concepts of adolescents, adults, and experts: Implications for curricular improvements. *Journal of Research in Science Teaching, 31*(6), 657–678.

Lin, C.-Y., & Hu, R. (2003). Students' understanding of energy flow and matter cycling in the context of the food chain, photosynthesis, and respiration. *International Journal of Science Education, 25*(12), 1529–1544.

Liu, J., Dietz, T., Carpenter, S. R., Alberti, M., Folke, C., Moran, E., Pell, A. N., Deadman, P., Kratz, T., Lubchenco, J., Ostrom, E., Ouyang, Z., Provencher, W., Redman, C. L., Schneider, S. H., & Taylor, W. W. (2007). Complexity of coupled human and natural systems. *Science, 317*(5844), 1513–1516.

Long Term Ecological Research Network [LTER]. (2007). *Integrative science for society and environment: A plan for research, education, and cyber-infrastructure in the US long-term ecological research network.* Retrieved from http://www.Lternet.edu/decadalplan/

Loverude, M. E., Kautz, C. H., & Heron, P. R. L. (2002). Student understanding of the first law of thermodynamics: Relating work to the adiabatic compression of an ideal gas. *American Journal of Physics, 70*(2), 137–148.

McKeown, M. G. (1991). Learning word meanings from definitions: Problems and potential. In P. Schwanenflugel (Ed.), *The psychology of word meanings* (pp. 137–156). Hillsdale: Lawrence Erlbaum Associates.

Millar, R. (2005). Teaching about energy (Research Paper 2005/11). York: Department of Educational Studies, University of York. Retrieved October 30, 2012, from www.york.ac.uk/education/research/research-paper

Mohan, L., Chen, J., & Anderson, C. W. (2009). Developing a multi-year learning progression for carbon cycling in socio-ecological systems. *Journal of Research in Science Teaching, 46*(6), 675–698.

Morris, R. J. (1972). Lavoisier and the caloric theory. *The British Society for the History of Science, 6*(1), 1–38.

National Research Council. (1996). *National Science Education Standards.* Washington, DC: National Academy Press.

National Research Council. (2011). *A framework for K-12 science education: Practices, crosscutting concepts, and core ideas.* Washington, DC: The National Academies Press.

NGSS Consortium of Lead States. (2013). *Next generation science standards: For states, by states.* Washington, DC: Achieve, Inc.

Nordine, J., Krajcik, J., & Fortus, D. (2011). Transforming energy instruction in middle school to support integrated understanding and future learning. *Science Education, 95*(4), 670–699.

Pinker, S. (2007). *The stuff of thought.* New York: Penguin Group.

Pinto, R., Couso, D., & Gutierrez, R. (2005). Using research on teachers' transformations of innovations to inform teacher education: The case of energy degradation. *Science Education, 89*(1), 38–55.

Schmid, G. B. (1982). Energy and its carriers. *Physics Education, 17*(6), 212–218.

Shanks, N. (2001). *Animals and science: A guide to the debates.* Santa Barbara: ABC-Clio.

Talmy, L. (1988). Force dynamics in language and cognition. *Cognitive Science, 12*, 49–100.

Trumper, R. (1998). A longitudinal study of physics students' conceptions on energy in pre-service training for high school teachers. *Journal of Science Education and Technology, 7*(4), 311–318.

Watts, D. (1983). Some alternative views of energy. *Physics Education, 18*(5), 213–217.

Chapter 10
Contextual Dimensions of the Energy Concept and Implications for Energy Teaching and Learning

Xiufeng Liu and Mihwa Park

10.1 Introduction

The role of contexts in student understanding of scientific concepts has been recognized for a long time. Nevertheless, learning is situated and rooted in culture (Chaiklin and Lave 1993; Lave and Wenger 1991; Vygotsky 1978). Therefore, it should be expected that there is a confounding effect of contexts on students' performances of solving scientific problems. For example, Park and Lee (2004) reported that students prefer to solve everyday problems despite that they perceive everyday problems to be harder than non-everyday problems. Specifically on the concept of energy, Liu and Ruiz (2008) found that, controlling for energy content (i.e., energy activities, energy sources and forms, energy transfer, energy degradation and energy conservation) and cognitive demands (i.e., simple understanding, reasoning and applications), there was a negative correlation between students' performance on large-scale standardized test questions on energy and the context of the questions. Specifically, change from everyday context to non-everyday context results in a statistically significant decrease in students' percent-correct answers on the energy questions.

However, context can be a complex construct when considering teaching and learning about energy; an examination of various contexts related to the energy concept is necessary. Here, *context* is defined as backgrounds (e.g., personal, historical, social, political) in which energy knowledge and understandings are specified in content standards, taught in the classroom and assessed on standardized and other tests. In the PISA 2006 assessment framework (Bybee et al. 2009), contexts are conceptualized as various issues citizens confront such as health, natural resources, environment, and hazards, and can exhibit at personal, social and

X. Liu (✉) • M. Park
Department of Learning & Instruction, University at Buffalo, State University of New York, Buffalo, NY, USA
e-mail: xliu5@buffalo.edu

R.F. Chen et al. (eds.), *Teaching and Learning of Energy in K-12 Education*,
DOI 10.1007/978-3-319-05017-1_10,

global levels. In this chapter, we examine the cultural, social and political contexts related to the energy concept. We then discuss implications of the above contexts for energy teaching and learning.

10.2 Cultural Context of the Energy Concept

Energy conservation was a significant discovery in the history of science (Kuhn 1959). However, the significance was not just limited to science itself; it also had a profound implication to culture. Smith (1998) argues that the doctrines (i.e., energy conservation and dissipation) were also promoted as a new approach to cosmology. This is apparent in how the first and second laws of energy, i.e., the new doctrines, have been stated; they are: (a) the energy of the universe is constant; and (b) the entropy of the universe tends to a maximum. Maxwell once stated that:

> The scientific importance of the principle of the conservation of energy does not depend merely on its accuracy as a statement of fact, nor even on the remarkable conclusions which may be deduced from it, but on the fertility of the methods found on this principle . . . It gives us a scheme by which we may arrange the facts of any physical science as instances of the transformation of energy from one form to another. It also indicates that in the study of any new phenomenon our first inquiry must be: How can this phenomenon be explained as a transformation of energy? What is the original form of the energy? What is the final form? and What are the conditions of transformation (cited in Smith 1998, p. 126).

The science of energy as a new cosmology was a significant cultural revolution in the mid-nineteenth century to present a new "scientific naturalism" (Smith 1998). The intention of the Scottish scientists led by William Thomson, along with their British, French and German allies, to promote energy science as a new cosmology was comparable to the effort Thomas Huxley and his like-minded scientists in promoting Darwin's evolution as a new cosmology. Fundamental to this new cosmology is the belief that all events in the universe are irreversible and no matter how and what changes take place, there is one quantity, i.e., energy, remains unchanged. Granted this cosmology based on energy doctrines is not all positive. Particularly, the energy dissipation doctrine may be perceived as being pessimistic, as Bertrand Russell wrote in *A free Man's Worship*, "all the labours of the ages, all the devotion, all the inspirations, all the noonday brightness of human genius, are destined to extinction in the vast death of the solar system, and that the whole temple of Man's achievement must inevitably be buried beneath the debris of a universe in ruins" (cited in Smith 1998, p. 314).

Since the 1870s when energy doctrines became commonly accepted, the culture implications of the energy concept have been further elaborated. In a seminal article by an American Anthropologist, Leslie White (1943) stated that "everything in the universe may be described in terms of energy (p. 335)". This applies not only to natural world and living systems, but also civilizations or cultures of mankind. He defined "Culture is an organization of phenomena-material objects, bodily acts, ideas, and sentiments-which consists of or is dependent upon the use of symbols (p. 335)".

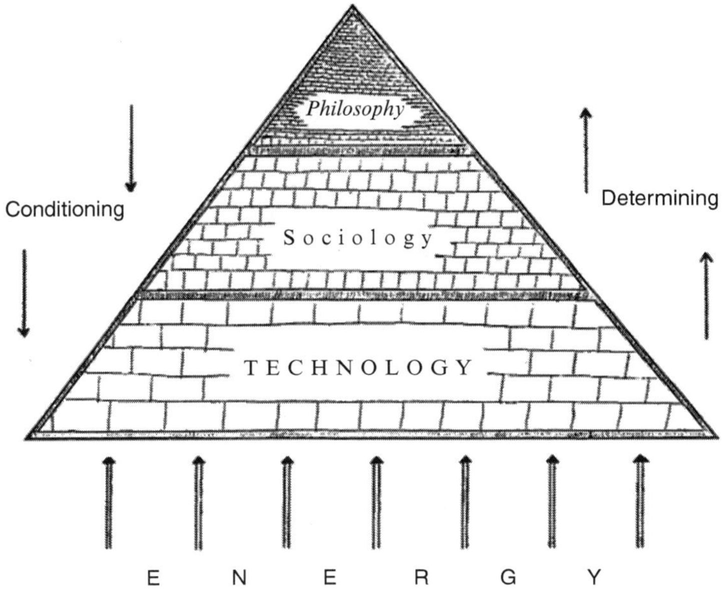

Fig. 10.1 The pyramid of culture (White 1969)

White further elaborated his theory of culture in terms of energy in a later book entitled *The Science of Culture* (White 1969). He stated that culture consists of three systems: the technological, the sociological, and the ideological/philosophical, and that energy is foundational to all the systems. The technological system is comprised of all the physical, mechanical, biological, and chemical instruments that are available for the purpose of manipulating matter; the sociological system consists of the various interpersonal relationships between members of a culture in terms of collective as well as individual patterns of behavior, psychology, and modes of social conduct; and the ideological system encompasses the philosophy, artistic forms, patterns of logic, and epistemologies peculiar to a given society. The three systems interact to form a culture. Figure 10.1 shows the interrelationship among the three systems.

While the role of energy and its foundation to culture by White (1943, 1969) may be too wishful (Wilk 2002) because it is not necessarily true that the more energy is harnessed and consumed, the more responsible human behaviors will become, which in turn the more advanced the civilization will be. Different cultural systems (e.g., the American suburban culture and the north European urban culture) are based on quite different energy demand and consumption patterns. Energy can never be a single determining factor of a cultural system; energy production and use must also give consideration to the sustainability of the environment (Dooley 2006).

A search of the term "energy" in the Thesaurus.com (http://thesaurus.com/browse/energy) can demonstrate how intertwined energy is in culture. Energy can be defined in culture as a person's spirit and vigor, something done, or strong desire for success, and each of these definitions is associated with many synonyms such as activity, animation, happening, force, commodity, and so on. There can be various dimensions of cultures of energy (Strauss et al. 2012) including the connection between resource flows and social relationships in energy systems; cultural transformation and notions of progress and collapse; the blurring of technology and magic; social tensions that accompany energy contraction; and sociocultural changes required in affluent societies to reduce dependence on fossil fuels. Today, pursuing a culture of energy toward sustainable development is becoming more and more desirable. How energy shapes a culture which in turn determines energy uses can have important implications for energy teaching and learning.

10.3 Social Context of the Energy concept

According to Devine-Wright (2007), energy as a social construct can be demonstrated in the following conceptions: (a) energy as commodities, (b) energy as ecological resources, (c) energy as social necessities, and (d) energy as strategy materials. Each of the above conceptions is associated with particular groups or stakeholders. For example, energy industries are the primary holders of the energy as commodities conception. A major assumption of this conception is that energy is a market phenomenon and should be left to the determining mechanism of the market economy, thus any social and political interference will be undesirable. On the other hand, the conception of energy as social necessities is mostly associated with civil right activists who advocate equality and social justice among all citizens. A major assumption of this conception is that the current energy systems are unfair and unjust because they are in favor of those in power. Energy as a social construct is to acknowledge that there can different social constructions of energy, and different constructions of energy will have different implications in terms of future energy production, distribution and uses. For example, energy as commodity construction will likely help maintain the current centralized energy production and distribution technologies while energy as social necessities construction will likely help promote decentralized and individual household based energy technologies. Devine-Wright (2007) claims that the decentralized energy use and production technologies call for energy citizenship because they should be in control of their own energy uses and production.

Energy as a social construct involves not only the scientific knowledge and understanding of the energy concept and technology but also the individual and social actions associated with energy. The connection between knowledge and action in relation to energy is usually assumed to be linear, i.e., the more knowledge and understanding an individual has about energy, the more likely the person will act more responsible in terms of sustainability. Using wind technology as an example, Aitken (2010) demonstrates the deficiencies of the above linearity assumption. Aitken argues that knowledge and understanding are not sufficient for responsible actions; we must accept the legitimacy of individuals' cultures and values in making personal decisions. A large body of literature has examined the various social and behavioral aspects of energy use (Lutzenhiser 1993). No simple relationships exist among various aspects; personal knowledge, habit, geographical location, social economic status, life-style, and so on, all have an impact on individuals' patterns of energy use. Culture can also have a major effect on energy use patterns (Wilhite et al. 1996).

Through a national survey, Bittle et al. (2009) found that people are willing to change their behavior in many ways, but they don't want to be forced into it. They also found that the public's knowledge level is low on energy, with significant numbers who do not know some basic facts about energy (e.g., difference between smog and global warming; difference between fossil fuel and non-fossil fuel). The above findings suggest that, changing energy behaviors can be complex; improving people's knowledge and understanding alone would not be sufficient. We must consider energy use and conservation in a broad social context.

10.4 Political Context of the Energy Concept

Humans use energy from many sources to maintain lives and growth. Such sources include the sun, human energy, animal power, fossil fuels, and renewable energy. With the increase of world population and improvement of living standards, the dependency on energy has been increasing. According to Pimentel and Pimentel (2008), for example, the US imports about 63 % of its oil at a cost of $120 billion per year and this number is expected to increase to 95 % by 2020. For the entire world, it is estimated that there are only 40 years of oil and natural gas resources left. The need to develop sustainable and renewable energy sources is becoming more and more urgent; energy has become an urgent social issue.

Because of the essential role energy plays in personal life and national economy, energy is always associated with politics. "Waiting in line at gas stations during the winter of 1973–1974 as a result of an oil embargo was a new experience for most Americans. On the night of 11 November 1975 New York City and the Northeast were in darkness because of an electrical blackout. In the summers of 1970s consumers were faced with a series of brownouts. In the winter of 1977 natural gas and oil supplies were insufficient to meet the demand. In 1979 political waves in Iran and the middle east brought odd/even gas selling days to the states on both coasts of the" (Allen 1980, p. 6). The energy crisis in the 1970s depicted in the above quote, although seems to be a remote event now, has changed American and world politics forever.

Since 1970s, various US government administrations, both democratic and republican, have sought for energy independence. In 1977, the Carter administration announced its general plan and called the energy crisis "the moral equivalent of war". Over the long term, the plan sought renewable and "inexhaustible" sources of energy to sustain economic growth and a high quality of life for all Americans. However, in the 1980s, renewable sources of energy remained too costly in order to reach commercial scale. During the1990s, environmental concerns dominated the US energy-policy debates and natural gas became an increasingly popular alternative to oil and coal. During the 2000s, national energy policy focused on increasing domestic production under a Republican administration, followed by energy efficiency and low-carbon sources under President Obama (CQ Press 2011).

Despite almost four decades of efforts by various US governmental administrations, achieving the US energy independence remains a remote goal (Brown et al. 2006). Further, the US now faces a host of new challenges that threaten the nation's economy, security, and lifestyle. Bittle et al. (2009) identified three challenges facing the US related to energy: (a) economics: While the oil price spike of 2008 faded in the global financial crisis of 2009, most analysts say prices will keep going up over the long run as countries like China and India require more fuel for their booming economies; (b) Oil dependence: The United States imports about 60 % of the oil it needs, and a significant amount of it comes from more problematic nations, leaving the US vulnerable to supply disruptions and unstable or even hostile regimes; and (c) Experts warn that it is no longer a question of whether world temperatures increase as a result of global warming; it's a matter of how much.

Dealing with the above challenges requires major government policies at both the national and local levels. The contrast in energy policy between the democrats and republicans is never clearer. The Obama administration has been pressing for more federal investment in renewable energy. However, congressional Republicans have been advocating for increased development of domestic oil and natural gas and other carbon based energy sources. The debate on energy policy in the US will continue at the federal level for the foreseeable future. This is because energy is a backbone of the US and world economies. At the state level, debate on hydro-fracking has just begun and will intensify in the next few years.

Energy consumption is also a contentious topic in international politics. While the developed nations often blame the energy shortage in the world for unsustainable population growth in developing world, they are also blamed by developing world for their un-proportional share of energy consumptions (Stern et al. 1997). While the connection between human activities and global climate change has been supported by overwhelming evidence in science, this issue has always been politicized in the US. Only 49 % of the US public believes that the Earth is getting warmer because of human activity (Pew Research Center 2009).

Major government energy policies bear significant economic and social consequences. Take biofuel as an example. In 2007, the US congress enacted and president Obama signed into law the Renewable Fuel Standard (RFS2) which mandates biofuel consumption in the US from the 2008 level of 9 billion gallons to 36 billion gallons by 2022 with minimum consumption levels for each year. In order to ensure the achievement of the above biofuel consumption goal, the legislation provides tax subsidies and market price supports along with other incentive programs to encourage biofuel production and consumption. According to an analysis of the legislation process of this major energy policy, Holleman (2012) found that the entire decision-making process was one-sided with voices overwhelming from the biofuel industry; voices from environmental and social groups were largely absent. Thus, the consequences of this legislation to environment and world food supplies were overlooked. For example, besides the potential ecological degradation of the major role corn plays in food production globally (in animal feed, sweeteners, starch, masa harina, etc.), the legislation has resulted in significant price increase in food on the world market, resulting food riots and exacerbation of impoverishment in some countries (Holleman 2012).

10.5 Implications of Contexts of the Energy Concept

We have examined various contexts of the energy concept in the above sections. We now discuss their implications for energy teaching and learning in K-12.

Project 2061s *Science for All Americans* (AAAS 1990) proposes four criteria for deciding what people should know in order to become scientifically literate:

(a) Utility: Can the content significantly enhance an individual's personal decision-making and employment?
(b) Social responsibility: Is the content likely to help individuals to make social and political decisions on matter related to science and technology?
(c) Intrinsic value: Is the content fundamental to human history or pervasive in our culture?
(d) Philosophical value: Does the content contribute to people's thinking about the world, such as where do we come from, why things are happening in particular ways.

Teaching and learning about energy in K-12 obviously meets the first criterion, because an understanding of energy can enhance individuals' ability to make personal decisions and increase economic productivity. However, meeting the other three criteria requires that teaching and learning about energy in K-12 must attend to contextual dimensions discussed above.

10.5.1 Energy as a Scientific Worldview and Cultural Construct

The social cultural context of the energy concept suggests that we need to approach energy from civics, history, economics, sociology, and psychology, in addition to science, math, engineering, and technology. The most recent conceptual framework for the next generation science standards (NRC 2012) includes energy as both a cross-cutting concept and a core disciplinary concept. This is a good start, but it can go further. Energy as a cross-cutting concept in the *conceptual framework* refers to how energy as a quantity within a physical system remains constant. If we consider the system broadly to refer to all systems, such as natural, living, ecological, social-cultural, to name a few, then energy as a cross-cutting concept presents a worldview. This worldview of energy will guide how we will approach any system from an energy perspective: we would ask questions not only in terms of forms and transfer of energy, but also the ownership and control of the energy source and its transportation, energy production industry, energy consumption, history of energy consumption, personal beliefs and preferences in energy uses, and so on. Essentially, this approach is to go beyond learning energy as a scientific concept, but also a technological means and cultural construct.

The *Benchmark* (AAAS 1990) includes a chapter on Human Society and states specific knowledge and understandings students from K-12 should develop in order to become scientifically literate. Specifically, the document identifies the following aspects of the human society: cultural effects of behavior, group behavior, social change, social trade-offs, political and economic systems, social conflict, and global interference. The above aspects are broad; they can be applicable to energy. That is, in each of the above aspect, energy can be considered an important element. Studying each of the aspects from an energy perspective, e.g., the role of energy in social conflict, can help promote an energy worldview.

Similarly, the content standards of the *National Science Education Standards* (NRC 1996) include a standard on science in personal and social perspectives. Specifically, this standard includes the following perspectives: (a) personal health, (b) characteristics and changes in populations, (c) types of resources changes in environment, and (d) science and technology in local challenges. Again, although these perspectives are not specific to energy, they can be related to energy. Developing students these perspectives can help promote a worldview of energy.

A challenge to teaching energy as a worldview and cultural construct is the requirement of multidisciplinary expertise and the lack of relevant curriculum materials. We believe this challenge can be overcome by taking a thematic approach (Fredericks et al. 1993). A thematic unit is multidisciplinary and multidimensional; this approach can integrate all aspects of energy (e.g., scientific, technological, social, historical, cultural, personal) under one common theme. For example, a theme on green transportation would afford a potential to incorporate forms of energy and energy transfer, energy sources, history and its evolution of auto industry, people's life style, and environmental concerns. When examining these aspects, the energy conservation and degradation can be used as guiding principles.

10.5.2 Energy as Civic Literacy

Socially and politically relevant energy teaching and learning means that energy is a civic literacy; it is necessary for students to become productive citizens in national energy policy debate and actions. It is expected that students' understanding of the fundamental aspects of the energy concept to impact their attitude toward specific energy policies and even inform their actions toward particular energy initiatives.

According to an education guide by the US Department of Energy (DoE 2012) in collaboration with various federal and non-governmental organizations including the American Association for the Advancement of Sciences (AAAS), energy literacy is defined as "an understanding of the nature and role of energy in the universe and in our lives. Energy literacy is also the ability to apply this understanding to answer questions and solve problems (p. 4)". The guide further elaborates that, an energy-literate person:

- Can trace energy flows and think in terms of energy systems;
- Knows how much energy he or she uses, for what, and where the energy comes from;
- Can assess the credibility of information about energy;
- Can communicate about energy and energy use in meaningful ways;
- Is able to make informed energy and energy use decisions based on an understanding of impacts and consequences; and
- Continues to learn about energy throughout his or her life;

The DoE energy education guide (DoE 2012) further identities seven principles; they are:

1. Energy is a physical quantity that follows precise natural laws;
2. Physical processes on earth are the result of energy flow through the earth system;
3. Biological processes depend on energy flow through the earth system;
4. Various sources of energy can be used to power human activities, and often this energy must be transferred from source to destination;
5. Energy decisions are influenced by economic, political, environmental, and social factors;
6. The amount of energy used by human society depends on many factors;
7. The quality of life of individuals and societies is affected by energy choices.

Essentially, energy literacy assumes that increased knowledge on energy would lead to informed decisions and actions on energy. However, research conducted so far has not consistently provided support to this assumption. In a study on middle and high school students' conceptual understanding of the energy concept and its relationship with students' attitudes toward and behaviors about energy conservation (DeWaters and Powers 2011), it was found that students were in general concerned about energy problems (affective subscale mean of 73 % of the maximum attainable score), yet relatively low in cognitive (42 % correct) and behavioral (65 % of the maximum) scores, suggesting that students may lack the knowledge and skills they need to effectively contribute toward solutions. However, while there was an increase in cognitive knowledge scores from middle school to high school, there was a significant drop in energy conservation behavior from middle school to high school. Inter-correlations indicate that energy-related behaviors were more strongly related to affect than to knowledge.

Studies on the general public reported similar findings as the above for school children. Bang et al. (2000) found that (a) consumers who were more concerned about the environment were not significantly more knowledgeable about renewable energy than consumers who are less concerned about the environment; (b) consumers with a higher level of concern for the environment were considerably more likely to be willing to pay a premium to use renewable energy than consumers who indicated somewhat less concern about the environment; (c) the level of knowledge about renewable energy was not significantly related to consumers' beliefs about the positive effects of using renewable energy; (d) consumers with more strongly valenced beliefs about the positive consequences of using renewable energy were significantly more likely to indicate that they would be willing to pay a premium to use renewable energy than those with weaker beliefs; and (e) more knowledgeable consumers were found to be significantly more likely to be willing to pay a premium for renewable energy than consumers with relatively less knowledge about renewable energy. The above results suggest that consumers' environmental concern and beliefs about renewable energy are more emotionally charged than knowledge-based.

Energy literacy suffers the same limitations as various conceptions of scientific literacy in the past. Specifically, following Layton et al. (1993) and Liu (2009), energy literacy is based on the following assumptions: (a) deficit elimination and (b) one-way transport. In terms of the deficit elimination assumption, energy literacy assumes that students and the general public lack energy literacy, thus need to correct this deficiency. This deficit assumption ignores the fact that students and the general public do have a wide range of informal knowledge and experiences about energy. The one-way transport assumption assumes that energy literacy is achieved through activities conducted by the knowledgeable to the less knowledgeable. This assumption gives scientists and the scientific community an un-questionable status, which is problematic because what scientists value about energy may not necessarily be what the general public values.

Energy as civic literacy goes beyond energy literacy. We can not just develop energy knowledge and understanding; in order to educate energy citizens, active participation and actions related to energy use and conservation are necessary. First and foremost, we need to make energy teaching and learning relevant to students. As stated by a group of prominent international science educators (Linder et al. 2011):

> Citizens' lives are increasingly influenced by science and technology at both the personal and societal levels. Yet the manner and nature of these influences are still largely unaddressed in school science. Few students complete a schooling in science that has addressed the many ways their lives are now influenced by science and technology. Such influences are deeply human in nature and include the production of the food we eat, its distribution, and its nutritional quality, our uses of transportation, how we communicate, the conditions and tools of our work environments, our health and how illness is treated, and the quality of our air and water.
>
> Science education is not contributing as it could to understanding and addressing such global issues as feeding the World's Population, Ensuring Adequate Suppliers of Water, Climate Change, and Eradication of Disease in which we all have a responsibility to play a role. Students are not made aware of how the solution of any of these will require applications of science and technology, along with appropriate and committed social, economic, and political action. As long as their school science is not equipping them to be scientifically literate citizens about these issues and the role that science and technology

must play, there is little hope that these great issues will be given the political priority and the public support or rejection that they may need (pp. 2–3).

Energy as civic literacy is consistent with Hodson's (2003) four domains of scientific literacy: (a) learning science and technology, (b) learning about science and technology, (c) doing science and technology, and (d) engaging in sociopolitical actions. It is the last domain in energy civic literacy that is missing in the DoE energy literacy. Programs like activist science and technology education (Bencze and Carter 2011) can be adopted for energy civic literacy development.

In conclusion, considering various contexts of energy would require teaching and learning about energy to go beyond the canonical form of knowledge. Energy is not a static, nor an isolated concept; it is strongly rooted in society. Energy interacts with culture and politics; teaching and learning about energy should promote an energy worldview and energy civic literacy.

References

Aitken, M. (2010). Why we still don't understand the social aspects of wind power: A critique of key assumptions within the literature. *Energy Policy, 38*, 1834–1841.

Allen, R. F. (1980). *Energy education: Goals and practices*. Bloomington: Phi Delta Kappa Educational Foundation.

American Association for the Advancement of Science [AAAS]. (1990). *Benchmarks for science literacy*. New York: Oxford University Press.

Bang, H.-K., Ellinger, A. E., Hadjimarcou, J., & Traichal, P. A. (2000). Consumer concern, knowledge, belief, and attitude toward renewable energy: An application of the reasoned action theory. *Psychology & Marketing, 17*(6), 449–468.

Bencze, L., & Carter, L. (2011). Globalizing students acting for the common good. *Journal of Research in Science Teaching, 48*(6), 648–669.

Bittle, S., Rochkind, J., & Ott, A. (2009). *The energy learning curve: Coming from different starting points, the public sees similar solutions. A report from public agenda*. http://www.publicagenda.org/media/the-energy-learning-curve. Accessed 21 Sept 2013.

Brown, M. A., Sovacool, B. K., & Hirsh, R. F. (2006). Assessing U.S. energy policy. *Daedalus, 135*(3), 5–11.

Bybee, R., McCrae, B., & Laurie, R. (2009). PISA 2006: An assessment of scientific literacy. *Journal of Research in Science Teaching, 46*(8), 865–883.

Chaiklin, S., & Lave, J. (1993). *Understanding practice: Perspectives on activity and context*. Cambridge: Cambridge University Press.

Department of Energy (DoE). (2012). *Energy literacy: Essential principles and fundamental concepts for energy education*. Washington, DC: The US Department of Energy.

Devine-Wright, P. (2007). Energy citizenship: Psychological aspects of evolution in sustainable energy technologies. In J. Murphy (Ed.), *Governing technology for sustainability* (pp. 63–86). London: Earthscan.

DeWaters, J. E., & Powers, S. E. (2011). Energy literacy of secondary students in New York State (USA): a measure of knowledge, affect, and behavior. *Energy Policy,* 1–12. doi:10.1016/j.enpol.2010.12.049.

Dooley, B. (2006). *Energy and culture: Perspectives on the power to work*. Hampshire: Ashgate Publishing.

Fredericks, A. D., Meinbach, A. M., & Rothlein, L. (1993). *Thematic unites: An integrated approach to teaching science and social studies*. New York: HarperCollins College Publishers.

Hodson, D. (2003). Time for action: Science education for an alternative future. *International Journal of Science Education, 25*(6), 645–670.
Holleman, H. (2012). Energy policy and environmental possibilities: Biofuels and key protagonists of ecological change. *Rural Sociology, 77*(2), 280–307.
Kuhn, T. S. (1959). Energy conservation as an example of simultaneous discovery. In M. Clagett (Ed.), *Critical problems in the history of science* (pp. 321–356). Madison: University of Wisconsin Press.
Lave, J., & Wenger, E. (1991). *Situated learning: Legitimate peripheral participation*. Cambridge: Cambridge University Press.
Layton, D., Jenkins, E., Macgill, S., & Davey, A. (1993). *Inarticulate science? Perspectives on the public understanding of science and some implications for science education*. Driffield: Studies in Education Ltd., University of Leeds.
Linder, C., Ostman, L., Roberts, D., Wickman, P., Erickson, G., & MacKinnon, A. (Eds.). (2011). *Exploring the landscape of scientific literacy*. New York: Routledge.
Liu, X. (2009). Beyond science literacy: Science and the public. *International Journal of Environmental & Science Education, 4*(3), 301–311.
Liu, X., & Ruiz, M. (2008). Using data mining to predict K-12 students' performance on large-scale assessment items related to energy. *Journal of Research in Science Teaching, 45*(5), 554–573.
Lutzenhiser, L. (1993). Social and behavioral aspects of energy use. *Annual Review of Energy and the Environment, 18*, 247–289.
National Research Council [NRC]. (1996). *National science education standards*. Washington, DC: National Academy of Sciences.
National Research Council [NRC]. (2012). *A framework for K-12 science education: Practices, crosscutting concepts, and core ideas*. Washington, DC: National Academy of Sciences.
Park, J., & Lee, L. (2004). Analysing cognitive or non-cognitive factors involved in the process of physics problem-solving in an everyday context. *International Journal of Science Education, 26*(13), 1577–1595.
Pew Research Center. (2009). *Scientific achievements less prominent than a decade ago: Public praises science; scientists fault public, media*. Washington, DC: The Author.
Pimentel, D., & Pimentel, M. H. (2008). *Food, energy, and society*. Boca Raton: Taylor & Francis Group.
Press, C. Q. (2011). Energy policy: Should the US use more clean-energy sources? *CQ Researcher, 21*(20), 457–480. Thousand Oaks: Sage publications.
Smith, C. (1998). *The science of energy: A cultural history of energy physics in Victorian Britain*. Chicago: The University of Chicago Press.
Stern, P. C., Dietz, T., Ruttan, V. W., Socolow, R. H., & Sweeney, J. L. (1997). *Environmentally significant consumption: Research directions*. Washington, DC: National Academies Press.
Strauss, S., Rupp, S., & Love, T. (Eds.). (2012). *Culture of energy: Power, practices, and technologies*. Walnut Creek: Left Coast Press.
Vygotsky, L. (1978). *Minds in society: The development of higher psychological processes*. Cambridge, MA: Harvard University Press.
White, L. A. (1943). Energy and the evolution of culture. *American Anthropologist, 45*(3), 335–356.
White, L. A. (1969). *The science of culture*. New York: Farrao, Straus, and Giroux.
Wilhite, H., Nakagami, H., Masuda, T., Yamaga, Y., & Haneda, H. (1996). A cross-cultural analysis of household energy use behavior in Japan and Norway. *Energy Policy, 24*(9), 795–803.
Wilk, R. R. (2002). Culture and energy consumption. In L. Orr, R. Bent, & R. Baker (Eds.), *Energy: Science, policy, and the pursuit of sustainability* (pp. 109–130). Washington, DC: Island Press.

Chapter 11
Towards a Research-Informed Teaching Sequence for Energy

Robin Millar

11.1 Introduction

Teaching energy ideas poses a greater challenge to science teachers and educators than other science topics. For most topics, there is broad agreement about what would constitute an appropriate understanding of the topic at different stages of the education process. For energy, this is not the case. Much of what has been written in science education books and journals about the teaching and learning of ideas about energy is not about learners' views and understandings, or the effectiveness of different teaching approaches and interventions, but about what should be taught and the language that teachers and textbooks ought to use. This is often critical of the scientific accuracy of textbook accounts and of common teaching approaches. The ensuing debates do not seem to have led to consensus within the science education community about the goals of energy teaching. Yet without clear and generally agreed learning goals, empirical studies of teaching and learning are of limited value in advancing knowledge of how students' ideas and understandings typically develop, or for improving practice.

One reason for the difficulty in deciding what to say about energy at school level is that the scientific idea of energy is very abstract. It is, for example, impossible to say in simple language what energy is, or means. Another problem is that the word 'energy' has entered everyday discourse, with a meaning that is related to, but rather different from, the scientific one. A further complication for the curriculum planner is that the idea of energy is used in somewhat different ways across the sciences, so that the things that are said about it differ from physics to chemistry and biology.

Section 11.2 discusses the difference between the everyday and scientific discourses of energy, and how science educators have tried to bridge the gap between these by developing an intermediate educational energy discourse. This then leads

R. Millar (✉)
Department of Education, University of York, York, UK
e-mail: robin.millar@york.ac.uk

R.F. Chen et al. (eds.), *Teaching and Learning of Energy in K-12 Education*,
DOI 10.1007/978-3-319-05017-1_11,

into a discussion of the main issues that have been raised by science educators about this educational energy discourse. Any proposal for teaching about energy needs to take these issues into account, and ideally to explain how they have been taken into account. Section 11.4 (and the Appendix) then present and discuss a teaching sequence for the topic of energy up to age 16. A teaching sequence sets out the content of a domain in an order that is designed to facilitate the development of understanding, identifying the age or stage at which specific ideas might be introduced. It is based on an analysis of the content of the domain, but also draws on what is known about students' typical learning difficulties, and it is grounded in a view of why an understanding of ideas in the domain are of value to students. It makes no claim to represent the pathway that a 'typical' student follows in developing understanding. The chapter concludes by looking at the importance of textbooks and, more particularly, of assessment instruments in testing and refining any proposed teaching sequence of this sort.

11.2 Discourses of Energy

The earliest use in English of the word 'energy' (according to the Oxford English Dictionary) was around 1600, when it meant 'force or vigour of expression'. In the intervening four centuries, the word 'energy' has passed into everyday spoken and written discourse. We can say that someone is 'full of energy', or 'has no energy'. Advertisements claim that certain foods or drinks 'give you instant energy', or 'an energy boost'. In news reports and other kinds of public information, energy is something we 'use' and 'consume'. We buy it from the 'energy utilities'. We are advised to insulate our homes, switch to new types of central heating boiler or car engine or light bulbs, in order to 'save' (or not to 'waste') energy. Governments publish data annually on 'energy use' and 'energy consumption' in different sectors of the economy (industry, transport, domestic, and so on), and debates rumble on about how we can meet our future 'energy needs'. In this discourse, energy is a commodity or resource. It comes in different forms, and from different places. We buy it and use it. When it's used, it's gone.

This is very different from the scientific meaning of the word 'energy'. In science, energy is the name for a property or attribute of an object, or a system (a group of objects that interact with each other). It is a quantitative property; you can measure how much of it an object or system has gained or lost between two instants. By 'measure', I mean put a number on the amount of it. This is not done directly – we do not have an 'energymeter' – but indirectly by measuring other primary quantities and then using an agreed formula. In this respect, energy is no different from many other physical quantities that we measure, such as speed, density, resistivity, thermal conductivity, and so on. The reason – indeed the only reason – why energy is useful is because it is conserved. If something loses some energy, something else must have gained it. The total amount at the end of any event or process is the same as it was at the beginning. In his celebrated *Lectures on Physics*, Feynman puts it like this:

> There is a fact, or if you wish a law, governing all natural phenomena that are known to date. There is no exception to this law – it is exact so far as is known. The law is called the conservation of energy. It says that there is a certain quantity, which we call energy, that does not change in the manifold changes which nature undergoes. That is a most abstract idea, because it is a mathematical principle; it says that there is a numerical quantity, which does not change when something happens. It is not a description of a mechanism, or anything concrete; it is just a strange fact that we can calculate some number and when we finish watching nature go through her tricks and calculate the number again, it is the same. (Feynman et al. 1963, pp. 4–1)

A key point here is that energy is not a description of a mechanism. It cannot provide the basis for a causal explanation of anything. It does not explain why anything happens or how it happens. All it can do is tell us that certain events are possible and others are not, or draw our attention to an outcome of an event that we might otherwise have overlooked, in order to 'balance the books'. To give one famous example of this, the requirement that energy (and momentum) must be conserved led Pauli to 'discover' the neutrino.

To try to bridge the very substantial gap between the everyday and scientific discourses of energy, many science educators, since the major curriculum reforms of the 1960s, have adopted an 'intermediate' form of energy discourse. This treats energy as something that can be stored in different places and can flow from place to place. In such a model, energy is implicitly portrayed as a quasi-material entity, invisible and intangible, which can change its form as it goes. The model is almost always introduced qualitatively, and students are asked to apply it to a wide range of everyday and simple laboratory events and processes. A fluid model is, of course, likely to work well for any conserved quantity. If there is less of it in one place, there is more in another – so it looks as if something has moved, or flowed. Energy is not conserved because it is 'like a fluid' – rather it behaves like a fluid because it is conserved (Ogborn n.d.). Most of us are easily drawn into seeing and talking about energy as though it was a fluid.

11.3 Issues and Disputes

The educational discourse of energy has, however, been subjected to persistent criticism, on a range of grounds. In this section, I will review the main issues that have been raised, to set the context for a proposal about the teaching of energy to students in the 5–16 age range.

11.3.1 Defining Terms

Warren (1982, 1991) argues that the treatment of energy as though it were a quasi-material substance is fundamentally flawed, in introducing terms without proper definition, and failing to acknowledge and to convey the abstract mathematical character of energy. He also believes it creates an obstacle to future learning. Warren argues that energy should be defined as 'the capacity to do work', and hence that ideas like force and work should be taught first. This definition of energy is, however, disputed on the grounds that the energy of a hot object cannot all be used to do work. The meaning of 'capacity' is therefore at best unclear. Physics has also moved on from this kind of conception of energy. The work of the early twentieth century German mathematician Emmy Noether showed that energy is the quantity that must be conserved if the laws of physics are invariant in time. Post-Einstein, energy is that which gravitates. The vast majority of the energy around us is the rest mass of objects; the energy changes that we notice are just ripples on the surface of a vast ocean of rest mass energy. These are clearly not ideas we would use to introduce learners to the scientific idea of energy, but they might convince us that we cannot approach the idea of energy via a 'correct' definition. As Feynman wrote in 1963: 'It is important to realize that in physics today, we have no knowledge what energy *is*' (Feynman et al. 1963, pp. 4–1). Fifty years on, this remains the case.

Another kind of objection to Warren's view is that approaching energy via force and work makes students think the idea applies only to mechanical situations, rather than seeing it as a very general and over-arching concept that applies to all physical processes. Everyday talk about energy is more likely to be in the context of fuels and food (the number of 'calories' different foods contain) than situations involving forces – and science teachers are likely to want to talk about energy in chemical and biological contexts, where forces are not an obvious or prominent feature, well before the ideas of force and work are taught. You might say that chemistry and biology want to begin where physics, on its own, would prefer to end up.

Warren (1991) recognised that a consequence of his view is that energy should not be taught until the upper secondary school, after ideas like force and work have been taught and learned. This, however, is a minority view amongst science educators and not an acceptable solution to most people – as a look at curriculum requirements and textbooks from many countries quickly shows. For most science educators, energy is too important, in personal and social contexts, to delay all discussion of it until students have reached the upper secondary stage. The everyday discourse of energy, although it differs in significant ways from the scientific view, is not without meaning and involves ideas and understandings that are practically useful and socially important. There are important ideas about energy provided by foods, and about the present and future supply and use of energy resources, that school students need to understand and become more able to use and apply. Gaining a functional understanding of these is an essential part of 'scientific literacy'. Many science educators would also argue that the school science programme should also start students on the road towards an understanding of the scientific view of energy, on the grounds of its cultural significance and its importance within the scientific disciplines. So we need to begin somewhere to build on, and refine, the everyday understanding of energy and help students to move towards the scientific understanding. Many would therefore agree with Duit (1987) that a model of energy as a quasi-material substance is an acceptable, indeed potentially valuable, way of approaching a difficult idea.

Table 11.1 Alternative accounts: energy or mechanism

Situation	A possible account in terms of ... 'Forms of energy'	The underlying mechanism
A battery powered vehicle	Chemical energy in the battery is transformed into electrical energy which is carried by the wires to the motor. It is transformed to kinetic energy in the motor, making the buggy move	The battery causes an electric current in the coil of the motor, which makes the coil rotate. This is then used to make the buggy move
A light bulb	The bulb lights because electrical energy flows from the battery to the bulb	The battery pushes a stream of electric charges through the bulb filament. This makes it get very hot

11.3.2 Forms of Energy

Treating energy as a quasi-material substance has led to the idea that it can appear in different forms, and can change its form as it moves from place to place. Many textbooks and teaching schemes include lists of these forms of energy: kinetic, chemical, heat, electrical, and so on. There is not complete agreement, however, on the list of forms, or on the names to use for some of them. Taber (1989) points out that textbooks offer different lists of forms of energy, and propose different names for the same form (for example, *movement energy*, *moving energy* and *motion energy* as simplifications of *kinetic energy*). The use of the term 'potential' is particularly variable, with some textbook authors choosing to replace it with the apparently simpler 'stored' (but not applying this also to chemicals or moving objects, where energy could equally well be said to be stored).

The 'forms of energy' approach has been criticised as teaching students to apply a set of labels which add little to understanding of processes. Table 11.1 provides two typical illustrations of this. The accounts in the left-hand column may appear to be explanations but are not. As the extract from Feynman above makes clear, energy is not a description of a mechanism or cause. The accounts in the right-hand column are more useful because they begin to point towards mechanisms and open up the possibility of probing more deeply towards a fuller explanation. Indeed the opening sentence of the first example is not merely uninformative, but is misleading. There is no object, or group of objects, in an electric circuit that has an amount of 'electrical energy' that is in principle measurable. Modelling energy as a quasi-material substance is leading here to the invention of quantities that have no observable referents in the real world. Whatever the statements in the left-hand column are, therefore, they are not science.

A 'forms of energy' analysis of multi-step processes or events can also readily lead to the identification of variables that have no bearing on the outcome (Millar 2005, 2011). Consider, for example, the arrangement shown in Fig. 11.1, where a battery runs a motor, which turns a pulley wheel and raises a load.

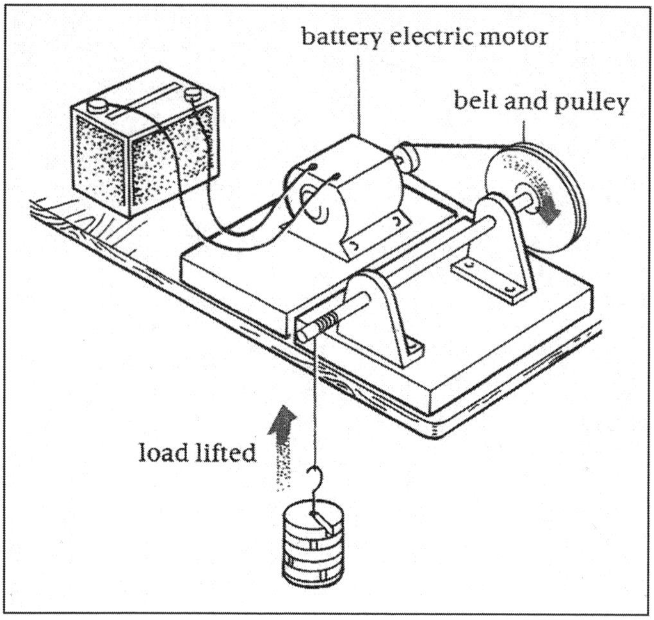

Fig. 11.1 Using a motor and pulley to raise a load (From Avison 1984)

A typical analysis of this event in energy terms in a school textbook might look like this:

chemical energy →	electrical energy →	kinetic energy	→ potential energy +	thermal energy
(in battery)	(in wires)	(in moving parts of motor and belt)	(in load)	(in motor, pulley)

The problem with 'electrical energy' has already been noted. But the kinetic energy stage is also problematic. We could certainly identify objects that are moving, and therefore have kinetic energy. But the *amount* of kinetic energy they have is irrelevant to a scientific analysis and explanation of the overall event. If we replaced the belt by a lighter one, but of the same strength, it would have less kinetic energy – but everything would continue as before. The kinetic energy of the moving parts is not a useful quantity to know. There is no need to bother about it. A potentially more useful analysis would simply focus on where energy is stored at the beginning and end of the event; energy ideas do not illuminate or explain the mechanisms involved.

Ellse (1988) is critical of the 'forms of energy' approach to teaching energy ideas, on the grounds that it focuses attention in the wrong place, on the 'form' of the energy at different points, rather than on the processes by which energy is transferred from one object, or place, to another. He argues that a focus on the latter is simpler, and leads to more useful and more important insights. He proposes that we should not use any labels for forms of energy, but simply talk about 'energy' being 'transferred' from place to place, rather than 'transformed' or 'converted' from one form to another. This can, indeed, be applied quite easily to many situations, but it fits some quite common ones rather less comfortably. For example, when an object gains speed by falling or sliding down an incline, it seems more natural to think of energy being stored in a different form, rather than being 'transferred' from one place to another. 'Forms' labels can also be useful to clarify what we mean to say. For instance, a battery sitting on a high shelf has energy by virtue of the chemicals it contains, and by virtue of its elevated position. Labels like 'chemical' and 'gravitational' would be useful for expressing this, and making clear which of them we want to talk about.

Kaper and Goedhart (2002a, b) discuss the usefulness of 'forms of energy' as 'an intermediary language on the road to thermodynamics' (p. 81). Their conclusion is ambivalent about the merits of portraying energy as a quasi-material substance, but they consider that forms can be useful in developing students' ideas about energy and do not create an obstacle to a fuller understanding of thermodynamics, if each corresponds to a formula that students will learn at a more advanced stage for calculating changes in the amount of energy stored.

11.3.3 Energy as a Cause

Another criticism of the way energy ideas are commonly taught is that energy is often portrayed as the cause of events and is used to offer explanations. Examples of this have already been shown in Table 11.1. Whilst research (Trumper 1990, 1993; Watts 1983) shows that many students think of energy as the cause of events, Ogborn (1986) points out that it is incorrect to portray energy as 'the go of things' – as what makes them happen. Causal statements like 'A ball keeps moving because it has kinetic energy' or 'Petrol makes a car go because it contains energy' misrepresent the scientific idea of energy. But whilst we should try to avoid the 'because', a store of energy *is* often needed to make things happen. And things that can release a lot of stored energy can make more happen than things that can release only a little. Ogborn (n.d.) acknowledges that these are 'halfway-useful half-truths' (p. 6). In introductory teaching about energy, it would be difficult to avoid implying that a store of energy of some kind is often the cause of a process or event – even if we might later want to explore this more deeply with some students, and try to reconcile it with the fact that energy is conserved.

For the deep point here is that a conserved quantity cannot explain why any process runs in one direction rather than the reverse. Something else is required to explain this. In thermodynamics, this 'something else' is entropy, or free energy. At school level, Ogborn (1990) suggests introducing the idea that changes are driven by a difference of some kind. So a difference in temperature, or in height in a gravitational field, can cause a change to occur. As a process runs, the difference that drives it gets less, or is used up. There is widespread support amongst science educators (for example, Duit 1981, 1986; Solomon 1982, 1992) for the underlying principle here – that the teaching of energy must include ideas about the dissipation of energy as well as its conservation, if students are to be helped to move their understanding on from the everyday discourse of energy 'use' and 'consumption'.

11.3.4 The Problem of Heat

In addition to the general issues about energy teaching outlined in the previous section, there is a strand of discussion in the science education literature about thermal phenomena, and in particular the use of the term 'heat'. As with 'energy', the everyday use of the term 'heat' is rather different from, and significantly less precise than, its scientific meaning. Heat is sometimes used almost interchangeably with temperature ('The heat outside today is unbearable.'; 'It's cold in here – turn up the heat.') In science textbooks, as several authors have pointed out (Mak and Young 1987; Summers 1983; Warren 1972, 1976), 'heat' is often used in ways that are inconsistent and, from a scientific perspective, incorrect. In particular, it is used to mean both the energy that a hot object has by virtue of its higher temperature, and the energy that is transferred spontaneously from a hotter object to a cooler one due to their temperature difference. If heat is added to an object, its temperature rises – and if it loses heat, its temperature falls. This is essentially the caloric theory of heat, developed in the second half of the eighteenth century and used by scientists until a more complete understanding of thermodynamics was articulated by Clausius around 1850. It breaks down in situations where work is done, either mechanically or electrically, to raise the temperature of an object without any interaction with another object at a higher temperature. In classical thermodynamics, the energy stored in a hot object is called its *internal energy*, and *heat* is energy that is transferred spontaneously from an object at a higher temperature to one at a lower temperature. The two quantities are not the same; two terms are needed.

Some people have argued, however, that it is unnecessary to introduce the term 'heat' at all, and better simply to talk about 'energy' being transferred due to a difference in temperature – which we call the process of 'heating' (Heath 1974, 1976; Summers 1983). Bringing in the idea of heat is, they argue, an unnecessary complication. So rather than reasoning that 'internal energy in the hot object becomes heat which then becomes internal energy in the cold object', we would simply say that 'energy is transferred from the hot body to the cold body; this process is called heating'. But in everyday language, 'heating' something means raising its temperature (or perhaps, in some situations, changing its state), however this is achieved. Terms like 'heat' (both as a noun and as a verb: 'to heat') and 'heating' are deeply embedded in everyday language, and consequently in our thinking. So avoiding them is difficult, if not impossible, when talking about thermal processes.

Recently, the term 'thermal energy' has become more common in textbooks, perhaps in response to the argument that the word 'heat' should be avoided. But this is not a solution. No single term can serve for both 'internal energy' and 'heat' (as defined in classical thermodynamics); two terms are needed. Most texts that use the term 'thermal energy' do not make clear which of the two they mean, or use it inconsistently to mean sometimes one and sometimes the other. The change required is a change of model, not just of terminology.

11.4 Tiptoeing Through the Minefield

Having summarised the main issues that have been raised about the teaching of energy, I want now to turn to the question of what might be done to address these and encourage a way of teaching about energy that is less open to the kinds of objections outlined above. Proposing a teaching approach feels a somewhat risky undertaking, as the section title implies. Perhaps a metaphor of trying to dodge sniper fire would be more apposite. But, to make progress, it seems necessary to offer proposals that can be discussed and improved, rather than simply to rehearse the issues and difficulties.

I should make clear at the outset that I do not see the goal as 'teaching the correct scientific understanding of energy'. There are understandings of energy in science at various levels of sophistication. The fact that these co-exist, and that many scientists talk and write about energy using aspects of everyday energy discourse and of the intermediate educational discourse of energy, especially in communications intended for non-expert audiences, is a reminder that scientific ideas and models are better judged on the criterion of usefulness for a particular purpose, rather than of correctness or truth. We cannot in any case expect learners to make the transition from an everyday understanding of anything to the accepted scientific explanation in a single step. We may first have to teach ideas that apply only to a restricted set of cases, in order later to develop these further towards wider applicability. And we will certainly have to use in our teaching some terms from everyday discourse, even where these are ill-defined or carry a meaning that is different from the same term in scientific discourse, because progress in understanding has to build on learners' prior knowledge and ideas.

All of this, of course, requires judgment and so people may legitimately differ on the simplifications and compromises that are acceptable. Two general principles that many would agree are important for a teaching sequence on any science topic are that:

- It should begin from ideas and contexts with which learners are familiar, and build on these rather than seeking to replace them.
- It should develop ideas and understandings that students can value, because they offer an intellectually satisfying insight into the behaviour of the natural world, and/or because they are practically useful in situations in which they may find themselves.

Applying these to the specific topic of energy, this might imply that:

- It should aim to refine and improve students' understanding of the everyday discourse of energy, because this involves ideas and actions that are of critical importance in personal and social contexts, as well as helping students to move towards an understanding of the scientific discourse and to begin to appreciate the elegance and potential usefulness of analysing events and processes from an energy perspective.

Drawing on the discussion above of issues concerning the teaching of energy, we might then also want to argue that:

- It should present energy as a quantity that is in principle measurable from as early a stage as possible, to reflect the fact that energy (in science) is essentially a quantitative construct. (This does not mean introducing equations and formulae, but rather using everyday sources of quantitative energy information, like food labels and domestic energy bills.)
- It should introduce ideas about the dissipation of energy alongside ideas about conservation, in order to make the latter more intelligible in relation to everyday energy ideas and observations.
- It should avoid using energy ideas to provide (apparent) explanations of the mechanisms or processes underlying events.
- It should exercise some care over the use of labels for 'forms' or 'types' of energy, restricting these to the different ways in which energy can be stored, and separating these from the different ways in which energy can be transferred from one store to another.

A teaching sequence for energy based on these criteria is presented in the Appendix. It has been developed, as one strand of a progression framework for school physics up to age 16, to inform and underpin a major curriculum development project for the lower secondary school in England (students aged 12–14) (see www.york.ac.uk/education/projects/yorkscience). Many of its features are in agreement with the overview of energy in the *Framework for K-12 Science Education* published by the US National Research Council (2012). This, for example, also takes the view that 'the idea that there are different forms of energy ... is misleading' and that it is 'misleading to call sound or light a form of energy' (p. 122).

The proposed starting point of the sequence is exploring and extending the ideas students bring from everyday life about the fuels we use for heating and for making things move, and about food as a fuel for humans and other animals. This might begin in the upper years of primary school. Students might also be taught about the origins of fossil fuels, and hence that they are a fixed and finite resource and the implications of this. Developing an understanding of how energy is released through a process of breaking and making bonds in chemical reactions will, of course, have to come much later.

Fuels make a good entry point because they are indeed 'used up' and not conserved, in line with our intuitions. Work on fuels at lower secondary level provides a context for introducing some numbers and doing some simple calculations, to convey the sense that energy is a quantitative variable. Energy values are marked on the packaging of every foodstuff, and the energy available can be compared with data on the amounts needed to do different jobs. Again in an everyday context, power ratings on domestic electrical appliances can be used to explore of the rate at which these require energy to be supplied, and hence the cost of operating them. It is important to realise that heating is relatively expensive! The basic notion of efficiency (getting a job done using as little fuel as possible) can also be introduced. This approach is, in effect, treating amounts, and units, of energy and power (MJ, kW, kWh) as 'socially defined' – defined by custom and practice rather than by a formal definition. To find out the power rating of a kettle you look at its baseplate. You come to an understanding of what 'several kW' means by using things like kettles. Similarly, fuel bills are given in kilowatt-hours (kWh) or megajoules (MJ). Here cost becomes part of getting the meaning; costs are made up of price × amount, so the energy is an amount. In practice, many terms (such as length, mass, volume, force, temperature) are first 'defined' through social custom and practice, with a formal definition coming later, if at all.

To make progress in applying and using energy ideas, students need to differentiate between the ideas of heat (in its everyday usage, meaning 'internal energy') and temperature. Extensive research indicates that this is a significant challenge for many learners (Linn and Songer 1991; Tiberghien 1984), as we might anticipate in the light of scientists' protracted struggle to separate these ideas in the seventeenth and eighteenth centuries (Wiser and Carey 1983). Whilst some science educators have criticised the teaching of what is in effect (though rarely in name) the caloric theory of heat, an understanding of this would represent a significant conceptual step for many learners. Any class discussion of thermal processes will involve the word 'heat', introduced by the students even if the teacher manages to avoid using it. I am not anyhow persuaded that it is necessary, or desirable from an educational perspective, to try to exclude it. Learning must build on prior knowledge. Ideas that can be used in explaining and predicting are valuable, even if we recognise that they apply only to a restricted range of phenomena (here, ones involving only solids and liquids where a negligible amount of work is done on or by the external environment). Better to treat this as a useful model (not as 'the truth') for the moment, and plan to revisit it later with some students at least, when it is possible to show its limitations and recognise why a better model is needed. The sequence in the Appendix proposes doing this with 15–16 year-old students who are on a track that leads towards the more advanced study of science.

The sequence suggests that students might also in the lower secondary years get some opportunities to investigate simple mechanical machines (levers, pulley systems, gears), perhaps in a topic that is ostensibly unrelated to energy. The key observation, that you can increase the force applied but at the expense of reducing the distance it moves, prepares the ground for the idea that the quantity 'force × distance' is a measure of the amount of energy transferred by a force – and that this cannot ever be 'multiplied' by a machine.

The core of the proposed sequence is in the age 13–14 section, where the idea that change is caused by a difference is introduced, along with a framework for looking at common events and processes in terms of 'energy stores' and the 'pathways' by which energy is moved from store to store. The language of 'forms of energy' is deeply embedded in the discourse of scientists, as well as that of science teachers. So a proposal to get rid of it is unlikely to succeed. A more feasible aim is to try to modify it, to create a somewhat clearer and more coherent framework for talking about events and processes in energy terms. To this end, the idea of using labels for the different ways in which energy can be stored ('energy stores'), rather than for 'forms of energy', has been proposed by several authors (Boohan 2007; Lawrence 2007; Millar 2000, 2011; Papadouris and Constantinou 2011; Tiberghien 2000) and by projects such as the Institute of Physics *SPT 11–14* project (www.talkphysics.org), and the Gatsby Science Enhancement Project (www.nationalstemcentre.org.uk/sep). The common energy stores are:

- kinetic
- chemical
- internal
- gravitational
- magnetic
- electrostatic
- elastic.

At a later stage, the term 'potential' may have to be introduced and attached to some of these, but at this initial stage it is more likely to be a source of confusion than of illumination. The label 'internal' may be useful in pointing towards the idea that the energy of a hot body is stored in the random motions, and relative positions, of its sub-microscopic particles, though some might prefer 'thermal'. This is an acceptable alternative, as is 'strain' instead of 'elastic'. When first introducing the idea of energy stores, it is probably better to treat gravitational energy as stored in the raised object. The idea that it is stored in the configuration of two (or more) objects in the field between them can come later, and may be more easily approached through magnetic and electrostatic situations.

The list of energy stores above does not include 'electrical energy', 'light energy' or 'sound energy' – because these are not 'stores'. Rather, they are 'pathways' by which energy can be moved from one store to another. The US National Research Council (2012) *Framework for K-12 Science Education* takes essentially the same approach and uses similar language. At an introductory level, we might introduce four such pathways. Energy can be moved (or transferred):

- mechanically (by a force acting over a distance);
- electrically (by charges being moved through a potential difference);
- by heating (as a result of a temperature difference);
- by radiation (both electromagnetic and mechanical).

By using labels only for the different ways in which energy can be stored, and clearly separating stores and pathways, it becomes possible to tell a somewhat clearer and more consistent story about events and processes from an energy perspective. It is not entirely unproblematic. For example, a space through which light is travelling contains a large number of photons, each of which has a measurable amount of energy; but the total energy stored at any instant in the collection of photons in a given region of space is rarely a variable of much importance or interest. Also, the decision to classify an energy transfer by infrared radiation as 'by radiation' rather than 'by heating' is simply because it may be easier for young learners to see it in this way. The argument is not that a framework based on the ideas of 'energy stores' and 'energy pathways' is perfect, but that it is 'good enough', and significantly better than frameworks that have been widely used in the past. An advantage of the 'stores and pathways' framework is that each of the 'energy stores' corresponds to an equation that can be used to calculate changes in the amount of energy that an object or system has – and so prepares the ground for a more quantitative treatment of energy ideas at a later stage.

The move from 'forms of energy' to 'energy stores' is not simply a cosmetic change. It focuses attention on the beginning and end of a process, and shifts attention away from intermediate stages and mechanisms, about which energy ideas have almost nothing useful to say. It is crucially important to choose examples for analysis and discussion that have a clearly specified beginning and end (or to make clear the moments that are being taken as the beginning and end), as energy is essentially a mathematical 'book-keeping' quantity that can be used to compare the states of a group of interacting objects at two moments in time, and not a mechanism. And it is important to make clear the boundaries of the system being considered; for example, when discussing the operation of a mains-powered electrical device, do we include the power station and its fuel within the system, or draw the boundary closer and have an energy flow into the system? It may be easier initially to consider systems which are closed, with no need for an analysis to include energy flows in or out.

Without going into details of possible teaching methods and approaches, it may be worth mentioning the use of diagrammatical representations of change processes in terms of energy transfers, in the Nuffield *Energy and Change* project (Boohan and Ogborn 1996a, b), the Institute of Physics *SPT 11–14* materials (Lawrence 2007), and the teaching intervention developed by Papadouris and Constantinou (2011). These are potentially useful tools for communicating abstract ideas about stores and pathways more effectively, and helping students to appreciate that energy ideas can be applied across a very wide range of phenomena.

The teaching sequence in the Appendix then proposes that, with scientific literacy in mind, all 15–16 year-olds should be helped to develop a fuller understanding of the main methods of electricity generation and of technologies (such as heat pumps) that are becoming more commonly used. And they should have opportunities to discuss and debate important issues around the current and future use of energy resources that will impact on their lives, and on which they should hold informed views. For those students who wish to hold open the option of more advanced study of physics, a more quantitative understanding can be built on the qualitative model introduced earlier, ideas about the particulate nature of matter can be used to explain some of the mechanisms of energy storage and transfer, and students can begin to develop a better model of thermodynamic processes, recognising why two quantities (internal energy and heat) are needed and understanding why they are not identical. The particulate model of matter is also useful here for explaining aspects of energy dissipation, and why it is difficult to use the energy of random motion of particles in an object.

11.5 Testing a Teaching Sequence

Peer review is one important way to test and refine any proposed teaching sequence. Another is to try to write an exposition of the ideas, following the proposed sequence, in words that would be suitable for learners of the target ages envisaged. This could be a 'textbook account', but does not need to be as polished or detailed as this to achieve its purpose – which is to test the sequence and to help identify issues of order and terminology that inevitably arise when you try to 'tell the story' to a learner. The act of writing an account is an important 'experimental test' of the hypothesis that any outline teaching sequence represents. We might expect the hypothesis to require some revision in the light of such a test.

The most powerful way, however, to test any teaching sequence is to specify precisely what we would expect students to be able to *do* if they had successfully learned the things we want them to learn at each stage. Duncan and Hmelo-Silver (2009) point out that the 'current characterization of [learning progressions] emerged from work in the field of assessment, in particular the development of assessment systems designed to track student progress' (p. 606).

In a discussion of the teaching of basic ideas in electricity, Mulhall et al. (2001) highlight the role of assessment by asking 'what, in detail, do we expect students to learn when we talk of "conceptual understanding" in electricity?' (p. 583). Their view is that 'we [the science education community] do not have even the beginnings of systemic answers' (ibid.). They go on to say that 'some justified response to [this question] is a necessary, if not sufficient, condition for any helpful advances in the thinking about and practice of teaching electricity' (ibid.). I agree – and I think the point applies with particular force to the teaching and learning of energy ideas.

Articles and proposals about the teaching of energy, for instance, frequently talk about helping students to gain 'an understanding of the energy concept', and even about 'a full understanding of the energy concept' (as though anyone had, or could claim to have, such a thing). But what is 'the energy concept', and what is 'an understanding' of it? This, I would argue, can only be defined operationally.

We must write specific tasks and questions that, in our view, would provide evidence that an individual student had, or did not have, the 'understanding' we have in mind. This is not, I stress, a technical matter of writing good questions to test an understanding that we can specify clearly in other ways. The meaning – indeed the only meaning – of the phrase 'an understanding of energy' is the ability to give a certain kind of answer to some specific questions or tasks.

Because of the central role that assessment plays in defining our intentions, curriculum development should involve a constant iteration between statements of intended learning outcomes, lesson activities that we think will help students attain these outcomes, and instruments that might provide evidence that a given student had (or had not) attained each outcome. It is a mistake to leave the development of assessment instruments to the end. Developers learn though the process of creating the instruments that can provide evidence of the intended learning. This often forces them to review and perhaps revise their statements about what is to be taught or learned. The result is a clearer specification of the intended learning outcomes, and a better alignment between the stated outcomes and the instruments used to obtain evidence of learning. The assessment instruments also become the most effective way of communicating the intended teaching sequence to teachers and other key actors, such as textbook writers and the writers of questions for high-stakes tests and examinations. (The argument outlined in this paragraph is developed more fully in Millar (2013)).

One reason why the extensive discussion of the teaching of energy over the past half-century has resulted in so little real change in practice is the lack of emphasis on assessment of understanding. To make progress, we need to develop assessment instruments that ask students to do things that we deem worthwhile, and which support and encourage the teaching approach that we want to promote. This is what the project for which this teaching sequence was developed is seeking to do. It is a large task and it is unlikely that we will succeed completely. But it seems the right place to begin.

Acknowledgements My understanding of energy and my ideas about the teaching of energy have been challenged and extended by conversations with Jon Ogborn over many years, and by Jon's writings on the subject. The influence of these may be apparent to readers of this chapter. In particular the idea that, in practice, amounts and units of energy and power are 'socially defined' – understood through social practices rather than from formal definitions – comes from Jon's comments to me on an earlier draft of this chapter.

The development of the teaching sequence in the Appendix has involved discussions with my colleagues at York, Elizabeth Swinbank and Mary Whitehouse, and with Jon Ogborn and Charles Tracy. Their comments on previous drafts have helped to clarify and improve the sequence.

Figure 11.1 is sourced from John Avison's *The World of Physics*, published in 1984 by Nelson. It is reproduced with the permission of Nelson Thornes Limited, Cheltenham, UK.

Appendix: A Proposed Teaching Sequence for the Topic of Energy to Age 16

Student age	Content
Up to age 8	–
9–11	Machines and animals (including humans) use up energy resources to do useful jobs or to heat something
	Fuels (wood, coal, oil, gas, etc.) are important energy resources; food[a] is the fuel for animals
	Wind, sunlight, moving water (flowing rivers, tidal movements, waves) are also important energy resources
	Energy is often supplied by electricity; electricity has to be generated using another energy resource (a primary energy resource)
	Some energy resources are continuously available or can be replaced at the same rate as they are used (renewable); others (such as fossil fuels) cannot
	The origin of fossil fuels
11–12	Amounts of energy needed for different jobs, and supplied by different fuels and foods, can be measured (in joules, J)
	The amount of energy available determines whether a particular job can or cannot be done
	Ratings in watts (W) on electrical appliances indicate the rate at which energy has to be supplied to them to operate them[b]
	Typical power ratings of common domestic mains appliances. Jobs involving heating are relatively expensive in energy terms. Domestic fuel bills; possible savings from insulation, changing lighting, etc.
	Different ways of doing the same job may require different amounts of energy. A process or device is more efficient than another, if it needs less energy do the same job
	Distinguishing between heat (internal energy) and temperature
	Energy (heat) moves spontaneously from an object at a higher temperature to one at a lower temperature, driven by the difference in temperature
	If energy (heat) is transferred to/from an object, its temperature goes up/down (in situations where there is no change of state)
	Thermal conduction
	Thermal insulators: reducing the rate at which energy (heat) is transferred due to a temperature difference
	Convection
	Maintaining a constant temperature: energy in = energy out[c]
	Simple machines: inclined plane, lever, pulley systems, gears. Increasing the force, but at the expense of reducing the distance it moves[d]
13–14	Explaining change: Spontaneous changes are caused by a difference of some kind (e.g. of concentration, temperature, height); a change driven by a difference tends to destroy that difference: batteries run down; objects come to the same temperature

(continued)

(continued)

Student age	Content
	When objects interact, if some lose/gain energy, others gain/lose energy
	Identifying objects, or groups of objects, that have gained (or lost) energy[e] because they:
	Are moving faster (or slower) than they were
	Have reacted chemically
	Are hotter (or cooler) than they were
	Have been raised (or lowered) in a gravity field[f]
	Are magnets (or electric charges) which have been moved apart/together in a magnetic (or electric) field
	Are springy and have been distorted (stretched, compressed, bent), or allowed to spring back from a distorted state
	Convenient labels for these different types of energy store: kinetic, chemical, internal, gravitational, magnetic, electrostatic, elastic
	Energy release in chemical reactions: energy is stored in the set of reagents, not in one of them (e.g. fuel + oxygen; the chemicals in a battery)
	Relating internal energy to the particulate model of matter
	Analysing energy changes in simple process and events in terms of:
	Where energy has come from and gone to (energy stores)
	Identifying objects, or groups of objects, that have gained (lost) energy, and saying how we know they have more (less) energy
	Identifying how the energy was transferred from the initial energy store(s) to the final energy store(s) (energy pathways): mechanically (by a force), electrically (by an electric current), by heating, by radiation
	Efficiency: the fraction of the energy supplied that goes to the intended final energy stores, or that goes by the intended energy pathway
	Heating by friction: a common source of inefficiency
	In any process or event, the total amount of energy of all the objects involved is the same at the beginning and end: energy is conserved. But energy also tends to get more spread out amongst the objects[g] (dissipated)
15–16 (all students)	Electricity generation: thermal power stations, other power stations using electromagnetic induction, photovoltaic cells
	Fissile materials as energy stores (nuclear energy stores)
	How thermal power stations work (in outline terms); efficiency
	Nuclear fusion: source of the Sun's energy; the promise of fusion reactors
	Refrigerators and heat pumps : using energy to create and maintain a temperature difference
	National and global energy resources: patterns and trends in usage. Ways of increasing efficiency of processes and activities. Possible future scenarios, and the issues they raise

(continued)

(continued)

Student age	Content
15–16 (students who may wish to continue the academic study of science beyond age 16)[h]	Transferring energy mechanically: work = force × distance Analysing simple machines: force input/output, energy input/output; efficiency Calculating changes in kinetic energy ($E_k = ½ m v^2$) and gravitational potential energy ($E_g = m g h$). Analysis of simple situations in terms of loss/gain of gravitational potential energy and gain/loss of kinetic energy Recognise that gravitational potential energy is not stored in a raised object, but in a system of two (or more) masses that are interacting gravitationally. Calculating changes in internal energy: thermal capacity; energy of melting/evaporation (latent heat). $E_i = c m \Delta t$; $E_i = l m$ Electrical work: done when a charge is moved within an electric field Power (in watts, W) of a device or appliance as a measure of the rate at which work is done by a battery (or power supply) to operate it[i] Energy transfers in electric circuits: power = current × potential difference; energy = current × potential difference × time Release of energy in chemical reactions, in terms of the energy needed to break bonds in reagents and the energy released when (different) new bonds form Two ways of raising the temperature of an object: by a transfer of heat (as a result of interaction with another object at a higher temperature); or by doing work (mechanically, or electrically). The distinction between heat (energy transferred as a consequence of a temperature difference) and internal energy. Simple thermal processes involving gases All of the energy in some energy stores can be used to do useful jobs; but it is never possible to use all of the internal energy of hot objects to do useful jobs. Energy in disordered (random) motion of particles is difficult to recover and use

Notes:

[a]More specifically, carbohydrate foods

[b]At the stated operating voltage, of course, but no need to emphasise this at this stage

[c]In the absence of any complicating factors such as changes of state, or adiabatic expansion

[d]This might be taught in the context of forces rather than energy. The ideas are not developed at this stage to include a consideration of work input/output, but this prepares the ground for this at age 15–16 for those who may wish to study science further

[e]The emphasis is on gains and losses, as it is *changes* in the amount of energy stored that we can measure, not absolute amounts

[f]At this stage, it is acceptable to see the raised object as the energy store. This can be developed later to a fuller understanding that the store is two or more interacting masses

[g]'The surroundings' are being regarded here as one of 'the objects involved'

[h]Note that this is *in addition to* the material for all 15–16 year old students above

[i]The idea that a power rating indicates the rate at which energy has to be supplied comes much earlier. The development here is that power can be defined in terms of work done

References

Avison, J. (1984). *The world of physics*. Nelson: Walton-on-Thames.

Boohan, R. (2007). *Innovations in practical work: Making energy real*. London: Gatsby Science Enhancement Programme.

Boohan, R., & Ogborn, J. (1996a). *Energy and change. Introducing a new approach; Activities for the classroom; Background stories for teachers*. Hatfield: Association for Science Education. Retrieved October 30, 2012, from http://stem.org.uk/rx7rg

Boohan, R., & Ogborn, J. (1996b). Differences, energy and change: A simple approach through pictures. *School Science Review, 78*(283), 13–20.

Duit, R. (1981). Understanding energy as a conserved quantity. *European Journal of Science Education, 3*(3), 291–301.

Duit, R. (1986). In search of an energy concept. In R. Driver & R. Millar (Eds.), *Energy matters* (pp. 67–102). Leeds: Centre for Studies in Science and Mathematics Education, University of Leeds.

Duit, R. (1987). Should energy be introduced as something quasi-material? *International Journal of Science Education, 9*(2), 139–145.

Duncan, R. G., & Hmelo-Silver, C. E. (2009). Learning progressions: Aligning curriculum, instruction and assessment. *Journal of Research in Science Teaching, 46*(6), 606–609.

Ellse, M. (1988). Transferring, not transforming energy. *School Science Review, 69*(248), 427–437.

Feynman, R. P., Leighton, R. B., & Sands, M. L. (1963). *The Feynman lectures on physics* (Vol. 1). Menlo Park: Addison Wesley.

Heath, N. E. (1974). Heating. *Physics Education, 9*(7), 490–491.

Heath, N. E. (1976). Heating. *Physics Education, 11*(6), 389.

Kaper, W., & Goedhart, M. (2002a). 'Forms of energy', an intermediary language on the road to thermodynamics? Part I. *International Journal of Science Education, 24*(1), 81–95.

Kaper, W., & Goedhart, M. (2002b). 'Forms of energy', an intermediary language on the road to thermodynamics? Part II. *International Journal of Science Education, 24*(2), 119–137.

Lawrence, I. (2007). Teaching energy: Thoughts from the SPT11–14 project. *Physics Education, 42*(4), 402–409.

Linn, M. C., & Songer, N. B. (1991). Teaching thermodynamics to middle school students: What are appropriate cognitive demands? *Journal of Research in Science Teaching, 28*(10), 885–918.

Mak, S. Y., & Young, K. (1987). Misconceptions in the teaching of heat. *School Science Review, 68*(244), 464–470.

Millar, R. (2000). Energy. In D. Sang (Ed.), *Teaching secondary physics* (pp. 1–43). London: John Murray.

Millar, R. (2005). *Teaching about energy* (Research Paper 2005/11). York: Department of Educational Studies, University of York. Retrieved October 30, 2012, from www.york.ac.uk/education/research/research-paper

Millar, R. (2011). Energy. In D. Sang (Ed.), *Teaching secondary physics* (2nd ed., pp. 1–48). London: Hodder Education.

Millar, R. (2013). Improving science education: Why assessment matters. In D. Corrigan, R. Gunstone, & A. Jones (Eds.), *Valuing assessment in science education: Pedagogy, curriculum, policy* (pp. 55–68). Dordrecht: Springer.

Mulhall, P., McKittrick, B., & Gunstone, R. (2001). A perspective on the resolution of confusions in the teaching of electricity. *Research in Science Education, 31*(4), 575–587.

National Research Council. (2012). *A framework for K-12 science education: Practices, cross-cutting concepts, and core ideas* (Core idea PS3 energy, pp. 120–130). Washington, DC: The National Academies Press.

Ogborn, J. (1986). Energy and fuel – The meaning of the 'go' of things. *School Science Review, 68*(242), 30–35.

Ogborn, J. (1990). Energy, change, difference and danger. *School Science Review, 72*(259), 81–85.

Ogborn, J. (n.d.). *Fundamentals*. Retrieved October 30, 2012, from www.nuffieldfoundation.org/practical-physics/fundamentals

Papadouris, N., & Constantinou, C. P. (2011). A philosophically informed teaching proposal on the topic of energy for students aged 11–14. *Science & Education, 20*(10), 961–979.

Solomon, J. (1982). How children learn about energy, or does the First Law come first? *School Science Review, 63*(224), 415–422.

Solomon, J. (1992). *Getting to know about energy – In school and society*. London: Falmer Press.

Summers, M. K. (1983). Teaching heat – An analysis of misconceptions. *School Science Review, 64*(229), 670–676.

Taber, K. (1989). Energy – By many other names. *School Science Review, 70*(252), 57–62.

Tiberghien, A. (1984). Critical review on the research aimed at elucidating the sense that the notions of temperature and heat have for students aged 10 to 16 years. In *Research on Physics Education. Proceedings of the first international workshop* (pp. 75–90). Paris: Editions du CNRS.

Tiberghien, A. (2000). Designing teaching situations in the secondary school. In R. Millar, J. Leach, & J. Osborne (Eds.), *Improving science education. The contribution of research* (pp. 27–47). Buckingham: Open University Press.

Trumper, R. (1990). Being constructive: An alternative approach to the teaching of the energy concept. Part 1. *International Journal of Science Education, 12*(4), 343–354.

Trumper, R. (1993). Children's energy concepts: A cross-age study. *International Journal of Science Education, 15*(2), 139–148.

Warren, J. W. (1972). The teaching of the concept of heat. *Physics Education, 7*(1), 41–44.

Warren, J. W. (1976). Teaching thermodynamics. *Physics Education, 11*(6), 388–389.

Warren, J. W. (1982). The nature of energy. *European Journal of Science Education, 4*(3), 295–297.

Warren, J. W. (1991). The teaching of energy. *Physics Education, 26*(1), 8–9.

Watts, D. M. (1983). Some alternative views of energy. *Physics Education, 18*(5), 213–217.

Wiser, M., & Carey, S. (1983). When heat and temperature were one. In D. Gentner & A. L. Stevens (Eds.), *Mental models* (pp. 267–297). Hillsdale: Lawrence Erlbaum.

Chapter 12
Distinctive Features and Underlying Rationale of a Philosophically-Informed Approach for Energy Teaching

Nicos Papadouris and Constantinos P. Constantinou

12.1 Introduction

Understanding energy is recognized as a major learning objective of school science, starting from the primary school (AAAS 1993). This is reflected in the recently published framework for K-12 science education standards (NRC 2012), in which energy is identified as one of four core concepts for science learning. Despite this wide recognition, teaching about energy is a challenging task (Driver and Millar 1986; Doménech et al. 2007; Solomon 1992). Even though existing research has contributed important findings about students' initial ideas and conceptual difficulties (Driver and Warrington 1985; Duit 1984; Kesidou and Duit 1993; Lawson and McDermott 1987; Solomon 1992) and also some useful insights with respect to possible teaching approaches (Arons 1990; Boohan and Ogborn 1996; Doménech et al. 2007; Nordine et al. 2011; Schmid 1982; Van Heuvelen and Zou 2001) there is still a need for further research on the development of teaching innovations and the improvement of our understanding of issues relevant to teaching and learning about energy.

In this chapter, we begin with an analysis of the epistemological barriers that tend to hamper attempts to introduce the concept of energy. We then propose a novel teaching approach, for middle school, that could contribute toward addressing this instructional challenge. We describe the rationale underlying this teaching approach, we provide an overview of the structure of a corresponding activity sequence that we have developed and we highlight its distinctive features. We conclude with a brief discussion of possible directions for further research in this area.

N. Papadouris (✉) • C.P. Constantinou
Learning in Science Group, University of Cyprus, Nicosia, Cyprus
e-mail: npapa@ucy.ac.cy

R.F. Chen et al. (eds.), *Teaching and Learning of Energy in K-12 Education*, 207
DOI 10.1007/978-3-319-05017-1_12, © Springer International Publishing Switzerland 2014

12.2 Epistemological Barriers Associated with Teaching and Learning About Energy

Teaching about energy constitutes a significant instructional challenge, which is particularly salient in the case of the elementary and middle school grades. This challenge is directly related to the abstract nature of energy as a construct. Striking an appropriate balance between developmental appropriateness and students' available conceptual resources, on the one hand, and epistemological coherence and content validity, on the other, presents an inherently difficult task. Next, we review three characteristics of energy that tend to perplex this task.

12.2.1 Energy Is a Crosscutting Construct

Energy is a crosscutting construct in the sense that it transcends individual, disparate domains such as mechanics, electric circuits, or magnetism. It is a unifying construct (Arons 1999; Holton and Brush 2001) that runs through all these individual domains. Thus, in teaching physics it would not be productive to fragment instruction of energy to individual fields, which, incidentally, is commonplace in conventional teaching. Instead, any attempt to build an epistemologically consistent definition of energy needs to reveal this crosscutting and unifying nature. This imposes a significant challenge to be addressed in any attempt to introduce and elaborate this construct in school science.

12.2.2 Energy Cannot Be Easily Attached to Kinesthetic Content

It is possible to help students develop insights about various concepts of science by anchoring them to experiences linked to familiar situations. For instance, the concepts of time, temperature or force could be readily associated with ideas of growth/development, hotness and push/pull, respectively. Thus, it should be possible to devise teaching proposals that help students develop qualitative, experientially-informed notions of those physical quantities. On the contrary, energy, in its most general sense, i.e., independent of its individual forms, cannot be depicted by means of tangible, concrete representations that could be directly linked with kinesthetic experiences. Energy is an abstract, purely mathematical construct that derives its utility from its conservation, which enables quantitative analysis of the behavior of physical systems (Feynman et al. 1965). This abstract nature of energy tends to hamper attempts to help students develop a satisfactory, intuitively-appealing conceptualization of this construct.

The reader might be tempted to conclude that we are suggesting that students do not possess productive experiential resources, even kinesthetic ones, which could be drawn and built upon in teaching about energy. This is not the case (Hammer et al. 2012). It is indeed possible to help students emerge with intuitive understandings about (and experiential insights into) at least specific aspects of energy. For example, it could be possible to help them appreciate both velocity and mass as relevant parameters that could influence the amount of kinetic energy possessed by a moving object (e.g., through exploring the extent of damage that it could cause by colliding with another object). In a similar manner, students' experiences with the fact that a battery connected to a circuit eventually goes flat, could serve a useful role for motivating the introduction of the features of energy conservation and degradation. However, what we do argue is that while it is possible to help students develop intuitive insights about individual forms of energy it is not possible to help them develop intuitive understanding about energy, in its more general expression, as a crosscutting construct.

12.2.3 Energy Does Not Lend Itself to an Operational Definition

Another epistemological barrier, which is directly connected to the inability to help students emerge with experiential insights into the nature of energy, involves the fact that it is not possible to construct an operational definition for energy. Seeking, formulating and refining operational definitions are essential aspects of science as an enterprise (Bridgman 1945). Additionally, it is also vitally important for science teaching. Developing an operational definition for a concept serves as a very valuable intermediate step towards coherent understanding of that concept (McDermott et al. 1996). It is important for facilitating the differentiation between relevant, although distinct concepts. For instance, it allows differentiating between mass and volume, velocity and position, time interval and time instant, or power, force and energy. Also, operational definitions are particularly powerful in terms of enabling students to generalize beyond the phenomena or specific examples they have encountered during instruction. Thus, from an instructional perspective, helping students formulate operational definitions constitutes a powerful teaching strategy that could foster functional conceptual understanding. This strategy, however, does not seem applicable to energy. It is not possible to develop a workable operational definition for energy that remains true to its epistemological structure and complexity. While it is possible to formulate operational definitions for individual forms of energy,[1] it is not feasible to do so for energy in its general

[1]For instance an operational definition that students could be guided to develop for measuring changes in thermal energy would be the following: the amount of energy that causes the temperature of a gram of water to increase (or decrease) by 1 °C is equal (or equivalent) to 1 calorie.

sense, as an overarching construct (Sexl 1981; Vokos 2010). Thus, attempts to help students develop a simple definition of energy are not likely to be successful. This is illustrated, for example, in the case of the definition of energy as the ability to do work, which is commonplace in the mainstream science textbooks. This definition is, at best, inaccurate, due to its inconsistency with the second law of thermodynamics, which essentially states that not all energy can be used to do work (Hicks 1983; Lehrman 1973). Additionally, it does not seem to be very informative or intelligible from the perspective of students. Given that it posits that mechanical work has already been defined, it might not be obvious why it would be useful to have a different construct representing the capacity to do work. Another issue that further exposes the problematic nature of this definition is that it seems tautological: it essentially defines energy in terms of an energy transfer process.

12.3 A Philosophically-Informed Teaching Proposal About Energy in Middle School

12.3.1 Energy as a Theoretical Framework Rather Than a Physical Quantity: An Alternative Perspective for Elaborating Energy

Conventional teaching about energy has usually relied on purely conceptually oriented approaches. These are exemplified by the *energy as the ability to do work* approach. Underlying this tendency towards conceptually-oriented approaches is the assumption that energy has to be introduced as a physical quantity. However, attempts to introduce energy under this assumption do not seem likely to effectively address the relevant epistemological barriers. For instance they cannot address the question *"what is energy?"* in a manner that is both satisfactory for students and acceptable to physicists.

We take the perspective that any attempt to introduce energy in school science needs to help students address the fundamental epistemological question "what is energy, why is it useful and how do we use it?" in an effective manner. This could have a significant impact on the extent to which students are likely to develop coherent understanding of energy and effectively integrate it in their learning pathway. We believe that a potentially productive way of addressing this question in a manner that avoids the epistemological barriers analyzed earlier, involves disengaging from the assumption that energy has to be dealt with as a physical quantity and shifting, instead, towards a philosophically-oriented perspective. In particular, rather than pursuing a definition of energy as a physical quantity we suggest that the question "what is energy?" could be more usefully situated in a philosophically-oriented context, in which students bootstrap their understandings of energy as they develop and enlarge a *theoretical framework* for it, as the epistemic needs arise. Specifically, one could begin with the idea that, in science, we formulate theoretical frameworks so as to account for observations and phenomena and that this involves inventing and elaborating interpretive frameworks for the operation of these phenomena. Once this idea has been sufficiently elaborated, energy could then be introduced as a theoretical framework that has been invented in science so as to enable the unified analysis of the operation of diverse *physical systems*. The emphasis could then be shifted to the gradual elaboration of this framework, through the introduction of the various features of energy (i.e., transfer, form conversion, conservation and degradation) and its application for the qualitative analysis of systems. Below we seek to demonstrate this rationale in the context of a teaching module that we have specifically developed to embody this teaching proposal.[2]

12.3.2 Overview of the Structure for a Teaching-Learning Sequence

The teaching materials, which are targeted at middle school students, consist of three main parts. The *first*, engages students in explicit and sustained epistemic discourse about certain aspects of the Nature of Science (NOS). The most important of these include (i) the distinction between observations and inference, (ii) the idea that in science we often build theoretical frameworks in order to describe, interpret and predict phenomena, and (iii) the idea that it takes human creativity to invent such theoretical frameworks. These ideas are promoted through three different types of activity, which vary according to the context that is attached to the epistemic discourse. Specifically, students are initially engaged in a couple of activities that are not connected to the disciplinary knowledge of science. Lederman and Abd-El-Khalick (1998) discuss various examples of such activities. One of these activities, which has been included in the teaching/learning materials, presents students with a drawing that illustrates a specific pattern of two different marks and asks them to, firstly, record observations about this pattern (e.g., there are two different sets of black marks that are of different size and shape) and, secondly, come up with possible interpretations as to what that pattern could be indicating (e.g., it could be that the marks were left by two birds that were heading towards a specific spot). This specific activity was intended to engage students in discussion about the distinction between observation and interpretation. The advantage provided by such non-integrated activities is that they provide a very simple context for the initial exposition of students to epistemic discourse.

The second type of activity, transfers the epistemic discourse to science-related topics. For this, we draw on case studies from the history of science, namely Lavoisier's caloric theory and Aristotle's theory of violent and natural motion. Students are presented with brief narratives describing the essential ideas involved

[2] A more thorough description of the underlying rationale and the structure of the teaching materials is given in Constantinou and Papadouris (2012) and Papadouris and Constantinou (2011).

in each of these theories and providing an overview of the range of phenomena they were intended to account for. In each case, students are engaged with identifying in the narratives specific examples of observations that fall within the scope of each theory and the corresponding interpretation. For instance, in the case of the narrative about Lavoisier's caloric theory, students are guided, through specially designed probes, to identify relevant observations (e.g., when two samples of water at different temperatures get mixed the temperature of the mixture attains a temperature, which is always intermediate between the initial temperatures of the two samples) and also how the caloric theory accounted for those (e.g., the amount of caloric in an object determines its temperature; when the two samples of water are mixed, there is a flow of caloric from the sample that is at a higher temperature to the sample at lower temperature; this flow stops as soon as the difference in the temperatures of the two samples cancels out).

Finally, in the third type of activity, epistemic discourse is contextualized in students' own inquiry activities. Specifically, while engaged in system analysis using energy (see description of third section presented next) students are consistently asked to reflect on the epistemological underpinnings of their own activities (e.g. distinction between identifying observable changes in physical systems and pursuing energy-based interpretations for these changes).

The ideas elaborated in this first part of the teaching/learning materials serve to build a philosophically-informed, working framework that students consider as an example of the philosophical ideas they are engaged with during the introduction of energy as an invented construct. These ideas are also revisited repeatedly throughout the instructional materials so as to inform the subsequent elaboration of the various features of energy.

The *second* part of the teaching/learning materials engages students in the identification of changes occurring in physical systems. This is intended to provide students with an appreciation of the notions of *system* and *change.* The learning materials include a variety of systems, spanning various branches of physics, which depict a broad range of changes (e.g., temperature increase/decrease, moving objects coming to halt, objects accelerating from rest, electric bulbs starting to glow etc.). In addition to identifying changes, students are also engaged with the formulation of interpretations for those changes. Initially, they are asked to account for individual changes independently whereas, in the next instance, they are asked to come up with a single interpretation that could account for all these changes. This is indeed a difficult question, which students typically fall short of addressing in an effective manner. However, getting students to engage in this question serves a useful role in the activity sequence in that it provides a context for eliciting the value of a single unifying framework and, hence, prepares the ground for the introduction of energy as a construct that could serve this specific purpose. In particular, energy is introduced at this stage as a theoretical framework that has been invented in science so as to offer a unifying perspective for the analysis of changes occurring in systems drawn from phenomenologically disparate domains.

The *third* part of the teaching/learning materials involves the gradual elaboration of energy as a framework for system analysis, through the introduction of its main features, namely *transfer, transformation, conservation* and *degradation*. These features are introduced in a gradual manner in the context of analyzing pre-selected physical systems. Care is taken to highlight how each of these contributes to the interpretive and predictive power of the energy framework. Energy transfer and transformation are introduced as useful features that could be drawn upon to provide qualitative mechanistic descriptions of the operation of physical systems. One important idea that is consistently promoted throughout the learning materials refers to the distinction between energy transfer processes (e.g., work, heat, sound, light etc.) and forms of energy (kinetic energy, elastic potential energy, internal energy etc.). In this section students are also guided to develop the *energy chain* as a model for depicting these descriptions in a graphical manner. Energy chains consist of arrangements of rectangles (denoting forms of energy) and arrows (denoting energy transfer processes) purporting to describe the operation of physical systems in terms of the corresponding energy changes (transfers and form conversions). These arrangements, essentially, involve diagrammatic depictions of the energy story underlying the operation of systems. Figure 12.1 provides three examples of energy chains associated with the operation of three specific systems.

Energy conservation and degradation are introduced at a subsequent stage and their epistemic utility is directly linked to their facility to provide predictions about changes in physical systems: energy conservation could facilitate the formulation of predictions for changes that are impossible to occur (those that violate the energy conservation principle) whereas energy degradation enables predictions about changes that are very likely to occur (because of the tendency of energy to degrade in quality). For instance, in a simple system involving a soft elastic ball that falls downwards after it had been released from a certain height, one could predict that it could not rebound to a height greater than the original height from which it was released because this would have violated the energy conservation principle. In a similar manner, one could take it a step further to predict that, in real settings, the ball will actually rebound to lower and lower heights, because of the tendency of energy to degrade in quality through heat transfer, including to the environment.

12.3.3 *Key Features of the Teaching Materials*

12.3.3.1 Continual Interplay Between Conceptual Elaboration and Epistemic Discourse

The difficulty to portray energy by means of tangible representations has led to substantial debate as to whether it is feasible (or useful) to help students address the question "what is energy?". One of the positions that had been expressed in this respect suggested that teaching about energy should be postponed until students are able to appreciate its abstract quantitative nature; attempts to introduce energy at an earlier stage run the risk of conveying invalid representations of energy as a material substance (Warren 1986). An opposite position that has been expressed is

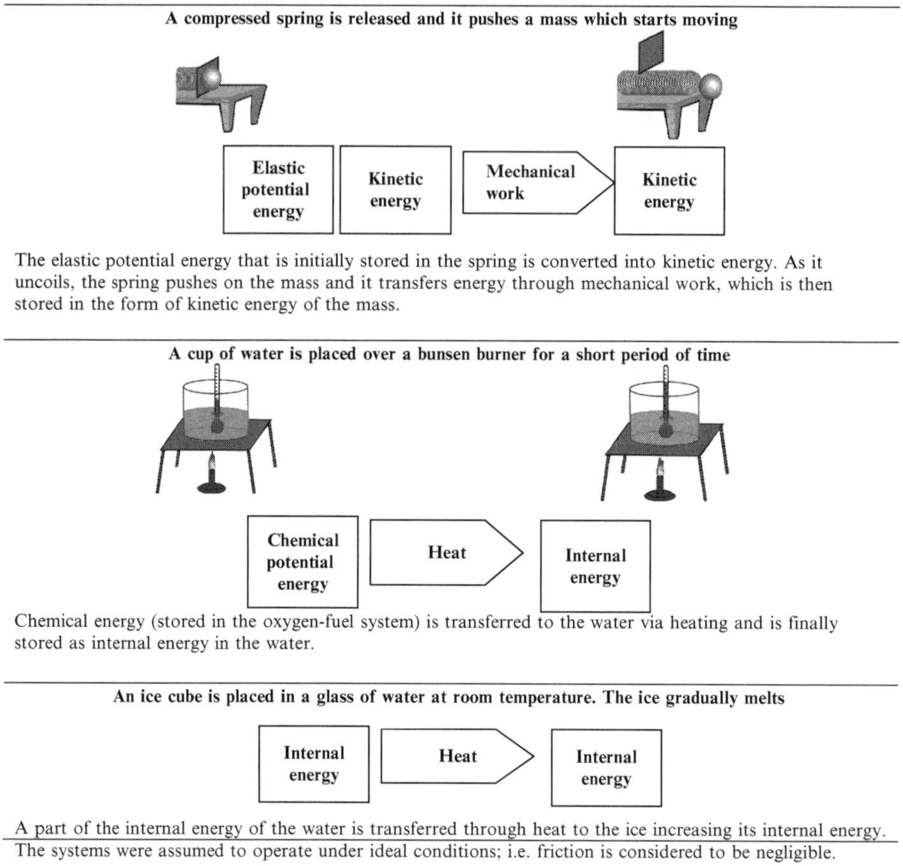

Fig. 12.1 Examples of energy chains

that it should be useful to help students develop energy as a substance-like entity (Duit 1987). This would provide students with access to a productive representation of energy at an early stage. A third position, departs from the premise that given the difficulties associated with defining energy, it should be fine to totally bypass any definition of energy or any attempt to address its nature and focus, instead, on how this construct is used for analyzing the operation of systems (Nordine et al. 2011).

The present teaching proposal takes a different perspective on this. It departs from the premise that despite the difficulties associated with addressing the question "what is energy?", it is still vitally important to tackle this issue and help students, from an early age, develop a qualitative notion of energy. Additionally, rather than seeking to address this question by taking a purely conceptual approach, which is typically beset with various epistemological obstacles, as discussed earlier, we take the perspective that it could be more productive to situate the teaching elaboration of energy in a philosophically-informed context. Underlying this approach is the assumption that engaging students in discourse about fundamental ideas of NOS could provide a useful framing for the introduction and elaboration of energy.

One important feature of this philosophically-oriented approach is the bi-directional interaction between epistemic discourse and conceptual elaboration. Specifically, the epistemic ideas about NOS serve to provide the backdrop against which energy is introduced as a theoretical framework that has been invented in science so as to facilitate the unified analysis of the operation of physical systems. On the other hand, the elaboration of the theoretical framework of energy provides a context for illustrating, and further exploring, these same epistemic ideas. For instance, the features of energy are introduced in a gradual manner and emphasis is placed on illustrating the contribution of each additional feature to the usefulness of the theoretical framework of energy, in terms of its facility to serve as a tool for analyzing (interpreting/predicting) changes occurring in physical systems.

This interplay between epistemic discourse and conceptual elaboration is an overarching feature of the teaching proposal, which is also evident in most of the remaining features of the teaching proposal discussed next.

12.3.3.2 Emphasis on the Crosscutting Nature of Energy

As mentioned earlier, the crosscutting nature of energy tends to perplex attempts to elaborate energy as a physical quantity. This complexity is usually bypassed in conventional science teaching through the introduction of energy in individual domains, which are dealt with in an isolated manner. However, as discussed earlier, dismissing the crosscutting nature of energy is liable to yield an epistemologically misleading account of energy. The proposed philosophically-informed approach allows us to actually capitalize on, rather than ignore, this fundamental characteristic of energy. Specifically, energy is introduced as a response to the quest for a construct that could offer unifying interpretations for changes occurring in physical systems regardless of the domain they are drawn from. Students are explicitly guided to appreciate that while it is feasible to analyze the operation of a physical system without any reference to energy, through concepts drawn from the corresponding fields (e.g., force or momentum for mechanical systems; electric current, voltage or resistance for electric systems), energy becomes a powerful framework for unifying the analysis of changes occurring in different systems.

We chose to focus on systems drawn from different domains of physics while totally excluding chemical or biological systems. One could argue that this appears inconsistent with the explicit objective of the teaching proposal to reveal, and elaborate on, the crosscutting nature of energy. We do acknowledge this limitation and we recognize that the teaching proposal falls short of highlighting the facility of energy, as a crosscutting theme, to intersect and unify the analysis in all sciences. There is significant diversity in how energy is conceptualized in the physical, chemical, life and environmental sciences, which we do not address by limiting our work with young students in the context of physical systems. However, we still believe that highlighting the unifying power of energy, within physics, and helping students emerge with an appreciation of the epistemic value of energy as a unifying framework for analyzing physical systems, is still productive and valuable from an instructional perspective. It enables students to develop a coherent response to the question "why is the energy construct useful in science?" at a fairly early stage, without having to rely on advanced science knowledge. Even this limited account of the crosscutting nature of energy could serve as a starting point which could be further elaborated at subsequent stages with a due emphasis on coherence.

12.3.3.3 Emphasis on Integrating the Features of Energy into a Coherent Whole

In conventional teaching the various features of energy are introduced in a rather declarative manner. The idea that energy manifests itself in various forms and converts from one form to another is often reduced to the mere presentation of technical terms, corresponding to different forms of energy, and the demonstration of various instances of form conversion in the context of the operation of specific systems (e.g., electric systems). Incidentally, this has often been used as an argument against the inclusion of this idea in conventional teaching, on the grounds that this technical terminology is not accompanied by significant gains in terms of conceptual understanding (Ellse 1988). At a subsequent stage, students are introduced to the mathematic formulae for the calculation of the energy stored in specific forms, usually forms involved in the operation of mechanical systems (e.g., kinetic energy, elastic potential energy and gravitational potential energy) and they are guided to apply those in solving simple quantitative *energy problems*. Energy conservation is imposed as a fundamental aspect of nature and the emphasis is usually placed on the value of this principle for solving quantitative problems. The feature of energy transfer is usually taken for granted, probably due to its intuitive basis, whereas degradation does not typically receive consistent attention.

In the proposed teaching approach we seek to help students develop and integrate the various features of energy into a coherent whole. Towards this end, we largely draw on epistemic discourse emphasizing the role of energy as a framework for analyzing physical systems. In this context, we seek to help students appreciate how the various features of energy contribute to the analysis of systems. These features are introduced in a gradual manner and students are guided to appreciate how each contributes to the interpretive and predictive capacity of the theoretical framework of energy. We begin with the feature of energy transfer. We motivate the position that when energy transfers from one part of a system to another it causes some of its measureable attributes to change and, in this context, this feature is introduced as a mechanism for interpreting changes occurring in physical systems. At a subsequent stage, students are guided to appreciate certain limitations associated with energy transfer being the sole energy-based mechanism they can employ to account for changes. Specifically, we guide students to recognize that exclusive reliance on energy transfer tends to lead to somehow vague interpretations that preclude accounts that are specific to individual systems. For instance, the generic statement "because of the energy that was transferred" could be drawn upon invariably to account for changes in a wide variety of physical systems. Additionally, we also guide them to appreciate that there are changes that do not involve any profound energy transfer and, hence, could not be easily accounted for by means of energy transfer alone. Examples of such changes are depicted by a horizontally positioned spring that oscillates or a contracting elastic band.[3] Getting students to identify and elaborate these weaknesses helps elicit the need for an additional feature. Energy transformation is introduced at this stage as a means to alleviate these shortcomings and improve the interpretive capacity of the theoretical framework of energy. Thus, for example, the oscillation of the horizontal spring or the contraction of the elastic band could be accounted for as an instance of form conversion (between elastic and kinetic energy).

In introducing the features of energy conservation and degradation, care is taken to derive them as plausible consequences of the application of the theoretical framework of energy to the analysis of a broad range of systems. For instance, in the case of energy conservation, we seek to help students recognize that the decrease in the quantity of energy stored in a specific part of a (closed) system is always coupled with a corresponding increase of the energy stored in other parts of the system. Students are guided to appreciate this idea in the context of the energy chains they construct. For instance, any energy chain starts and finishes with a rectangle, which denotes a form of stored energy. This is consistent with the idea that, prior to any process, energy is stored in some form and that after the process is over energy again gets stored in some form, probably in other parts of the system. This is also consistent with the idea that the amount of energy associated with the rectangle that appears first in the energy chain decreases whereas there is a corresponding increase in the amount of energy associated with the rectangle that appears at the end of the energy chain.

In a similar manner, in the case of energy degradation students are guided to emerge with experiential insights into the idea that in the operation of any real process there is always an unavoidable (often undesired) increase of the temperature of some of its parts which in turn leads to energy transfer to the surrounding air through heat. We also guide students to reflect on the relative "quality" of the stored energy in terms of the ease with which it can be utilized to bring about desired changes and appreciate the gradual transfer of energy to the surrounding air in the form of internal energy as an indication of reduction in the quality of energy.

In addition to rendering conservation and degradation plausible and consistent with the theoretical framework of energy, emphasis is also placed on revealing their epistemic role in enhancing the predictive capacity of the theoretical framework of energy. In particular, students are guided to appreciate that energy conservation confines the possible configurations that can be attained by a system. Thus, it allows formulating predictions for changes that are impossible to occur (those that would

[3] Even though, both systems involve instances of energy transfer (e.g., through heat or sound) these relate to changes in the system other than the oscillation or the contraction.

violate the energy conservation principle). On the other hand, the contribution of energy degradation in the predictive capacity of the theoretical framework of energy lies on the identification of changes that are highly probable to occur because of the tendency of energy to degrade in quality.

12.3.3.4 Distinguishing Between States and Processes

Conventionally, teaching about energy does not pay much, or consistent, attention to highlighting the important distinction between forms of energy and energy transfer processes. The outcome of this is evidenced in students' tendency to use terms referring to specific forms of energy and corresponding energy transfer processes in an interchangeable manner. Perhaps the most widely documented instance of this confusion is heat (Ellse 1988; Romer 2001). Research evidence suggests that students fail to differentiate between heat (a process that changes the amount of energy stored in a system or an object) and thermal or internal energy (forms of energy stored in objects associated with their temperature). This distinction is not trivial as it hampers students' ability to derive appropriate energy-based accounts for the operation of systems. This is well documented in the research literature, which shows, for instance, that students (even at the college level) tend to believe that heat is the only possible way of increasing the temperature of an object and, for that matter, its thermal energy, neglecting that this same outcome could also emerge as a result of energy transfer through other processes such as mechanical work (Loverude et al. 2002).

Throughout the teaching materials we seek to make this distinction in a systematic manner. Specifically, care is taken to introduce forms of energy and energy transfer processes as distinct components of the theoretical framework of energy. While analyzing systems throughout the activity sequence students are systematically asked to explicitly identify the forms of energy that happen to change as part of the operation of the system of interest and the corresponding energy transfer processes that mediate between these changes. This distinction is also promoted through the energy chain model, in that forms of energy and energy transfer processes are depicted through different shapes (rectangles and arrows, respectively). Also, throughout the teaching materials students are consistently engaged in elaborating on the energy chains they develop and justifying their selection with respect to the shape they had incorporated for different elements of their energy chains.

12.4 Research Agenda

Our current research concentrates on two areas. The first relates to the empirical investigation of what students can achieve through their interaction with learning materials designed to embody the features of the proposed teaching approach.

Indeed, a seemingly fruitful direction for research on teaching and learning about energy involves experimenting with the enactment of teaching innovations in classroom settings so as to address the question: *"what could students achieve, in terms of learning gains about energy, when interacting with specially designed learning environments?"* This could allow exploring what students can potentially achieve under specified instructional conditions, which, in turn, could bring about significant insights into teaching and learning about energy. We have undertaken to empirically investigate the potential of the teaching proposal through its enactment in classroom settings and the collection of data on students' learning outcomes. So far, the teaching/learning materials have been implemented in three intact classes (total of 64 twelve-year-old students) and the empirical findings that have emerged provide quite encouraging indications. For instance, the data suggest that after their interaction with the teaching/learning materials students became clearly better positioned to distinguish between observations and interpretations and to appreciate the role of human invention and creativity in formulating interpretations for phenomena. These seem to have held an instrumental role in helping students appreciate energy as an invented construct and associate it with the epistemic act of interpreting. For instance, prior to the instruction more than half of the students (57 %) rejected the idea that energy has been invented by scientists, on the grounds that invention is not a legitimate component of science since it tends to undermine its trustworthiness. The frequency of this idea underwent a substantial decrease (8 %) after the implementation of the teaching/learning materials. In addition, 55 % of the students (compared to none prior to the implementation) explicitly stated that energy has been invented by scientists because it facilitates the interpretation of phenomena. A more elaborate description of these findings can be found in Papadouris and Constantinou (2014). In a similar manner, in the case of students' ability to employ energy for the analysis of changes in physical systems, our data suggest that, to a large extent, students became able to identify the energy transfers during the operation of simple physical systems, associate forms of energy with specific parts of the corresponding systems, identify the processes that facilitate the energy changes, and synthesize these elements into coherent energy chains providing an interpretation of the operation of the system as a whole. In particular, a significant percentage of students (approximately 60 %) were able to provide coherent energy chains for the changes depicted in the physical systems that we used in our assessment tasks.[4] An additional encouraging finding relates to the correlation between students' performance in analyzing the changes in the different systems we used in our assessment tasks. Specifically, the quantitative treatment of the data yielded a relatively high (approximately 0.6), statistically significant measure of

---

[4]The systems involved changes whose analysis allowed focusing on clearly discernible initial and final stages, associated with specific stores of energy, while glossing over much of what happens in between. One example of such system involved striking a ball with a golf club causing it to accelerate from rest.

association. This provides an encouraging indication as to the coherence underlying students' ability to engage with energy-based analysis of system changes.

The second area that we seek to address in our research relates to the formulation of a proposal about a possible learning progression for energy. This would describe an evolution of how students could reason about the key features of the teaching proposal in a gradually more sophisticated manner. Toward this end we are currently working on the refinement of a variant of the teaching materials, targeted at high school students (ages 16–18), in which we seek to increase the depth of teaching elaboration and include additional aspects of energy. Perhaps the most important of these aspects involves the quantitative nature of energy. The current version of the teaching materials addresses energy mainly in a qualitative manner, even though it does include a few activities that involve a coarse, semi-quantitative analysis. For example, it engages students in drawing inferences (based on the energy chain for the operation of the system under analysis) as to the forms of final storage whose total amount would be equivalent to the amount of energy that was transferred from the initial source. We are currently engaged with the development of a more sophisticated version that would enable a more elaborate version of quantitative analysis of systems. This would be an important addition to the theoretical framework of energy since it would enable quantitative analysis of systems, which is perhaps the most significant contribution of energy in physics and science, more broadly (Feynman et al. 1965).

Acknowledgements The Cyprus Research Promotion Foundation has provided financial support for this work through projects EKTEMA (PENEK20/02) and EPIKOITE (ENISX/0504/15). We would also like to thank Prof. Stamatis Vokos and Prof. Demetris Koliopoulos for valuable and insightful comments on draft versions of this manuscript.

References

AAAS. (1993). *Benchmarks for science literacy*. New York: American Association for the Advancement of Science.

Arons, B. A. (1990). *A guide to introductory physics teaching*. New York: Wiley.

Arons, A. (1999). Development of energy concepts in introductory physics courses. *American Journal of Physics, 67*(12), 1063–1067.

Boohan, R., & Ogborn, J. (1996). Differences, energy and change: A simple approach through pictures. *School Science Review, 78*(283), 13–19.

Bridgman, P. W. (1945). Some general principles of operational analysis. *Psychological Review, 52*(5), 246–249.

Constantinou, C. P., & Papadouris, N. (2012). An epistemologically informed teaching and learning approach on the topic of energy for students aged 11–14: Rationale and empirical results from a pilot implementation. *Studies in Science Education, 48*(2), 161–186.

Doménech, L. J., Gil-Pérez, G., Gras-Martí, A., Guisasola, J., Martínez-Torregrosa, J., Salinas, J., Trumper, R., & Valdés, P. (2007). Teaching of energy issues: A debate proposal for a global reorientation. *Science & Education, 16*(1), 43–64.

Driver, R., & Millar, R. (Eds.). (1986). *Energy matters*. Leeds: University of Leeds.

Driver, R., & Warrington, L. (1985). Students' use of the principle of energy conservation in problem situations. *Physics Education, 20*, 171–176.
Duit, R. (1984). Learning the energy concept in school-empirical results from the Philippines and West Germany. *Physics Education, 19*, 59–66.
Duit, R. (1987). Should energy be illustrated as something quasi-material? *International Journal of Science Education, 9*, 139–145.
Ellse, M. (1988). Transferring not transforming energy. *School Science Review, 69*, 427–437.
Feynman, R. P., Leighton, R. B., & Sands, M. (1965). *The Feynman lectures on physics*. Reading: Addison-Wesley.
Hammer, D., Goldberg, F., & Fargason, S. (2012). Responsive teaching and the beginnings of energy in a third grade classroom. *Review of Science, Mathematics and ICT Education, 6*(1), 51–72.
Hicks, N. (1983). Energy is the capacity to do work – or is it? *The Physics Teacher, 21*(8), 529–530.
Holton, G., & Brush, G. S. (2001). *Physics, the human adventure: From Copernicus to Einstein and beyond* (3rd ed.). New Brunswick: Rutgers University Press.
Kesidou, S., & Duit, R. (1993). Students' conceptions of the second law of thermodynamics – An interpretive study. *Journal of Research in Science Teaching, 30*(1), 85–106.
Lawson, R. A., & McDermott, L. C. (1987). Student understanding of the work-energy and impulse-momentum theorems. *American Journal of Physics, 55*(9), 811–817.
Lederman, N. G., & Abd-El-Khalick, F. (1998). Avoiding de-natured science: Activities that promote understandings of the nature of science. In W. F. McComas (Ed.), *The nature of science in science education: Rationales and strategies* (pp. 83–126). Dordrecht: Kluwer Academic.
Lehrman, R. (1973). Energy is not the ability to do work. *Physics Teacher, 11*(1), 15–18.
Loverude, E. M., Kautz, H. C., & Heron, P. R. L. (2002). Student understanding of the first law of thermodynamics: relating work to the adiabatic compression of an ideal gas. *American Journal of Physics, 70*(2), 137–148.
McDermott and the Physics Education Group at the University of Washington. (1996). *Physics by inquiry* (Vol. II). New York: Wiley.
Nordine, J., Krajcik, J., & Fortus, D. (2011). Transforming energy instruction in middle school to support integrated understanding and future learning. *Science Education, 95*(4), 577–770.
NRC. (2012). *A framework for K-12 science education: Practices, crosscutting concepts, and core ideas*. Washington, DC: The National Academy Press.
Papadouris, N., & Constantinou, C. P. (2011). A philosophically informed teaching proposal on the topic of energy for students aged 11–14. *Science & Education, 20*(10), 961–979.
Papadouris, N., & Constantinou, C. P. (2014). An exploratory investigation of 12-year-old students' ability to appreciate certain aspects of the nature of science through a specially designed approach in the context of energy. *International Journal of Science Education, 36*(5), 755–782.
Romer, H. R. (2001). Heat is not a noun. *American Journal of Physics, 69*, 107–109.
Schmid, G. B. (1982). Energy and its carriers. *Physics Education, 17*, 213–218.
Sexl, R. U. (1981). Some observations concerning the teaching of the energy concept. *European Journal of Science Education, 3*(3), 285–289.
Solomon, J. (1992). *Getting to know about energy: In school and in society*. London: Falmer Press.
Van Heuvelen, A., & Zou, X. (2001). Multiple representations of work-energy processes. *American Journal of Physics, 69*(2), 184–194.
Vokos, S. (2010). *Understanding forms of energy through testing novel cases*. Paper presented at the summer meeting of the American Association of Physics Teachers, Omaha, NE.
Warren, J. W. (1986). At what stage should energy be taught? *Physics Education, 21*(3), 154–156.

Chapter 13
Repairing Engineering Students' Misconceptions About Energy and Thermodynamics

Margot Vigeant, Michael Prince, and Katharyn Nottis

13.1 Introduction

Our primary concern is educating undergraduate engineers, specifically in the areas of thermodynamics and heat transfer. "Heat transfer" is used throughout this chapter to refer to the area within engineering focused primarily on movements of energy in systems that do not create mechanical work. This may be confusing to non-engineers, particularly since "heat" as a form of energy exists only within the transfer between two bodies of differing temperature. However, it is the term-of-art in engineering and a common course title. This course focuses on radiation, conduction, convection, and the design of devices for such transfers, chiefly heat exchangers (a steam-filled room heating radiator is an example of a heat exchanger). "Thermodynamics" as a field within engineering is concerned with energy transformations more holistically, encompassing the design of engines, refrigeration cycles, as well as the behavior of chemical mixtures and chemical reactions. *What* these students need to understand encompasses both the interrelationships and transformations between internal, potential, and kinetic energy, heat and work that enable cars, phones, satellites and skyscrapers. In a typical curriculum, an engineering graduate is taught to be conversant in the equations that describe these conversions, the Laws of Thermodynamics, as well as those that govern the rates at which these movements occur, part of Heat Transfer. Graduates use these equations to design, operate, and improve power plants, chemical plants, and engines that provide the infrastructure for modern society. This, for us, captures the first part of "*why*" students need to learn about energy.

M. Vigeant (✉) • M. Prince
Department of Chemical Engineering, Bucknell University, Lewisburg, PA, USA
e-mail: mvigeant@bucknell.edu

K. Nottis
Department of Education, Bucknell University, Lewisburg, PA, USA

R.F. Chen et al. (eds.), *Teaching and Learning of Energy in K-12 Education*, 223
DOI 10.1007/978-3-319-05017-1_13, © Springer International Publishing Switzerland 2014

The second part of "*why*" is broader. Even if a graduate never designs a chemical or power plant, even if they never touch the equations dealing with energy transformation again after their studies, they still require an excellent understanding of energy related concepts. For example, electric cars are hailed as "zero emissions" (Nissan 2012). Discussion of these vehicles in the popular press reveals that many people do not realize that the electricity needed to charge the car is typically generated by combustion; mass and energy are conserved and while the car is not (or is less of) a point-source for emissions, it is still an emissions source at the power plant. This is not to say such cars are not better; but that we should honestly engage with the issues. However, it is well documented that a technical understanding does not necessarily translate to a conceptual understanding of a given issue; misconceptions in the area of heat, energy, and temperature are well documented (Thomas et al. 1995; Carlton 2000; Jasien and Oberem 2002; Sozbilir 2003, 2004; Sozbilir and Bennett 2007). If engineering graduates cannot take a leadership role in honest evaluation of public energy policy based on solid technical and conceptual understanding, who will?

Students enter our classrooms, not as blank slates, but holding significant ideas about how the world works (Bransford et al. 2000). Even in advanced areas such as engineering thermodynamics, where one would think students' ability to form preconceptions would be limited, students' early life and educational experiences inform their understanding of concepts to as great or greater extent as their understanding of the relevant equations. For example, the same student who would correctly answer the question "Two surfaces in a windowless room are in contact only with each other and with quiescent air at 25 °C. The temperature of one surface is measured and found also to be at 25 °C. What is the temperature of the other surface?" may give an incorrect answer to the more experiential number-free question "Is the temperature of a tile floor higher, lower, or the same as a carpeted floor in the same room?" (Georgiou and Devi Sharma 2012).

Other work in this book focuses largely on initial construction of understanding. Our focus is on repair of misconceptions – misunderstandings about the physical world that go deeper than misunderstanding of factual knowledge but represent a flawed mental model of a fundamental process. Adding facts or heuristics to a flawed foundation will not address the underlying problem; to repair these misconceptions, students must rebuild their mental models in more conceptually accurate ways (Bransford et al. 2000).

While both procedural and conceptual knowledge are valuable, we have seen that at the university level, engagement of students' prior understanding is often neglected in favor of content-driven presentation of theories and equations. However, without the explicit engagement of prior understanding, students often layer new information over their previous incorrect understanding or entirely fail to connect new information at all, leaving them as procedural problem solvers rather than moving them along the trajectory towards expert-level problem solving. Expert-level problem solvers tend to organize their approach conceptually rather than procedurally, which is why we want to encourage students' understanding in this realm in addition to their procedural understanding (Bransford et al. 2000).

It is worth pointing out that persistent misconceptions in these areas are found both in the K-12 student population (Gönen and Kocakaya 2010; Jasien and Oberem 2002; Paik et al. 2007) and in the STEM undergraduate population (Prince et al. 2010, 2012a, b; Sozbilir and Bennett 2007; Yang et al. 2009). It would be a reasonable hypothesis that high school graduates who pursue engineering as undergraduates would be those for whom K-12 STEM instruction had been most successful and compelling, as evidenced by their high grades in math and science and subsequent enrollment in higher education. However, instruction built solely upon "teaching by telling" is not very effective at addressing misconceptions, and computational ability of the type that gets students into undergraduate engineering programs, does not guarantee that conceptual understanding was achieved.

It is the persistence of deep-seated misconceptions in the face of mathematical competence that drives our work. Engineers, called upon to design heat exchangers, will recognize which governing equations to use to describe and specify the system. However, in an equation-free context of societal decision-making, we need engineers who understand concepts as well as equations, both to represent their work appropriately to society at large but also to make competent personal decisions about energy consumption, production, and use.

Beyond engineers, there is a broader societal need for understanding of key ideas in both thermodynamics and heat transfer. The drive for conceptual understanding is even more critical for the general public, as they typically won't have a set of equations to fall back on. The first and second laws of thermodynamics – conservation of energy and increase of entropy – underlie all political and social energy discussions about 'energy independence', 'hydrogen economy', and 'green energy'. But as the thermodynamics is often bringing 'bad news', it is often downplayed as a negative 'point of view' by media coverage, as with the above example of the electric car.

Our work in particular seeks to repair engineering students' misconceptions about the energy-related areas of: the second law, the distinction between enthalpy and internal energy, the distinction between temperature and energy, and the distinction between factors impacting the rate of heat transfer and those impacting the amount of energy transferred. While typical engineering coursework is able to build computational competency in these areas, our work demonstrates that it is less successful at developing conceptual understanding. We will then discuss the inquiry-based activities, built upon the Workshop Physics model (Laws et al. 1999), and their success at repairing students understanding in these concept areas.

13.2 What Are the Challenges We Are Facing in Teaching Students About Energy?

In our study of engineering undergraduates' conceptual understanding of thermodynamics and heat transfer concepts, we developed a concept inventory assessment that captures students' understanding in these areas. Based on our work, engineering undergraduates enter thermodynamics and heat transfer courses with a concept inventory score of less than 50 % on energy-related concepts. After a university course-worth of instruction, students score about 60 % on these same concepts (see Table 13.3 below). This score is much lower than the students' typical grade within the courses, which is largely based upon students' ability to manipulate equations. Our challenge, therefore, is to maintain students' faculty with the equations of thermodynamics and heat transfer while building improved conceptual understanding. Complicating this challenge is the fact that students' preconceptions in these areas are often incorrect and resistant to change through simply 'telling'; something besides simply telling them the correct answer must be done to allow them to repair their understanding (Streveler et al. 2008).

In this section, we will present the common misconceptions that engineering students hold about the important energy-related areas listed above. We selected most of these areas for study because they were identified as both important and difficult to understand in a Delphi study (Streveler et al. 2003).

13.2.1 The Second Law

Burning propane in a gas grill rated to generate 50,000 BTU/h, won't be able to power at 20 hp lawn mower, even though the energy provided per hour by the grill is about the same as the power used by the mower to do its work. This will be true regardless of the sophistication of the system used to match the grill to the mower; although it can be improved by such technologies as fuel cells, it will still be the case that the mower will only be able to perform as work a fraction of the energy that the grill sends to its surroundings as heat. The second law of thermodynamics implies that the amount of work that may be generated by a given energy source is generally less than the total amount of energy that could be transferred by that source (Elliott and Lira 2010). This is perhaps the most important energy-related concept because it is the primary limiting feature in humanity's ability to harness energy to do useful work. When asked why a given system is not 100 % efficient at turning heat into work, a student might point to design factors such as friction and insulation, rather than the impossibility of this conversion based on the second law. Misconceptions about the second law and entropy have been documented elsewhere (Sozbilir and Bennett 2007; Kesidou and Duit 1993). This area is particularly important for society as a whole: the second law limits how far a car can go on a gallon of gas, helps explain why solar panels don't turn all incident light into electricity, and why running a process that produces hydrogen gas from water that is then returned to water in a fuel-cell car, does not in fact even produce as much work as went into the production of the hydrogen in the first place.

13.2.2 Temperature and Energy

Students regularly predict that small sparks of 1,000 °C would create a worse burn than would spilling a tablespoon of scalding water on their hands. The first system, a high temperature cinder with a mass of 0.01 g, will not be able to transfer as much energy to the student's hand as will the boiling water in the second system; the vast difference in mass trumps the difference in temperature. Students will often interpret the temperature of a system as the most direct indicator of the energy content of that system, neglecting other important factors such as the size of the system in question (Miller et al. 2006; Nottis et al. 2010; Prince et al. 2010; Streveler et al. 2008). An alternate way to consider this problem is as confusion between intensive variables (like temperature,) and extensive variables (those that depend on the size of the system under consideration, such as total mass). Misunderstandings about the relationship between temperature and energy have been documented in pre-college students and among scientists in addition to within the college population (Lewis and Linn 1994; Kesidou and Duit 1993).

13.2.3 Rate vs. Amount

If you would like to make your drink cold quickly, should you add 100 g of ice as chips or one big cube? How about if you want to make your drink really cold – will one of those options make the drink colder in the end? It turns out that the drink always reaches the same temperature, although the chipped ice, with its greater surface area for energy exchange, gets there more rapidly. Students often confuse factors impacting the rate at which heat is transferred from one substance to another with factors that impact how rapidly that transfer occurs. While this particular misconception area has not yet been extensively studied in the K-12 population, the authors' experience with children suggests this might be prevalent in that group as well; requests to put a stove or air conditioner on "high" in order to both change temperature "fast" and by "lots!" are fairly common.

13.2.4 Internal Energy and Enthalpy

When considering the energy-related state and processes of a given system, engineers typically consider quantities of and transformations between many different 'kinds' of energy. These are simplified for convenience into only those terms that have significant impact in an engineering context. This presents a challenge, as it means that in speaking of energy, the public, and members of STEM disciplines may have different ways to describe the same situation. In fact, even within the STEM disciplines, chemists may speak of energy differently than physicists, and chemical engineers do not use the same symbols or sign conventions as do mechanical engineers. This particular concept area deals with distinguishing to related but distinct measures of energy and is therefore unique on this list as being important for engineers and scientists, but not really that important for the broader community.

Enthalpy can be thought of as a shorthand notation that captures in one term both internal energy and flow-work. Flow-work is work done by a moving fluid as it pushes the fluid ahead of it out of the way (Koretsky 2004). Students' misconception in this area is that both internal energy and enthalpy terms are equivalent; or, stated another way, that flow work does not exist. Students tend to believe that any energy due to movement is completely described by accounting for kinetic energy, a term that can be significantly smaller than the flow-work for moving gasses. This concept is the most specialized of the concepts discussed here, and its confusion is evident less in everyday situations than in engineering calculations. Mistaken substitution of enthalpy for internal energy or vice versa could result in a calculation that over or under represents the energy change of a given system.

13.3 What Should Be Done to Meet These Challenges?

In the previous sections, we have defined the concept areas within energy that have been identified as particularly important and challenging for engineering students, as well as the common misconceptions that make learning these ideas a particular challenge. In this Section we describe our approach to repair of students' misconceptions and the success of the approach.

13.3.1 Inquiry-Based Approach

Laws et al., in Workshop Physics, suggest that inquiry-based activities are significantly more effective than lecture for repair of misconceptions in physics (Laws et al. 1999). In our work, we adapted their approach to create and test activities for the concept areas given above. Laws et al. cite several key aspects of inquiry-based activities, shown in Table 13.1.

Table 13.1 Elements of inquiry-based activity modules (Laws et al. 1999)

(a) "Use peer instruction and collaborative work
(b) Use activity-based guided-inquiry curricular materials
(c) Use a learning cycle beginning with predictions
(d) Emphasize conceptual understanding
(e) Let the physical world be the authority
(f) Evaluate student understanding
(g) Make appropriate use of technology
(h) Begin with the specific and move to the general"

In our implementation of this approach, students are presented with a physical situation or simulation quite similar to a question known to elicit misconceptions. While a physical activity was our design preference, interactive simulations were used when it was impractical or impossible to make the influence of key variables accessible. This approach to distinguishing when to use simulation and when experiment is appropriate and can be effective for instilling conceptual change (de Jong et al. 2013).

In the internal energy vs. enthalpy concept area, students are asked how the temperature of air emerging form a fan compares to the temperature of the air that entered the fan. Students record their prediction on a worksheet, and then engage in the experiment or simulation, making observations, 'playing' with the simulation/equipment, and answering questions as they work. Students are encouraged to 'play' with the activity and assure themselves that it is not a trick. In the case of the fan example, students may measure the temperature of air entering and exiting a hair-dryer (with the heating element 'off'), as well as air flow rate and the electrical energy entering the system. They "play" with it by altering the speed of the fan, measuring the temperature at multiple points in the air-stream, all verifying for themselves that, yes, the temperature actually is higher at the outlet than it was at the inlet. Finally, students are asked for a written reflection on their original prediction and an assessment of whether or not their original understanding was correct, and are encouraged to discuss this with their peers. In the case of the fan example, students discover that the temperature of the air emerging is higher than the air that entered; in part, this is because the motor grows warm. However, even were this not the case, they are able to determine that the temperature would still rise due to the work done by the fan on the air – energy is conserved!

In addition to incorporating the elements of Table 13.1, we designed activities to take no more than 20 min and require materials commonly available or that could be purchased for less than $20. The activities are intended to be used as experiments within the laboratory sequence accompanying college-level engineering courses on thermodynamics or heat transfer. Students might complete the activities during laboratory time, and then complete the questions and post-analysis as homework or as a lab report. In some universities, these courses do not have an accompanying laboratory, and are therefore completed in class, as in-class demonstrations, or as homework in the case of the activities that are simulations. Two inquiry-based activities were created for each concept area, and are summarized in Table 13.2 below.

13.3.2 Results and Discussion

In order to determine the extent of change in conceptual understanding in students' as the result of the inquiry-based activities, we administered concept inventories. These instruments are developed to be measures of conceptual understanding. A typical concept inventory is a multiple-choice assessment wherein the incorrect

Table 13.2 Summary of activities

Concept area	Activity	Type
Second law	Carnot engine: students control the temperature of the heat source and sink as well as the level of friction inside a virtual power plant, and track the resulting efficiency. The virtual power plant operates on the Carnot cycle, the most efficient process for turning heat into work	Simulation
	Cycle modeler: students control a piston-cylinder in which there is a fixed amount of an ideal gas. Students can change the pressure, temperature, or volume of that gas, or put the gas through an adiabatic (no heat exchange) step. By controlling a series of steps, students can try to create a cycle that produces work (an engine cycle) of their own design, and interactively attempt to create a system where all heat energy into the system is turned into work by the system. Through extensive experimentation, they reinforce the idea that such a system is not possible	Simulation
Internal energy vs. enthalpy	Hair-dryer: students measure the temperature of air entering and leaving the hair dryer and determine the source(s) of the observed temperature increase, which is, in part, due to the increased enthalpy of the outlet stream	Experiment
	Filling: students measure the sharp temperature increase that occurs when air is allowed to flow back into an evacuated container. Subsequent discussion and calculation reveals that this is because the flow-work exerted by the incoming air becomes a temperature increase in the closed system	Experiment
Temperature vs. energy	Liquid N_2: students add a small amount of boiling water or a large amount of ice water to identical cups of liquid nitrogen and observe which results in the most boil-off of the nitrogen	Experiment
	Adiabatic valve: students allow three gases to flow through a valve, expanding as they go, and predict the resulting temperature change	Simulation
Rate vs. amount	Ice-chips vs. Ice Cube: students add equal masses of ice in the form of chips and a single 'snowball' to identical cups of water and observe water temperature vs. time. Chips melt faster, but both cups reach the same final temperature	Experiment
	Hot blocks: students control the physical parameters of virtual metal blocks – surface area, density, temperature, as well as the number and size of blocks, added to equal masses of ice, allowing them to see which factors change how much heat is transferred and which factors contribute to how fast that transfer occurs	Simulation

Full activity packets available upon request from the corresponding author

33. A conventional electric table-top fan is placed in a large room and turned on, causing air to move as shown below (dotted line shows system boundary).

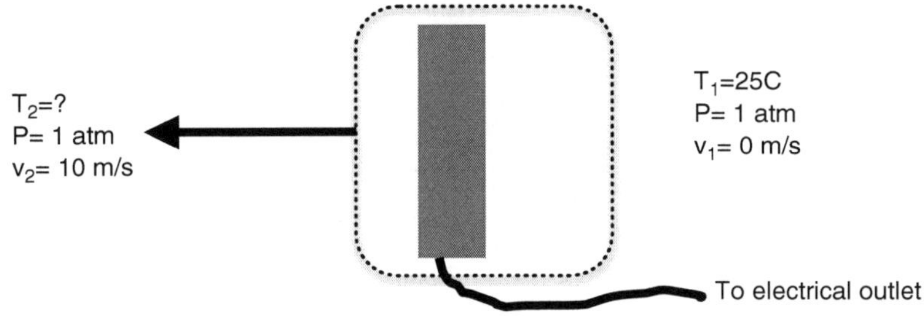

Air behind the fan is at room temperature (T_1=25°C), atmospheric pressure, and is not flowing. Air coming out of the fan is also at atmospheric pressure, but is at a higher net velocity (v_2=10 m/s). Recall that at room temperature, the kinetic theory of gasses tells us ideal gas molecules are typically moving at about 500 m/s. How do you expect the temperature of the air coming out of the fan to compare to the temperature of the air entering?

a) The temperature of exiting air will be slightly lower than of entering air because of convective cooling
b) The temperature of exiting air will be identical to that of entering air because there is no change in air pressure
c) The temperature of exiting air will be slightly higher than that of entering air because of its higher kinetic energy
d) The temperature of exiting air will be slightly higher than that of entering air because of the work added to the system by the fan

Fig. 13.1 A question on the "internal energy vs. enthalpy" concept. Most correct answer is "d"; most popular pre-test answer given by students is "a"

answers are distractors, items specifically written to be attractive to students holding well known misconceptions. For example, Fig. 13.1 shows a concept inventory question similar to the example activity described above.

The distractors in Fig. 13.1 were created based upon students' responses to a similar open-ended question. Option "a" is attractive to students who consider only their personal experience with fans; in discussing this answer with students, it's common for those who choose it to think briefly and immediately realize that there is no mechanism by which the air temperature is actually lowered by passage through the fan. Option "b" is what often directly follows for those who realize "a" must be incorrect; if one does not consider the energy balance (sum of all energetic inputs and outflows from the system, in this case, the fan and air passing through it), there appears to be no energy change. The most conceptually sophisticated distinction is between options "c" and "d". In both of these, students recognize that for energy to be conserved, the electrical work done on the system must result in some change. Option "c" is incorrect because it is not only the kinetic energy of the air that is changing (and measurements show that the kinetic energy change is actually quite

Table 13.3 Impact of activities on conceptual understanding

Concept area/measurement instrument	Control pre-	Control post-	Activities pre-	Activities post-
Entropy/CEIT	50.0 % (n = 271)	63.9 %* (n = 231)	49.6 % (n = 649)	68.0 %* (n = 205)
Internal energy vs. enthalpy/CEIT	26.5 % (n = 271)	38.5 %* (n = 231)	29.5 % (n = 649)	55.5 %* (n = 205)
Temperature vs. energy/HECI (Prince et al. 2012a, b)	53.6 % (n = 373)	56.4 %* (n = 344)	52.2 % (n = 463)	62.7 %* (n = 392)
Rate vs. amount/HECI (Prince et al. 2012a, b)	36.8 % (n = 373)	42.6 %* (n = 344)	33.3 % (n = 463)	63.5 %* (n = 392)

*Significant change at the $p < 0.01$ level; Note the large change in "n" from pre- to post- for entropy and internal energy is due to the data set being broken into subsets of students known to have completed *all* activities (shown) and students known to have completed only *some* of the activities (not shown). This group was omitted from this data set because we could not be assured they had completed the activities for the relevant concept areas at this time

small relative to the other energy inputs and outflows from this system). This leaves us with "d", in which is the most accurate because it encompasses not only the (small) kinetic energy term, but also internal energy and flow work. While this question may be answered correctly without reference to enthalpy directly, it cannot be answered correctly without recognition that kinetic energy is not sufficient to describe the energy flows in this system, which connects directly to the underlying misconception about the non-existence of flow work described in the Internal Energy vs. Enthalpy section above.

We developed, established the reliability of, and administered concept inventories in these concept areas to document the conceptual change as a result of using inquiry-based activities. The Heat and Energy Concept Inventory (HECI) (Prince et al. 2012a, b) and Concept Inventory for Engineering Thermodynamics (CIET) (Vigeant et al. 2011) were used. Each of these benefitted significantly from questions developed for the Thermal and Transport Concept Inventory (Miller et al. 2011). All three of these concept inventories are available to faculty online in full through the AIChE Concept Warehouse. For the 'control', the test was administered in the first 2 weeks of the relevant course, and again in the final 2 weeks, with typical instruction occurring during the course. For the 'activities' case, students participated in the inquiry-based activities described above, taking the same pre- and post- tests. The results reflect undergraduate engineering students responses from at least ten different institutions, both large and small, public and private, distributed throughout the United States. Number of students responding in each grouping shown in Table 13.3.

For each concept area, there are between 4 and 10 questions on the concept inventories. Somewhat less than half of these questions relate directly to phenomena observed in activities ("near transfer"), while the remainder ask about the same concepts in novel situations ("far transfer").

In every case, students score significantly higher after instruction than before, as shown by paired t-tests. However, using ANCOVA, students performing activities outperform students who did not have access to activities significantly in every case but for the "entropy" concept area, where the slight improvement shown is not statistically significant based on current data analysis. As described in more detail by (Prince et al. 2012a, b), in the heat transfer concepts, students using activities improved significantly more than their non-activity using peers for both near- and far- transfer questions, although their improvement for near-transfer questions was larger.

Overall, the inquiry-based activities result in significant improvements in engineering students' conceptual understanding in energy-related areas. While scores do not approach 100 %, this is a significant improvement in conceptual understanding as assessed at the end of the course that only requires the investment of 20–40 min of class time. This is particularly notable because these measurements are taken up to 3 months after the activities were initially completed. The immediate post-activity understanding of the specific situation presented within the activity is nearly 100 %, based upon the long-answer post-processing questions asked as the 'reflection' component of the activity. This drops to the ~60 % we see by the end of the semester. So the changes appear to be relatively long-lived for a portion of the population. Making these changes persistent for a greater fraction of the students is part of our next challenge.

The extent to which either traditional class or class plus activities impact students' understanding is varied by concept area in ways it is challenging to interpret. Chi states that misconceptions related to emergent processes, such as those relating to heat, are intrinsically more robust (Chi 2005). It is also possible that areas where conceptual understanding is based on significant life experience, such as Temperature/Energy and Rate/Amount, are more difficult to change with 'typical' classes because instruction challenges ideas that students have 'known' for a very long time; temperature and energy misconceptions are documented for elementary school students (Confrey 1990). A direct challenge to their misconceptions however, such as comes from the unexpected results from the experiments in Table 13.3, can have a more lasting impact on concepts than learning of the relevant equations (Treagust 2007). By contrast, the area of Enthalpy/Internal Energy typically requires a minimum of advanced high-school science in order for students to form either correct or incorrect ideas about the subject at all. Therefore, having less accumulated misunderstanding to correct, a greater level of success is possible with typical class and with inquiry-based activities.

The area of the Second Law/Entropy is more complicated. While it would seem that, like the Enthalpy/Internal Energy area, it requires significant STEM coursework to 'achieve' confusion, misunderstanding of this concept is widespread. The idea that systems wind-down and that one can't extract as work out the energy that one put in seems antithetical to the idea of working hard to get ahead. However, of all four concepts here, this one tends to receive the most direct and sustained attention during typical coursework, which may explain why the gain with the addition of activities is not significant as for the other areas.

Returning to the original question, "How should energy be taught?", we can endorse students' active engagement and inquiry as key elements in developing long-term understanding, as has been demonstrated in physics education (Hake 1998; Deslauriers et al. 2011). These particular activities for Temperature/Energy and Rate/Amount could easily be adopted at the high-school level. The current 2nd Law/Entropy activities rely upon students possessing an understanding of 'power cycle' that may or may not be present until after high school physics.

An ongoing challenge is expanding the number of heat transfer and thermodynamics courses using these activities. Hidden in the data is a resource problem; not all faculty who participated in the study used all of the activities, for a variety of reasons. In our current work, we are examining the impact these changes have on students' learning, and the extent to which it is possible to adapt activities such that they are easy for the instructor to implement while maintaining effectiveness for students. Preliminary results suggest that pressures of space and money inhibit widespread implementation of experimental activities; even small-scale activities require significant money and laboratory space when run in classes with 100 students or more. Further, with downward pressure on the number of credits students take, most courses of this type do not have dedicated laboratory sessions; many faculty find it unreasonable to implement a 20 min activity within a 52 min lecture. In our ongoing work, we will be assessing the impact of the actives when used as demonstrations or when implemented only as simulations, and comparing this to the more time- and material- intensive model we currently follow.

We believe it is vital for engineering students to understand, both procedurally and conceptually, concepts relating to the second law, enthalpy and internal energy, the distinctions between temperature and energy, and the distinctions between factors that make energy transfer more quickly and those that transfer more energy. It is clear that simply having the ability to appropriately manipulate the relevant equations is insufficient to cultivate conceptual understanding, particularly for situations that students perceive as part of the "real world" as opposed to as engineering classroom constructs. Our inquiry-based activities take a total of 20–40 min per concept area, and are able to shift students' conceptual understanding significantly upwards. While there is still room for growth, the payoff in understanding relative to the time invested is promising. We encourage our colleagues to consider blending procedural instruction with activities such as these, to help students solidify their conceptual understanding and thereby bridge the gap between equations and the real world.

Acknowledgements The authors acknowledge the National Science Foundation for funding support DUE#0442234 and #0717536; Ruth Streveler and Ron Miller for concept inventory development and activity input; and John Pershetti for activity input. Students contributing to this work include: Daniel Prudente, Jeff Detrich, Emily Eherenberger, and Jeff Stein for activity development and programming.

References

Bransford, J., Brown, A., & Cocking, R. (2000). *How people learn: Brain, mind, experience and school.* Washington, DC: Commission on Behavioral and Social Science and Education, National Research Council.

Carlton, K. (2000). Teaching about heat and temperature. *Physics Education, 35*(2), 101.

Chi, M. T. H. (2005). Commonsense conceptions of emergent processes: Why some misconceptions are robust. *Journal of the Learning Sciences, 14*(2), 161–199.

Confrey, J. (1990). A review of the research on student conceptions in mathematics, science, and programming. *Review of Research in Education, 16*, 3–56. Retrieved from, http://www.jstor.org/stable/10.2307/1167350

de Jong, T., Linn, M. C., & Zacharia, Z. C. (2013). Physical and virtual laboratories in science and engineering education. *Science, 340*(6130), 305–308. doi:10.1126/science.1230579.

Deslauriers, L., Schelew, E., & Wieman, C. (2011). Improved learning in a large-enrollment physics class. *Science, 332*, 862–864 (13 May).

Elliott, J. R., & Lira, C. T. (2010). *Introductory chemical engineering thermodynamics.* New York: Prentice Hall.

Georgiou, H., & Devi Sharma, M. (2012). University students' understanding of thermal physics in everyday contexts. *International Journal of Science and Mathematics Education, 10*(5), 1119–1142.

Gönen, S., & Kocakaya, S. (2010). A cross-age study: A cross-age study on the understanding of heat and temperature. *Eurasian Journal of Physics and Chemistry Education, 2*(1), 1–15.

Hake, R. R. (1998). Interactive-engagement versus traditional methods: A six-thousand-student survey of mechanics test data for introductory physics courses. *American Journal of Physics, 66*(1), 64–74.

Jasien, P. G., & Oberem, G. E. (2002). Understanding of elementary concepts in heat and temperature among college students and K-12 teachers. *Journal of Chemical Education, 79*(7), 889–895.

Kesidou, S., & Duit, R. (1993). Students' conceptions of the second law of thermodynamics–An interpretive study. *Journal of Research in Science Teaching, 30*(1), 85–106.

Koretsky, M. (2004). *Engineering and chemical thermodynamics.* New York: Wiley.

Laws, P., Sokoloff, D., & Thornton, R. (1999). Promoting active learning using the results of physics education research. *UniServe Science News*, 13.

Lewis, E. L., & Linn, M. C. (1994). Heat energy and temperature concepts of adolescents, adults, and experts: Implications for curricular improvements. *Journal of Research in Science Teaching, 31*(6), 657–677.

Miller, R. L., Streveler, R. A., Olds, B. M., Chi, M. T. H., Nelson, M. A., & Geist, M. R. (2006). *Misconceptions about rate processes: Preliminary evidence for the importance of emergent conceptual schemas in thermal and transport sciences.* Proceedings from presented at ASEE Annual Conference, Chicago.

Miller, R., Streveler, R., Olds, B., & Slotta, J. D. (2011). *Thermal and transport concept inventory.* Retrieved from, http://www.thermalinventory.com/surveys.html

Nissan. (2012). *Features: Power management (Nissan Leaf advertising materials).* Retrieved November 2, 2012, from, http://www.nissanusa.com/leaf-electric-car/key-features?next=ev_micro.root_nav.features

Nottis, K., Prince, M., & Vigeant, M. (2010). Building an understanding of heat transfer concepts in undergraduate chemical engineering courses. *US-China Education Review, 7*(2), 1–8.

Paik, S.–.H., Cho, B.-K., & Go, Y.-M. (2007). Korean 4- to 11-year-old student conceptions of heat and temperature. *Journal of Research in Science Teaching, 44*(2), 284–302. doi:10.1002/(ISSN)1098-2736.

Prince, M., Vigeant, M., & Nottis, K. (2010). Assessing misconceptions of undergraduate engineering students in the thermal sciences. *International Journal of Engineering Education, 26*(4), 880–890.

Prince, M., Vigeant, M., & Nottis, K. (2012a). *Using inquiry-based activities to repair student misconceptions related to heat, energy, and temperature*. Proceedings from Frontiers in Education, Seattle, WA.

Prince, M., Vigeant, M., & Nottis, K. (2012b). Assessing the prevalence and persistence of engineering students' misconceptions in heat transfer. *Journal of Engineering Education, 101*(3), 412–438.

Sozbilir, M. (2003). A review of selected literature on students' misconceptions of heat and temperature. *Bogazici University Journal of Education, 20*(1), 25–40.

Sozbilir, M. (2004). Students' ideas and misunderstandings of enthalpy and spontaneity: A review of selected researches. *Hacettepe University Journal of Education, 26*, 155–159.

Sozbilir, M., & Bennett, J. M. (2007). A study of Turkish chemistry undergraduates' understandings of entropy. *Journal of Chemical Education, 84*(7), 1204–1208. Jul 2007.

Streveler, R., Olds, B., Miller, R., & Nelson, M. (2003). *Using a Delphi study to identify the most difficult concepts for students to master in thermal and transport science*. Proceedings from presented at ASEE annual conference, Nashville, TN.

Streveler, R. T., Litzinger, T., Miller, R., & Steif, P. (2008). Learning conceptual knowledge in the engineering sciences: Overview and future research directions. *Journal of Engineering Education, 97*(3), 279–294.

Thomas, M. F., Malaquas, I., Valente, M., & Antunes, M. (1995). An attempt to overcome alternative conceptions related to heat and temperature. *Physics Education, 30*(1), 19–26.

Treagust, D. F. (2007). Research-based innovative units for enhancing student cognitive outcomes and interest in science. In *Contributions from science education research* (pp. 11–26). Springer. Retrieved from, http://link.springer.com/chapter/10.1007/978-1-4020-5032-9_2

Vigeant, M., Prince, M., & Nottis, K. (2011). *Engineering undergraduates' conceptual understanding of thermodynamics: Assessment and change after normal instruction*. Proceedings from Hawaii international conference on education, Honolulu, HI.

Yang, D., Streveler, R., Miller, R., & Santiago, A. (2009). *Repairing misconceptions: A case study with advanced engineering students on their use of schema training modules*. Proceedings from presented at ASEE annual conference, Austin, TX.

Part IV
Opportunities and Approaches for Teaching and Learning About Energy

The *Framework for K-12 Science Education Standards* (National Research Council 2012) outlined a bold vision for future energy instruction in elementary and secondary schools. By naming energy as both a disciplinary core idea and a crosscutting concept, the *Framework* emphasized the critical role that energy plays in making sense of the natural world and in providing an analytical lens that can cross disciplinary boundaries. The *Framework* also stresses the importance of weaving disciplinary core ideas and crosscutting concepts with science and engineering practices in order that students develop understandings that are useful in practice. Yet, traditional energy instruction typically fails to connect with the world beyond school walls and does not demonstrate how the energy-related ideas in one scientific discipline connect to the energy ideas in another. In this instructional environment, students commonly fail to build more sophisticated understandings of energy over time. While some large-scale cross-sectional studies suggest that students progress in their understanding with more exposure to school, this progress is slow and very few students develop a deep understanding of energy conservation by the time they finish secondary school. While some attribute this slow and difficult progress to the abstract nature of the energy concept and fundamental cognitive limitations associated with maturation, there exists compelling evidence that the problem is primarily instructional. A growing body of literature is demonstrating that non-traditional instructional methods (i.e., methods that emphasize conceptual development through scaffolds for student reflection and investigations to gather evidence for the role of energy in real-world systems) show promise for helping students build more sophisticated energy understandings much sooner than they tend to during traditional instruction (Lacy et al. 2014; Nordine et al. 2011; Trumper 1990). Additional research shows the importance of learning concepts using science and engineering practices (Duschl et al. 2007; McNeill et al. 2006). In this way, students develop understanding of the energy concept that they can use to solve problems, explain phenomena and learn more.

The chapters in this part contribute to the growing base of knowledge about how to design instruction that helps students develop robust understandings of energy that will enable them to reason conceptually about a wide range of phenomena and to use the energy concept in scientific settings to analyze the behavior of systems under study. While each chapter describes a different approach to energy instruction, all approaches build upon existing research into students' thinking about energy to develop strategies tuned to different ages and learners. In this way, the chapters provide images of how teachers can support students' energy learning over time and thus provide insight into the instructional approaches that will be necessary to meet the ambitious goals for energy learning laid out by the *Frameworks*.

The first chapter in the part is written by Sara Lacy, Roger Tobin, Marianne Wiser, and Sally Crissman and focuses on energy instruction in grades 3–5. While some have argued that elementary students are simply too young to learn about energy, Lacy, et al., argue that elementary students are capable of progressing toward a more sophisticated understanding of energy by learning to view systems through an "Energy Lens". This lens, which is developed over many years, prompts students to evaluate familiar systems by asking questions such as "What is the system of interest?", "Where does the energy come from? Where does it go?", and "What is the evidence for our answers?". The authors propose a preliminary energy learning progression for grades 3–5 that builds on students intuitive ideas about energy, and they provide *in situ* examples of how elementary students can grow in their ability to analyze simple systems from an energy perspective.

The second chapter also focuses on energy learning at the elementary level, but takes a different approach. Kristen Wendell analyzes a set of published elementary engineering activities and contends that there are many existing opportunities to explicitly teach elementary students to reason about systems using an energy perspective, though these opportunities are not always emphasized. Wendell argues that educators should capitalize on such opportunities to help students in developing an "applied knowledge of energy" that prompts them to become aware of the need for an energy input for systems/technologies, to evaluate energy storage options, and to consider strategies for energy transfer within and across systems. Though they have a different focus, these goals align with the questions that Lacy, et al., encourage students to consider in their Energy Lens.

Both elementary-focused chapters in this part emphasize the importance of student interactions with real systems and gathering evidence regarding the behavior of energy. In the third chapter, Angelica Stacy, Karen Chang, Janice Coonrod, and Jennifer Claesgens echo this instructional emphasis but focus on high school chemistry students. Stacy, et al., argue that chemistry instruction has too often relied upon rote (and often incorrect) energy simplifications, such as "chemical bonds store energy and release it when they are broken". They stress that chemistry instruction must guide students away from such tempting simplifications by engaging them in inquiry-oriented activities in which they build evidence for the role of energy in chemical processes over time. This chapter describes a set of connected activities from the *Living Chemistry* curriculum in which students investigate evermore sophisticated questions as they explore how an energy perspective can help them to predict and explain fire.

In the fourth chapter, Melanie Cooper, Michael Klymkowsky, and Nicole Becker also focus on teaching students to use energy ideas to analyze chemical systems, but emphasize that current chemistry instruction at the collegiate level often fails to help students understand how different approaches to energy are interconnected. By unpacking how energy is used in the atomic-molecular, macroscopic, quantum mechanical approaches to chemical systems analysis, the authors describe how these lenses connect with each other and discuss how many chemistry classes ignore these links. They describe how a new introductory chemistry curriculum, *Chemistry, Life, the Universe and Everything* (CLUE), structures energy instruction to develop the three approaches over time and in conjunction with each other.

The fifth chapter, written by Rui Wei, Lei Wang, and William Read, explores a commonly used approach to energy instruction – the use of metaphor. By analyzing the benefits and drawbacks of several common metaphors related to teaching entropy, the authors contend that metaphors can help students to understand nuanced energy-related principles, but that this approach comes at a cognitive price.

The sixth chapter also focuses on conceptual tools to reason about the energy in systems, but Lane Seeley, Stamatis Vokos, and Jim Minstrell focus on providing conceptual and pedagogical tools for teachers. They describe a series of professional development activities in which teachers track energy transfers and transformations in everyday systems. These activities include an "Energy Theater", in which teachers act out energy changes as a group, "Energy Cubes", where teachers track energy changes by moving and flipping marked plastic cubes, and "Energy Tracking Diagrams", in which teachers design their own representations for energy changes in phenomena. In the activities, teachers address nuanced phenomena with physical, biological, and chemical connections – e.g., the energy changes as a person *lowers* a ball with a constant speed. Because they focus on familiar energy-related questions that commonly flummox students and teachers alike, the authors demonstrate the power of their approach in helping students use an energy perspective to reason about the non-idealized phenomena that they will encounter in their lives outside of school.

While the chapters in this part represent a range of instructional approaches for a variety of learners, they have some common elements: a focus on explicit connections between ideas and activities, using a consistent analytical lens in a variety of contexts, engaging students in various science practices, and stressing that learners' energy ideas must be built gradually over time and through interactions with familiar phenomena. These elements align with the *Frameworks* and will no doubt form a cornerstone of successful K-12 energy instruction into the future.

References

Duschl, R. A., Schweingruber, H. A., & Shouse, A. W. (2007). *Taking science to school: Learning and teaching science in grades K-8*. Washington, DC: National Academies Press.

Lacy, S., Tobin, R., Wiser, M., & Crissman, S. (2014). Looking through the energy lens: A proposed learning progression for energy in grades 3–5. In B. Chen, A. Eisenkraft, D. Fortus, J. Krajcik, K. Neumann, J. Nordine, & A. Scheff (Eds.), *Teaching and learning of energy in K – 12 education*. New York: Springer.

McNeill, K. L., Lizotte, D., Krajcik, J. S., & Marx, R. W. (2006). Supporting students' construction of scientific explanations by fading scaffolds in instructional materials. *Journal of the Learning Sciences, 15*(2), 153–191.

National Research Council. (2012). *A framework for K-12 science education: Practices, crosscutting concepts, and core ideas*. Washington, DC: The National Academies Press.

Nordine, J., Krajcik, J., & Fortus, D. (2011). Transforming energy instruction in middle school to support integrated understanding and future learning. *Science Education, 95*(4), 670–699. doi:10.1002/sce.20423.

Trumper, R. (1990). Being constructive: An alternative approach to the teaching of the energy concept – Part one. *International Journal of Science Education, 12*, 343–354.

Chapter 14
Looking Through the Energy Lens: A Proposed Learning Progression for Energy in Grades 3–5

Sara Lacy, R.G. Tobin, Marianne Wiser, and Sally Crissman

14.1 Introduction

The *Framework for K-12 Science Education* (National Research Council [NRC] 2012) specifies that by the completion of elementary school students will have basic conceptual understanding of what energy is and how it behaves, including acceptance of the basic ideas of energy conservation. It is expected, for example, that it will make sense to students to use the same term, "energy," to discuss phenomena as varied as a moving ball, a battery, a cup of hot water, a stretched rubber band, and sunlight; and to understand the motion of a pendulum, or the operation of a slingshot, as involving the transfer of this elusive "stuff" from one object and/or one "form" to another.

Here we propose both a general framework for thinking about the goals of pre-college energy education, and a detailed learning progression for Grades 3–5 aimed at getting students to the required level of understanding by the end of the elementary grades. (Following the *Framework* (NRC 2012) we do not envision formal energy instruction in earlicr grades.) Our proposal is based on a careful consideration of the role of energy concepts in science and society and on prior research on children's understandings of energy, supplemented and shaped by our own interviews and exploratory interventions with students in 3rd and 5th grades and with teachers.

Learning progressions describe how knowledge in a domain (e.g., energy or matter) can evolve from young students' ideas (which radically differ from

S. Lacy (✉) • S. Crissman
TERC, Cambridge, MA 02140, USA
e-mail: Sara_Lacy@TERC.edu

R.G. Tobin
Department of Physics and Astronomy, Tufts University, Medford, MA 02155, USA

M. Wiser
Department of Psychology, Clark University, Worcester, MA 01610, USA

R.F. Chen et al. (eds.), *Teaching and Learning of Energy in K-12 Education*, DOI 10.1007/978-3-319-05017-1_14, © Springer International Publishing Switzerland 2014

scientists') to a deep and productive understanding of a scientific theory. They articulate *how* students can reach a scientific understanding through specific learning experiences that promote the progressive reconceptualization and integration of students' ideas. A learning progression provides a roadmap for developing curricula that foster "successively more sophisticated ways of thinking about a topic that can follow one another as children learn about and investigate a topic over a broad span of time" (NRC 2007). We believe that this approach is particularly well suited to the abstract nature of energy. In fact, a learning progression approach has shown promise in teaching energy in middle school (Nordine et al. 2010).

14.2 Foundational Ideas of the Learning Progression

This learning progression focuses on a network of *foundational ideas* about energy carried across grades 3–5 toward productive intermediate understandings (the 5th grade *Stepping Stone*) that can help students progress toward scientific understanding while ensuring that they keep making sense of phenomena. We have identified an "Energy Lens" and four foundational ideas (the "Energy Quartet") that are central to a scientific understanding of energy, essential for an informed citizen, and can progressively and meaningfully evolve, with instruction, from their precursors in childhood to principles endorsed by scientists. While our specific proposal and investigations here are confined to Grades 3–5, we would suggest that the Energy Lens and Energy Quartet could provide a useful organizing structure through all the pre-college years.

14.2.1 The Energy Lens

The principle of conservation is at the very heart of the concept of energy in science (Duit 1981; Elkana 1974; Goldring and Osborne 1994; Millar 2005). It is precisely the fact that energy is conserved that makes it useful. Like any conservation principle, it constrains the possible behaviors of a system, allowing us to say with confidence, across a vast range of natural phenomena, what *cannot* happen (no matter what it's made out of, a dropped ball will not bounce higher than its starting point) and what machines *cannot* be made (e.g., a car that runs on water), no matter how clever we are or how much money we spend. It can lead to scientific discoveries, such as the neutrino, or practical insights, like why you could heat your kitchen by leaving the oven door open but you can't cool it by leaving the refrigerator door open. But the concept of energy conservation is a peculiar tool. It typically does not tell us what *will* happen, or explain the causal mechanisms involved in a phenomenon. It usually focuses on initial and final states of a system, ignoring the messy processes in between, and is most helpful when those processes are at their messiest. Energy arguments can help us to conclude that plants require sunlight to live and grow, or to ask ourselves where the energy for an electric car comes from, but they do little to elucidate the details of photosynthesis or the design of electric cars.

If the principle of conservation of energy is taught without experience in why it is useful, how to reason with it, and what kinds of questions it can and cannot address, the principle is unlikely to be more than empty words, and indeed is unlikely to be believed.

Our proposal is that energy education focus on how scientists use what we call the "Energy Lens" to examine a broad range of phenomena, rather than on definitions, facts, or even equations (though all of those are necessary). "Energy" should be taught not as a discrete topic but as an analytical tool (Jin and Anderson 2012) or intellectual stance that provides partial but powerful insights into many topics in science and society. It differs radically from the force-dynamic, time-sequential, causal reasoning that is more naturally adopted by most students – and most adults (Driver and Warrington 1985; Jin and Anderson 2012). It often requires thinking in terms of systems rather than objects and willfully ignoring striking and fascinating details in favor of a focus on an abstract and invisible quantity. We have observed college physics students who will do a page of dynamic/kinematic calculations rather than two lines of energy analysis, because the causal relationship between force and motion makes more sense to them. Learning to look at phenomena through an "Energy Lens," therefore, is not a matter of a few weeks or even a few years; rather it must develop and be nurtured over a period of many years, and in many different contexts, including chemistry, biology and engineering as well as physical science. Indeed it took the scientific community hundreds of years to come to an understanding of energy (Elkana 1974).

Using the Energy Lens entails asking a set of questions that are appropriate to virtually any phenomenon in the physical world:

- What is the system of interest?
- What observable or measurable changes or other interesting behaviors are taking place?
- Where in the system are energy changes occurring?
- Where does the energy come from?
- Where does the energy go?
- What is the evidence for our answers?

We place the Energy Lens at the intellectual and pedagogical center of our proposal. We believe that through repeated exposure to these questions in various forms and contexts, students will acquire the habit of mind of asking – and tentatively answering – these questions about phenomena they encounter in their daily lives and throughout their science and engineering education. In addition, the Energy Lens provides teachers with an understanding of energy as a crosscutting concept and a straightforward way of bringing energy ideas into other curricular units. (Some teachers have told us that since being exposed to these ideas they see energy everywhere in their science curriculum and have begun to use it as a unifying theme, e.g., *"'Energy' can be viewed as an overarching concept that I can keep revisiting during the school year to connect content as we move through each curricular area."*)

14.2.2 The Energy Quartet

While the Energy Lens represents a powerful analytical stance for the analysis of natural phenomena and technological applications, its use depends on a firm grasp of four key aspects or strands of the energy concept, the "Energy Quartet"[1]:

1. The nature and manifestations ("forms") of energy;
2. Transfer and transformation of energy;
3. Dissipation and degradation of energy;
4. Conservation of energy.

The specific phenomena and experiences that students encounter are aimed at developing their understanding of these aspects of energy, while simultaneously building the habit of mind of looking at diverse phenomena through the Energy Lens.

From a scientific perspective the strands do not have equal status. Understanding energy conservation is the overarching goal, and so stands apart from and above the others. Dissipation and degradation could be viewed as a special case of transfer (to the environment) and transformation (into thermal energy). But our experience with students and teachers has persuaded us that they deserve their own category.

Most importantly, the strands are interconnected, interdependent and concurrent. As Goldring and Osborne (1994) point out, the nature of energy and the principle of conservation cannot be taught sequentially, but must be developed together, "holistically, gradually and over a period of time, with progressively deeper insight leading to the development of a more precise understanding." The idea that the energy stored in a stretched rubber band is the same kind of thing as the energy of a rolling ball makes sense only because one can be transferred to the other. Dissipation makes sense only if "heat" has already been recognized as a form of energy and it has been shown that other forms of energy can be converted into thermal energy. Conservation is believable only once dissipation is fully accepted; yet one is unlikely to accept, or even consider, the idea of dissipation without at least a tentative belief in energy conservation in other contexts. So these four strands must be interwoven and developed in parallel.

[1]We recognize that some of these terms are controversial, particularly "forms" and "transformations," and we discuss those issues later in this paper.

14.3 What Ideas Do Students Bring to Grade 3? What Should They Know by Grade 5?

The foundational ideas described above are, we believe, relevant for energy education throughout the pre-college curriculum. We turn now specifically to the elementary grades. The Grade 3–5 learning progression builds on students' initial ideas about energy (the Lower Anchor), and teaches students to think of phenomena in terms of energy from the start. The four thematic strands and the Energy Lens develop progressively from the "Lower Anchor" to the "5th grade Stepping Stone" that prepares students for further instruction and learning in middle and high school.

14.3.1 3rd Grade Lower Anchor

The Lower Anchor represents the set of ideas and cognitive resources that most children bring to 3rd grade, prior to any instruction about energy. It is derived from prior research on young children's understanding of energy and our own exploratory interviews. Very young children focus strongly on causal relationships (Leslie 1995; Rovee-Collier et al. 2001; Goswami 2008), so the non-causal perspective of the Energy Lens is likely to be alien.

The literature is largely concerned with middle school students, and does not always differentiate between younger and older students, whose ideas about energy are heavily influenced by instruction. To establish our Lower Anchor, it was important to determine what third-graders think about energy. To investigate children's ideas in more detail, and also to explore how readily they would accept or adopt new ideas, we conducted clinical interviews with eight pairs of 3rd grade students recruited from an urban school in Massachusetts with a high proportion of low-income and non-white families and English language learners. Each interview comprised two 30-min sessions and involved a range of hands-on activities including marble collisions on a horizontal track, a Newton's Cradle, a ball and spring, a balsa wood airplane propeller spun by a twisted elastic, and questions about what it means to "have energy."

Third graders told us that objects such as battery-powered devices and Slinkys have energy because they move by themselves, in unpredictable ways and/or "a lot." They believe that "energy" is the state of being in motion, or the ability to move. This makes "having energy" an inherent, either/or property of a certain class of objects that has the disposition to move in a certain way, as well the state of moving in such a way and the inner resources to do so. This corresponds to the anthropomorphic bias reported in the literature (e.g., Watts 1983; Trumper 1993); children see energy as an intrinsic property of living things and objects that move on their own. This is a correlate of a deeper concept: things that have energy are those with (apparent) agency and intentionality; they have inner resources to move and can be "recharged." In other words, they are the "animates" that children differentiate from "inanimates" starting in infancy (Leslie 1995; Spelke et al. 1995; Gergely et al. 1995; Gelman 2003; Hatano and Inagaki 1999). However, the third graders we interviewed gave no evidence of believing other energy frameworks reported in the literature, e.g., that energy is a causal agent, is in food or fuel, is a product, or is transferred. It is likely that those ideas develop later, as the result of instruction, exposure to media, and everyday language. Most third graders did say that batteries have energy. Batteries are perceived of as *enablers* (as described by Jin and Anderson 2012); switching on the battery allows the battery to "do its job," i.e., to give energy to a device in the sense of allowing the device to be in a state of energy.

We found that once the interview focused on energy and motion events, students readily adopted the view that all moving objects have motion energy and accepted speed as an indicator of the amount of energy. The interviewer could use language to implicitly and effectively convey "all inanimate objects can have motion energy" by saying, "Let's look at these colliding marbles. Do you think the blue one has as much energy after it collides with the red one?" This supports our view that language is an important pedagogical tool (Lemke 1990).

14.3.2 5th Grade Stepping Stone

The 5th grade Stepping Stone is the form of the Energy Quartet that students could understand by the end of 5th grade, and the contexts to which they could apply the Energy Lens. To define the stepping-stone we considered the Framework (NRC 2012), draft Next Generation Science Standards (Achieve Inc. 2012) and proposed learning progressions for energy (Jin and Anderson 2012; Nordine et al. 2010). We reviewed existing elementary and middle school curricula and discussed our ideas with a group of elementary school teachers. We also considered empirical evidence of whether and how students' ideas can progress when they engage in key learning experiences.

At the end of 5th grade, students' understanding of energy will still be fragmented and incomplete, but they will have a sound experiential and conceptual foundation on which to build a more solid understanding of the Energy Quartet and greater skill, versatility and sophistication in using the Energy Lens, through well-designed instruction in middle and high school. Students will understand that energy is a unitary entity and be able to recognize and characterize its most familiar manifestations, including energy of motion, stored energy of elastic deformation and batteries, stored gravitational energy, and thermal energy. They will be able to trace energy transfers and transformations in a variety of contexts, identifying and representing the gains and losses associated with various phenomena. They will recognize that energy gains in one object or system are consistently accompanied by energy losses in another and that magnitudes of gains and losses are correlated.

They will identify friction,[2] sound, and light as mechanisms for energy transport out of an object or system. They will include the possibility of "leakage" of energy into thermal energy in the environment and recognize that although such energy is not "gone," it is degraded in its usefulness for other purposes.

Since we are not, in these grades, attempting to quantify energy, quantitative conservation is not a goal. By the end of 5th grade, students should spontaneously look for energy losses when they see energy gains, and vice versa, and should expect that the magnitudes of those gains and losses are correlated. Without the means to quantify and compare amounts of energy in different objects and different forms, it is too early to claim that the gains and losses are equal, but students should be familiar with representations of energy (such as energy bars (described in Sect. 5.2) or energy cubes (Scherr et al. 2012)) that embody conservation, and be prepared to consider that possibility. We would not expect 5th graders to apply a strict conservation principle in most situations, particularly those involving subtle or undetectable forms, such as low-grade thermal energy in the environment. Further, we do not expect that students will fully distinguish between a form of energy and its indicator (e.g., thermal energy and temperature) though we expect they will recognize that other factors (e.g., the amount or kind of material) also influence the amount of energy.

The abilities and understandings included in the 5th grade Stepping Stone will provide the foundation for applying the Energy Lens to a variety of systems and phenomena, including those encountered outside the classroom. They will allow students to begin to make predictions and draw inferences based on energy considerations. Since the Energy Lens questions – where does the energy come from and where does it go? – implicitly contain the idea of conservation (energy cannot just appear or disappear), students who have become accustomed to asking and addressing those questions will be well prepared for a more formal treatment of energy conservation in subsequent instruction.

14.3.3 Construct Map

Table 14.1 shows our proposed Construct Map, which indicates possible levels of understanding for each strand from the 3rd grade anchor (L1) towards the 5th grade Stepping Stone (L4). It can serve as a basis for formative assessment. Students start with limited, accessible versions of the four strands and progressively build more general, richer, more complex integrated versions of them. We have

[2]More precisely, work done by the frictional force. At this grade level we do not think it is productive to focus on the distinction between force and work. "Friction" is used here to refer to a general phenomenon, aspects of which include the frictional force, energy dissipation, and frictional heating.

Table 14.1 Preliminary construct map for the learning progression for energy

Nature/manifestations ("forms")	Transfer and transformation	Dissipation/degradation	Conservation
L4: Verbal and nonverbal representations show awareness of unitary nature, and distinguish energy from indicators. Can explain thermal phenomena without invoking "cold energy." Describes gravitational energy and energy of foods/fuels as energy of systems, light and sound as energy carriers. May not fully distinguish food/fuels from associated energy or thermal energy from temperature	**L4:** Recognizes that energy gains in one object/system are consistently accompanied by energy losses in another. Describes gains and losses as transfers and transformations of the same "stuff." Represents and describes transfers and transformations in a range of situations, including those with multiple objects/forms	**L4:** Consistently includes thermal energy in environment when tracing energy transfers and transformations. Identifies frictional heating, sound and light as mechanisms for energy transport out of object/system. Understands that "lost" energy could exist but not be recoverable	**L4:** Spontaneously asks "Where does energy come from?" and "Where does energy go?" in new situations and includes transfer to/from environment in attempting to answer. Recognizes that magnitudes of energy gains in one object/system and losses in another are correlated. Considers possibility that total energy could be constant, but may not fully believe or consistently apply, especially to real world situations
L3: Describes energy as quantity or substance, and distinguishes energy from indicator in some but not all cases. Recognizes multiple manifestations exist simultaneously. Identifies stored gravitational energy and associates amount with height. Identifies thermal energy and associates amount with temperature. Understands that energy can be transferred without a transfer of matter in some contexts	**L3:** Can qualitatively track and represent energy increases and decreases in processes involving transformations within a single object/system as well as between objects/systems. Uses words like "give," "take" and "share" to describe energy changes. Shows increased ability to consider multiple objects/forms	**L3:** Shows understanding that in a large system, increases in thermal energy may not produce detectable temperature changes. Includes environment in system and in tracing energy flow in cases involving light, sound or obvious frictional heating. Understands that some forms of energy are less useful than others. May believe "lost" energy is fully recoverable	**L3:** Recognizes constancy of total energy in simple mechanical systems (e.g. pendulum) without dissipation. Recognizes "Where does the energy go?" as meaningful, and begins to include light, sound and frictional heating as possible answers, with guidance, but does not spontaneously ask question and/or thinks "it's just gone" in many situations

L2: Identifies energy of motion of inanimate objects and stored energy of elastic deformation (springs, elastics), batteries and capacitors, and associates amounts with appropriate indicators. Does not clearly understand that they are different manifestations of the same quantity or consistently differentiate energy from matter. Associates energy with specific objects	**L2:** Can qualitatively track and represent energy increases and decreases in processes involving transfer from one object to two or a few distinct objects, with or without transformations. Recognizes that gains/losses occur in combination, but does not show firm understanding of transfer and does not include multiple energy pathways	**L2:** Recognizes that other forms of energy can be transformed into thermal energy. Understands that light and heat can carry energy away from object. Begins to notice effects such as heating of objects and vibrations. Does not consider energy transfer to wider environment (e.g. air) or possibility of undetected losses	**L2:** Recognizes that energy has to "come from" somewhere to initiate a process, and connects gains and losses, but believes it disappears when visible process ends
L1: Identifies energy as inherent property of animate objects or objects that move on their own or in unusual ways	**L1:** Identifies only presence or absence of energy, not increases or decreases. Can describe processes in terms of gains/losses but does not show awareness of transfer	**L1:** Asserts that energy is just gone when motion stops. Attention is focused on discrete objects; systems/environment are not considered. No awareness of thermal energy	**L1:** Sees energy as inherent property of certain objects that can appear or disappear and is gone when not perceptible

developed this map based on our review of the relevant literature and our preliminary investigations, but like any such map it is a work in progress and will be revised iteratively as the learning progression is implemented and tested in the classroom.

14.4 The Proposed Learning Progression; Start with the Lower Anchor, Aim for the Stepping Stone

In this section we propose a pathway from the 3rd grade Lower Anchor to the 5th grade Stepping Stone. Our proposed learning progression indicates how students' understanding of each of the four strands will broaden and deepen over the course of three curriculum units and includes some productive contexts and experiences within each grade. Although the strands are described sequentially, we emphasize that they develop simultaneously and in parallel. The argument for the unitary nature of energy is inseparable from the ideas of transfer and transformation, for example, and the Energy Lens questions "Where does the energy come from?" and "Where does the energy go?" will be asked throughout. Across the strands there is a progression in the complexity of the phenomena studied and the sophistication of the representations and reasoning used.

14.4.1 Strand 1: Nature and Manifestations ("Forms") of Energy

A central challenge is to convey the idea of energy as a unitary quantity that is manifested in diverse ways. Fixating on a taxonomy of "forms," as many curricula do, can be misleading and confusing, and distract from the more important idea of unity. Some writers advocate abolishing the terms "forms" and "transformations" of energy altogether (Millar 2005; NRC 2012; Swackhamer 2005). Yet it is useful to name the diverse manifestations of energy, just as graphite and diamond have different names even though carbon is carbon. Furthermore, many familiar phenomena, such as a ball rolling down a ramp, are difficult to describe meaningfully at a grade-school level without the idea of transformation.[3] Rather than abandoning "forms" and "transformations" entirely, we propose to limit the number of categories, avoid sorting for its own sake, and construct activities that emphasize the underlying unity of these apparently disparate manifestations (Achieve Inc. 2012; DOE 2012; Jin and Anderson 2012; Nordine et al. 2010).

[3]The alternative is to describe the energy as being transferred between the gravitational field and the ball (NRC 2012; Swackhamer 2005) but we regard the idea of a field as too abstract to be meaningful in these early grades; in our experience it is difficult enough for college students.

In Grade 3, we begin with manifestations of energy that are familiar and accessible to young children – motion and batteries – and begin developing the ideas of energy amounts (none, less, more), transfer, and transformation. The energy stored in compressed springs and elastics is introduced through the ability to produce motion, such as spinning a propeller (Van Hook and Huziak-Clark 2008). The focus is on simple mechanical systems (colliding marbles, propellers, slingshots), in which the energy is associated with specific objects. All the activities deal with motion in a horizontal plane, intentionally deferring gravitational energy because it cannot (or at least should not) be identified with a specific object. We introduce the term "system" to describe a set of interacting objects (Driver and Warrington 1985; Jewett 2008).

The idea of a system becomes important with the introduction of gravitational energy in Grade 4. (It is also a crucial concept in the study of matter (TERC 2011) and a crosscutting concept in the *Framework* (NRC 2012)). Unlike the stored energy in a compressed spring, gravitational energy is a property not of the object itself, but of the system comprising the object and the earth as they interact through gravity[4] (NRC). Magnets will be used to scaffold this difficult idea, as a more accessible example of energy in a system of interacting objects that does not belong to either object by itself.

Energy in electrical circuits is also explored in Grade 4, allowing the introduction of light and sound as energy carriers[5] (NRC 2012). Heat is mentioned, but not explored in detail until Grade 5. Capacitors are used as sources of stored electrical energy, bringing several pedagogical advantages over batteries: They can easily be charged to various levels, including with a hand-cranked generator; there is a measurable indicator (voltage) for the amount of stored energy; and the stored energy can be exhausted in a few seconds or minutes.

Thermal energy, food, and fuels are introduced last, in Grade 5. We have had success using thermal phenomena as an entry point for energy investigations with adult teachers, but young children associate heat with energy only after some experience and instruction. (Giving up the idea of "cold energy" is challenging for both adults and children.) Understanding heat as energy is critical to the strands of transfer/transformation, dissipation/degradation and conservation, and understanding the relationship among food, fuels, and energy is central to many scientific and everyday applications (DOE 2012; Jin and Anderson 2012).

---

[4]Again, the alternative is to describe the energy as residing in the gravitational field (Swackhamer 2005). On a microscopic level, the energy in the compressed spring is due to interactions between the atoms and is not a property of the atoms themselves. So the distinction between the two cases is pedagogical rather than fundamental. Since the interacting atoms are all within the spring, we can describe the stored energy as a property of the spring, due to its compressed state, without reference to any external object.

[5]We do not think it is useful to treat electrical currents as carriers of energy (Millar 2005). It is not accurate to consider the energy as carried by the current, and such a description can contribute to misconceptions such as that the energy is kinetic energy of the electrons and that in a complete circuit the energy returns to the battery.

Accurately describing the energy associated with food and fuels is challenging (Jin and Anderson 2012; Millar 2005; Warren 1983). Since energy is released only when the food/fuel reacts (usually with oxygen) forming new chemical bonds, the energy is properly understood not as a property of the food/fuel itself but of the *system* comprising the food/fuel and oxygen. But a full understanding of the distinction requires an understanding of chemistry, chemical bonds and chemical reactions that very few students achieve even in high school (Jin and Anderson 2012). Further, the idea that energy resides in the food and fuel is pervasive, and in many contexts useful, or at least not harmful. It is far more important, in our view, that students understand that the food or fuel is not itself the energy.

14.4.2 Strand 2: Transfers and Transformations

In our interviews, 3rd grade students saw that decreases in one object's energy coincided with increases in another's (e.g. when a moving marble strikes a stationary one) but they did not conclude that energy had been transferred from one to the other. Adopting the model of energy as a kind of "stuff" that can be transferred is a productive, and perhaps even necessary step towards understanding the unity of energy and the principle of conservation (Millar 2005; Scherr et al. 2012; Swackhamer 2005). In our learning progression students observe, describe, and represent coordinated gains and losses in diverse systems of growing complexity. They begin in Grade 3 with examples of transfer from one object to another, and words such as "give" and "share" are introduced. They move from a view of energy as an "either/or" property to descriptions of amounts of energy as "none," "some," or "a lot."

Transfer and transformation become a central focus in Grade 4, with many examples of increasing complexity, including transformations within a single object or system (e.g., a falling ball), and situations in which energy is transferred to multiple objects and/or forms, as when an electrical circuit drives a propeller and a light bulb. They also begin to observe, and use representations to show, that gains and losses are correlated in magnitude (e.g., a capacitor charged to higher voltage can make a bulb burn brighter and longer). By the end of Grade 4 students should have the expectation that energy gains and losses occur in combination, and the habit of looking for the loss corresponding to an observed gain, and begin to think of those gains and losses as a transfer – that is, that the energy gained is the same "stuff" as the energy lost.

The emphasis in Grade 5 shifts primarily to thermal energy. Purely thermal phenomena, as when a hot object is placed in cool water, provide some of the clearest examples of transfer, but accepting "heat" as energy, rather than a distinct substance, also requires experiences with the transformation of other forms of energy into heat and of heat into other forms. It is also essential that students begin to recognize thermal energy as a ubiquitous byproduct of all the phenomena they study, and to include it in tracing energy transfers.[6]

14.4.3 Strand 3: Dissipation and Degradation

Any successful learning progression for energy must confront the apparent conflict between the scientific principle that energy is conserved and the everyday view of energy as something that is routinely "produced," "consumed," "wasted," and "used up." The statement that "Energy cannot be created or destroyed" stands in apparent conflict not only to the way energy is usually described, but to the universal experience that unless we keep providing them with energy, moving things stop, hot things cool off, and batteries (as well as living things) "die" (Goldring and Osborne 1994; Solomon 1985). The explanation that the energy is still present in the form of slight warming of the environment is not persuasive – unless one is already firmly committed to the principle of conservation – nor is it easily demonstrated. It is simpler, and common (for teachers as well as students), to conclude that conservation of energy applies only in idealized situations (e.g. in the absence of friction). Experiences in which mechanical or electrical energy is transformed into detectable heat – rubbing an eraser on a carpet, heating a resistor with a battery, burning a hole in paper by colliding steel spheres (Nordine et al. 2010) – are necessary, but they do not seem to transfer readily to the more common situation in which the thermal energy is too diffuse to produce a readily detectable temperature change.

This obstacle will not be overcome easily or quickly. The issue of "where does the energy go when it's gone" will need to be raised repeatedly and consistently over the course of years, in many contexts of varying complexity, and students (and teachers) must have multiple experiences that make visible the normally invisible thermal energy associated with dissipation, through the use of such tools as sensitive temperature probes and infrared photography. We hope to help students begin to understand dissipation by the end of 5th grade through such experiences, and by consistently raising the issue of "where the energy goes" while, in parallel, building the idea of energy conservation. But activities and instruction will need to continue to focus on the topic in the higher grades as well.

The question of "where does the energy go when it's gone" arises naturally in the Grade 3 activities, and children will be encouraged to speculate about it; in our exploratory interviews they noticed both vibrations and slight warming. In Grade 4

[6]We see no benefit to emphasizing or enforcing the thermodynamic distinction between heat and internal energy at this grade level, but we encourage the use of the term "thermal energy" to emphasize that "heat" is a manifestation or form of energy, rather than a distinct quantity or substance.

light and sound provide additional mechanisms by which energy can "leak out" of the system. In Grade 5 there will be activities in which small temperature changes are measured, and students become familiar with the idea of thermal energy in the environment as a sink for the "lost" energy, even if they do not yet fully accept the idea.

At the same time, the everyday idea that something is irreversibly lost when energy is "used" is correct, and we don't want to replace one misconception (that energy simply disappears) with another (that the dissipated energy could be recovered and reused). A full understanding would require grappling with the Second Law of Thermodynamics and the idea of entropy, but we will acknowledge, beginning in Grade 4, that the "lost" energy is "less useful" than other forms, so that although the total amount of energy does not change, the amount of "usable" energy is diminished in every process (DOE 2012; Goldring and Osborne 1994; Millar 2005; NRC 2012; Nordine et al. 2010).

14.4.4 Strand 4: Conservation

Scientists would not have invented the concept of energy, and we would not teach it, if it were not conserved; conservation is the whole point (Duit 1981; Elkana 1974). The Energy Lens makes no sense without conservation. So this strand is the core of the project, yet it is, deliberately, the one strand that is not explicitly taught, because it cannot be given real meaning until the other strands have been developed, and because a convincing case cannot really be made until one begins to quantify the various forms of energy. Instead, in these early grades we build, largely implicitly, a model of energy as a kind of "stuff," a body of evidence for such a model, and habits of mind that embody the model, such as asking "Where does the energy come from?" and "Where does the energy go?" and thinking of energy losses being accompanied by energy gains. The ideas of transfer and transformation will continue to be developed across the grades, and the language of the Energy Lens will be used consistently from the beginning – initially as prompts for speculation, but with a growing expectation of meaningful analysis and argumentation. The developing understandings will be reinforced as students are asked to use representations – both student-generated and provided – that will increasingly embody the model of energy as a unitary and conserved quantity (Scherr et al. 2012).

14.5 Exploratory Results from Classroom Research

As a partial test of the feasibility of our learning progression approach, we piloted activities with 3rd and 5th grade students. In each case we began with students' initial ideas (the Lower Anchor) and developed activities that we hoped would reveal students' conceptual difficulties and help them restructure their ideas toward the 5th grade Stepping Stone. Our findings are helping us to assess both the learning progression and the effectiveness of the learning experiences.

14.5.1 Grade 3 Teaching Experiments

We developed and administered a sequence of "teaching interventions" to investigate if students in Grade 3 could make progress toward the 5th grade stepping stone by using the Energy Lens. Initially, third graders typically think of energy as an all-or-none state, and believe that humans, but not inanimates, can give energy to objects. The idea that a stretched elastic or compressed spring has energy does not make sense to them. Our goal was for students to learn that energy can be transferred between moving objects, from a battery to a moving object, and between a moving object and a spring or elastic band. An interviewer met with pairs of students for four half-hour sessions over the course of 2 weeks. Eighteen 3rd graders participated, most of them from an urban school in Massachusetts serving low-income families.

The interviewer used a flexible script to interact with students over a series of tabletop activities. Students were encouraged to notice and correlate changes in speed and configuration (energy indicators), and to interpret them in terms of energy transfer. Since students easily associate energy with motion, we introduced a simplified version of the Energy Lens questions in the context of colliding marbles:

- Describe what happens. (E.g., Give me a play-by-play description of the changes in motion that you see.)
- Where are there energy changes? (When is there no energy of motion, some energy of motion, or lots of energy of motion? When do you see energy of motion increasing? Decreasing?)
- Where did the energy come from? (If a marble gained energy of motion, where did the energy come from?)
- Where does the energy go? (If one marble lost energy of motion when it hit the other marble, where did the energy go?)

Students were led to see speed as an energy indicator and to associate one marble's energy loss with the other's energy gain. The interviewer then asked students what they thought of the idea that one marble gave some of its energy to the other.

Given what students already know about batteries—they run things, they have energy, they "die"—we hypothesized that a battery-operated toy, in this case a battery-operated propeller, was a fruitful context for introducing the concept of stored energy, and for extending the practice of using the Energy Lens to associate gains and losses of energy (e.g., the energy gained by the propeller is lost by the battery). We also hypothesized that a battery-operated propeller could serve as a useful analogy for one operated by a stretched elastic. Drawing on the analogy between the two propellers, we asked students to consider whether a twisted elastic has stored energy, and encouraged them to apply the Energy Lens to the elastic

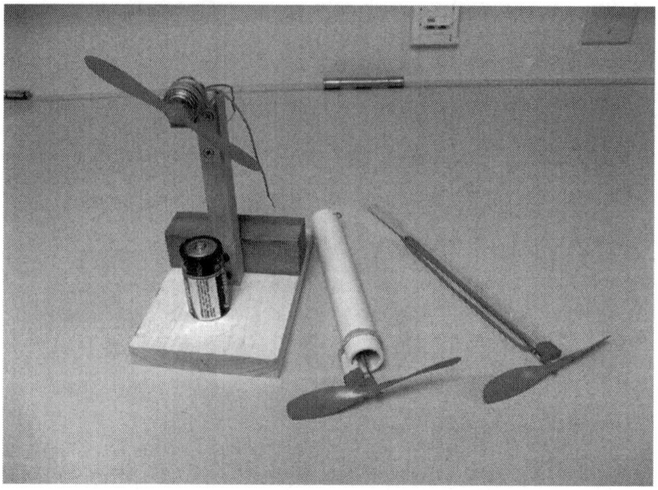

Fig. 14.1 Students explored propellers operated by battery and twisted elastics

operated propeller: the propeller blade begins to move; it gains energy of motion; the energy of motion comes from the twisted elastic. The elastic untwists; when the elastic is less twisted, it has less stored energy; the elastic gives motion energy to the propeller; when the elastic gives motion energy to the propeller it loses stored energy; loss of stored energy is associated with a gain of motion energy (Fig. 14.1).

The excerpt below from the Grade 3 teaching intervention *Can an elastic band store energy*? shows how students use what they have learned about energy stored in batteries, energy of motion, energy gains and losses, and energy transfer to investigate the propeller powered by a twisted elastic band. In a prior investigation, they have learned that stored energy lost from a battery is gained as energy of motion of the propeller.

A pair of students is given a propeller that is attached to an elastic hidden inside a tube: the elastic has been twisted and the propeller is held in place by a skewer that serves as a brake. The interviewer is guiding them to observe and describe what happens and to identify observable indicators of energy changes.

I = Interviewer, S = Student[7]

I: *What about this propeller. Do you think it has any energy?*

S: *No*

S: *Yes*

I: *How can you tell?*

S: *I think it does.*

I: *Can you explain why? It's not moving.*

---

[7]In transcribing the recorded interview we could not consistently distinguish the voices of the two students, so an utterance by either student is indicated by "S" without further identification.

S: *Because there's a rubber band inside it... Actually, no I don't think it has energy right now.*
S: *I don't think it has energy, because it's not moving. And usually when something has energy, it would be moving.*
I: *The battery wasn't moving, but you said it had energy.*
S: *It had stored energy.*
I: *Check this out. [Releases propeller and then stops it with the skewer brake.] Do you think it has energy now?*
S: *Yes*
I: *Why?*
S: *It has stored energy*
I: *How can you tell?*
S: *Because it has a rubber band. It's twisted up. And like, when you take the stick away, it'll go running away.*
I: *[Releases the propeller and lets it run until it stops.] What about now? Does it have stored energy now?*
S: *Not now because now the rubber band is, it's not twisted. Usually when the rubber band is twisted then it's gonna untwist.*
I: *[Shows the propeller without enclosing tube. The elastic is untwisted.] This is what it is like inside. Does this have stored energy?*
S: *Um. No.*
I: *Could you store energy in it?*
S: *Yes.*
I: *How would you do it?*
S: *[Tries to pull and twist the rubber band. Then twists the propeller, which in turn twists the rubber band.] It's working, it's working.*
I: *So now. Tell me why you think there's stored energy.*
S: *Whee.....[releasing the propeller and letting it spin].*
I: *[Winds the propeller.] Now is there energy stored in it?*
S: *Yes*
I: *If I let it go, do you see any energy?*
S: *Yep.*
I: *Did you see any energy pairs? What had no energy and then more energy?*
S: *First it didn't have energy. Before you twisted it.*
I: *You mean the rubber band?*
S: *Yeh. Now it has stored energy.*
I: *What about the red propeller. Does it have energy?*
S: *Not until this gave it energy.*
I: *What's getting energy if I let go?*
S: *This is [the propeller].*
I: *And what loses energy?*
S: *The rubber band loses energy and the propeller gains energy.*
I: *So it lost stored energy and it gained energy of motion.*
I: *Can you make it have more energy than it had the last time?*
S: *I keep twisting. I haven't stopped twisting it since I started.*

I: *How can you tell it has more energy?*
S: *Cause it will run longer and faster.*
I: *Is there any way you can tell by looking at the rubber band?*
S: *Yes. Because it's getting tighter and tighter.*
I: *Did you give it more energy?*
S: *Very.*
I: *What do you think is going to happen when you let go?*
S: *It's gonna go vroom.*
I: *So will it give the propeller more energy if it has more energy?*
S: *Yes, but this is gonna lose more energy.*
I: *I think you should let go now.*

We used a pre/posttest to understand if and how children's ideas progressed before and after the teaching interventions and to assess the promise of the activities. The pre/post was an interview structured around a series of tabletop activities; knocking down bowling pins with a rolling ball, launching a small pom-pom with an elastic-band slingshot, and using a battery-powered milk frother. Students did the activities in pairs, then watched slow-motion videos of the same phenomena and answered open-ended interview questions.

Students made progress from the pre- to posttest in their interpretation of all three events. In the pretest, students did not think that the bowling ball gave energy to the pins. ("*I don't think the ball is giving energy to the pins, I think it's just like making it be forced to fall down.*") In the posttest they were more likely to think that an object's energy can increase because it receives energy from another object; the majority of students believed both that the bowling ball gave energy to the pins and that the battery gave energy to the milk frother, i.e., that inanimate objects can give energy. (Detailed analyses of the test transcripts for all three events reveal that the strong association between humans and energy weakens but does not completely disappear.) In the pretest, students rejected the notion of elastic energy; most third graders told us that the slingshot did not have energy when it was stretched but not moving.

Interviewer: *As you pull it back, while you're pulling, is there any energy?*
Student: *Yes. Because it's moving and before it wasn't moving.*
Interviewer: *Now it's stretched all the way back, is there any energy?*
Student: *No, I don't think there's any energy because it's not moving.*

In the posttest, on the other hand, most students asserted that when energy is given to the slingshot (by stretching it), the slingshot has stored energy. In the posttest, students began to make the association that when energy is gained by one object, it is lost by another. Some students progressed from thinking that giving energy means, e.g., putting another object in a state of energy, to understanding that giving energy is transferring energy; they were more likely to say that the bowling ball slowed down because it gave some of *its* energy to the pins.

Student: *I saw Jimmy give energy to the ball and then the ball give energy to the pins to knock them down.*
Interviewer: *What happens to the ball at the very end?*
Student: *The ball stops, and it lost energy. The ball gained energy, and then the ball gave energy to the pins, and then the ball lost energy.*

In the posttest, some students said energy is transferred from the slingshot to the pompom. Others said that the slingshot gives energy to the pompom, but that is not why it loses energy, noting that the slingshot stops whether or not it launches a pompom.

Thus, we conclude that we are "on the right track." The sequence and goals of our teaching activities seem meaningful and productive: introduce the Energy Lens – beginning with the association of energy with observable indicators and the idea of energy as something that can be transferred between objects—in the context of collisions, extend it to battery-powered objects; and use the energy of batteries as an analogy for elastic energy.

14.5.2 Grade 5 Teaching Experiments

The focus of our work with 5th graders was to develop and test a sequence of activities that would help students think about phenomena involving thermal energy. We piloted two 1-hour sessions in 5th grade classrooms in one urban and one suburban school. Members of the research team led the activities in cooperation with the classroom teachers.

We have found that children of all ages and their teachers (Tobin et al. 2012) have difficulty considering the question "Where did the energy go," a key Energy Lens question that is essential for understanding energy conservation. A 5th grade learning progression target is that students become aware that a system can "lose" energy to its environment; that there are invisible forms of energy (such as low-grade thermal energy) that can make it appear that energy is lost; and that thermal energy is a "sink" for all energy. We know from prior research that not all children believe that heat is energy and that, in some situations, children think that *cold*[8] is transferred rather than heat – elements of the Lower Anchor. We hypothesized that experience with the Energy Lens in the context of motion phenomena, prior to the activities relating to heat, would pave the way for gaining a stronger sense that heat is energy and understanding the correlation between heat gain and heat loss in the context of thermal phenomena.

---

[8]This misconception is documented in the AAAS Project 2061 Science Assessment website: "When two objects at different temperatures are in contact with each other, thermal energy is transferred from the warmer object to the cooler object and "coldness" or "cold energy" is transferred from the cooler object to the warmer object." (AAAS n.d.)

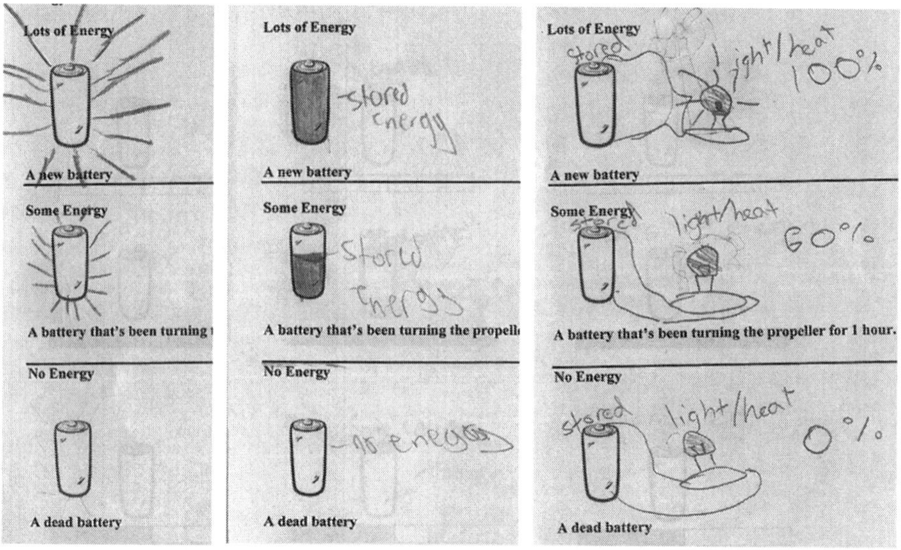

Fig. 14.2 Student representations of "no," "some," and "lots of" energy

We introduced the Energy Lens in the context of marble collisions, and taught students to look for and describe changes in terms of pairs of energy gains and losses in phenomena involving motion energy, "stored energy" of a battery, and "light energy" and "heat energy" of a light bulb. Students designed representations to show situations of no energy, some energy, and lots of energy. We began with an outline drawing of three batteries and asked them to come up with as many ways as they could think of to show "no," "some," or "lots of" energy (Fig. 14.2).

Next, using the same drawing of three batteries, we introduced "energy bars." We filled in energy bars to show "no energy," "some energy," and "lots of energy," and asked students to identify "Which battery has no energy? Some energy? Lots of energy?" (Fig. 14.3).

The next session aimed to extend their use of the Energy Lens to phenomena involving thermal energy, and to make the case that there is no such thing as "cold" energy. Students connected a hand-crank generator and a battery to a resistor and described what happened (heating of the resistor) in terms of energy gains and losses.

To gain insight into how they were thinking about energy transfers, we asked 5th grade students to use "energy bars" to represent energy gains and losses (a) when a resistor is connected to a battery and placed in a beaker of room temperature water and (b) when a room temperature copper cube is placed in a beaker of hot water. (See Fig. 14.4.) As students explained how they decided to fill in the energy bars, they revealed their understanding of energy gains and losses during heat transfer and, in some cases, their understanding of temperature as an indicator of energy and even dissipation. For example, "I gave the [room temperature] water two energy

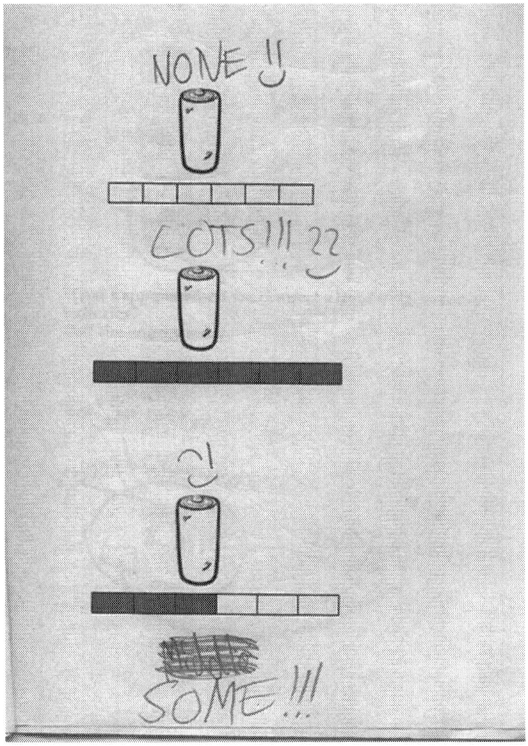

Fig. 14.3 Students label "energy bar" representations of "no," "some," and "lots of" energy

bars because I thought it had some energy before I put in the resistor. I thought the battery must be charged so I colored in all 6. After the resistor heated the water, I added 3 energy bars to the water that gained energy and took away 3 from the battery that lost energy." This student showed the water gaining the same number of bars as the battery lost, but we would not claim that this necessarily indicates an understanding of energy conservation. In general, we did find that students began to recognize that the magnitude of the resistor's energy loss was correlated with the water's energy gain. The energy bars represent a first, partial step in the direction of quantification and towards the idea that, when fully accounted for, the energy gains and losses are not merely correlated but in fact equal.

With such a limited intervention, and without the prior experiences that 5th Graders would have had in a fully implemented learning progression curriculum, we could not hope to fully evaluate whether our postulated Stepping Stone is reachable. We were able, however, to draw some tentative conclusions: By the end of these lessons, most students could describe changes in terms of (a) no, more, or less energy as a property of objects and systems, (b) different manifestations or forms of energy (and exhibited beginning understanding of energy in all its forms as a unitary thing), (c) energy transfer in terms of gains and losses in pairs or multiples, (d) energy transformations, e.g., motion energy to light and heat energy, heat energy to

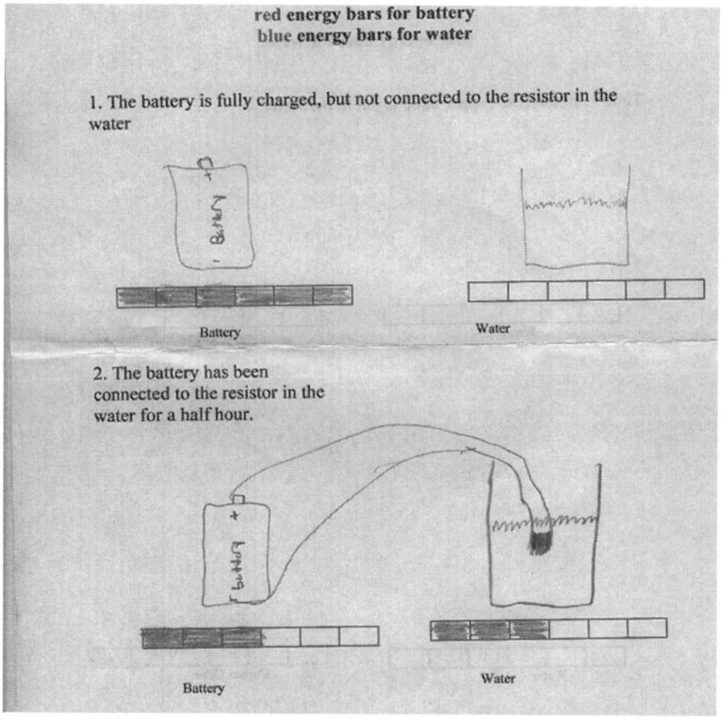

Fig. 14.4 A student uses "energy bars" to represent energy gains and losses when a resistor is connected to a battery and placed in a beaker of room temperature water

motion energy. Students explained that motion or stored energy can be transferred to an object such as a resistor and manifest itself as heat; heat energy can be transferred to an object and manifest itself as motion. Students were intrigued by the idea that "there is no such thing as cold energy."

A pre/post assessment involving a hands-on activity in which a hot stone was placed in cooler water was used to assess students' progress towards the learning goals. We asked students, "When the stone was in the water, its temperature went down and the temperature of the water went up. Can you explain what was happening?" In the posttest, 20 of the 37 students (compared to 4 of 37 in the pretest) were able to describe the phenomenon in terms of energy gains and losses.

Student #1 (post): *The stone is giving heat energy to the water and the stone losses heat energy and they act as a energy pair. The water gained heat energy while the stone loss heat energy. [sic]*

Before instruction, 80 % asserted the existence of "cold energy." A representative pre-test response was:

Student #2 (pre): *The stone transferred some of its heat energy to the water and the water transferes [sic] some of its cold energy to the stone.*

After instruction the percentage had dropped to 48 %.

Student #2 (post): *Some of the heat energy transferd [sic] to the water so the stone got less warm and the water got warmer.*

After such a limited intervention, however, this understanding is probably fragile and context-dependent. Many students hold on to the idea that "loss of heat energy" can also be described as "gain of cold energy."

The teaching activities also revealed conceptual difficulties that would be addressed in a longer, curricular intervention. The students did not readily look for more than one gain corresponding to a given loss (e.g., in the environment). As in Grade 3, the idea of gains and losses was more readily accessible than that of energy transfer. Students found it easy (and fun) to use graphical representations to show no, a little, or a lot of energy, and qualitatively or semi quantitatively show energy gains and losses. Using representations – both graphical (such as energy bars) and concrete (such as the energy cubes used by Scherr et al. 2012) – may provide a way for students to show that energy is not only gained and lost, but that the same energy is *transferred* from one system to another – that although energy is manifested in many ways, it is all the same "stuff" – as it can all be represented in the same way. Appropriate representational systems, moreover, may help students construct or assimilate – or at least consider – the idea of conservation (Scherr et al. 2012).

These preliminary results give us confidence that the learning goals we envision for the 5th grade, while ambitious, are both accessible and engaging for teachers and children in this age group.

14.6 Conclusion

We have outlined a learning progression for energy in grades 3–5 based on the idea that energy is a powerful analytical tool for thinking about a wide range of phenomena, rather than a discrete topic. Our proposed learning progression is structured around the parallel development of a network of interconnected and interdependent foundational ideas. It builds on young students' intuitive ideas (the Lower Anchor) and is sensitive to prevalent hurdles and misinterpretations revealed by prior research and our own preliminary investigations with children and teachers. Our learning progression is a hypothesis about how relevant instruction could progressively enrich, transform, and integrate students' knowledge toward a scientific understanding of energy. Exploratory interviews and teaching interventions provide preliminary evidence for the promise of this approach. The next step will be to use these ideas to develop the complex system of curriculum, assessment, and teacher professional development that will provide elementary students with the resources to develop a more sophisticated understanding of energy in middle school.

Acknowledgement This work was supported by National Science Foundation Awards #1020013 and #1020020.

References

AAAS. (n.d.). *AAAS Project 2061 Science assessment website*. Retrieved 2012, from, http://assessment.aaas.org/misconceptions/NGM016/248

Achieve Inc. (2012). *Next generation science standards, Draft May 2012*. Retrieved 2012, from, http://www.nextgenscience.org

Department of Energy (DOE). (2012). *Energy literacy: Essential principles and fundamental concepts for energy education*. Retrieved 2012, from, http://www.globalchange.gov

Driver, R., & Warrington, L. (1985). Students' use of the principle of energy conservation in problem situations. *Physics Education, 20*, 171–176.

Duit, R. (1981). Understanding energy as a conserved quantity: Remarks on the article by R.U. Sexl. *European Journal of Science Education, 3*(3), 291–301.

Elkana, Y. (1974). *The discovery of the conservation of energy*. Cambridge, MA: Harvard University Press.

Gelman, S. (2003). *The essential child*. New York: Oxford University Press.

Gergely, G., Nadasdy, Z., Csibra, G., & Biro, S. (1995). Taking the intentional stance at 12 months of age. *Cognition, 56*, 165–193.

Goldring, H., & Osborne, J. (1994). Students' difficulties with energy and related concepts. *Physics Education, 29*, 26–32.

Goswami, U. (2008). *Cognitive development: The learning brain*. New York: Psychology Press.

Hatano, G., & Inagaki, K. (1999). A developmental perspective on informal biology. In D. L. Medin & S. Atran (Eds.), *Folkbiology*. Cambridge, MA: The MIT Press.

Jewett, J. W. (2008). Energy and the confused student II: Systems. *Physics Teacher, 46*, 81–86.

Jin, H., & Anderson, C. W. (2012). A learning progression for energy in socio-ecological systems. *Journal of Research in Science Teaching, 49*(9), 1149–1180.

Lemke, J. L. (1990). *Talking science: Language, learning, and values*. Norwood: Ablex.

Leslie, A. M. (1995). A theory of agency. In D. Sperber, D. Premack, & A. J. Premack (Eds.), *Causal cognition: A multidisciplinary debate*. New York: Oxford University Press.

Millar, R. (2005). *Teaching about energy*. University of York Department of Educational Studies Research Paper 2005/11. Retrieved 2012, from, http://www.york.ac.uk/media/educationalstudies/documents/research/Paper11Teachingaboutenergy.pdf

National Research Council (NRC). (2007). *Taking science to school: Learning and teaching science in grade K-8*. Washington, DC: The National Academies Press.

National Research Council (NRC). (2012). *A framework for K-12 science education: Practices, crosscutting concepts, and core ideas*. Washington, DC: The National Academies Press.

Nordine, J., Krajcik, J., & Fortus, D. (2010). Transforming energy instruction in middle school to support integrated understanding and future learning. *Science Education, 95*, 670–695.

Rovee-Collier, C. K., Hayne, H., & Columbo, M. (2001). *The development of implicit and explicit memory*. Amsterdam: John Benjamin Publishers.

Scherr, R. E., Close, H. G., McKagan, S. B., Close, E. W., & Vokos, S. (2012). Representing energy dynamics in complex physical processes. *Physical Review ST-PER*, unpublished.

Solomon, J. (1985). Teaching the conservation of energy. *Physics Education, 20*, 165–170.

Spelke, E. S., Phillips, A., & Woodward, A. L. (1995). Infants' knowledge of object motion and human action. In D. Sperber, D. Premack, & A. J. Premack (Eds.), *Causal cognition: A multidisciplinary debate* (pp. 44–78). New York: Oxford University Press.

Swackhamer, G. (2005). *Cognitive resources for understanding energy*. Unpublished.

TERC. (2011). *The inquiry project: Bridging research and practice*. Retrieved from, http://inquiryproject.terc.edu.

Tobin, R. G., Crissman, S., Doubler, S., Gallagher, H., Goldstein, G., Lacy, S., Rogers, C. B., Schwartz, J., & Wagoner, P. (2012). Teaching teachers about energy: Lessons from an inquiry-based workshop for K-8 teachers. *Journal of Science Education and Technology, 21*, 631–639.

Trumper, R. (1993). Children's energy concepts: A cross-age study. *International Journal of Science Education, 15*(2), 139–148.

Van Hook, S. J., & Huziak-Clark, T. L. (2008). Lift, squeeze, stretch and twist: Research-based inquiry physics experiences of energy for kindergartners. *Journal of Elementary Science Education, 20*(3), 1–16.

Warren, J. W. (1983). Energy and its carriers: A critical analysis. *Physics Education, 18*, 209–212.

Watts, D. (1983). Some alternative views of energy. *Physics Education, 18*(5), 213–217.

Chapter 15
Opportunities for Reasoning About Energy Within Elementary School Engineering Experiences

Kristen Bethke Wendell

15.1 Introduction

As engineering emerges as an important discipline in pre-college education, most engineering experiences planned for the elementary school level – and many of those planned for the middle and high school levels – are emphasizing practices of engineering design. Much of recent elementary engineering curriculum development and research has focused on supporting and assessing young students' design abilities of planning, prototyping, testing, and following the "engineering design process." (e.g., Cunningham 2009; Dyehouse et al. 2011). For example, in the *Capturing the Wind* unit of the Engineering Is Elementary curriculum (Museum of Science, Boston 2012), children learn that the engineering design process involves the tasks of asking, imagining, planning, creating, and improving. They then apply that process as they construct miniature windmills. Despite this focus on design practices and processes, engineering design also involves attention to physical principles (Cross 2004) and the modeling of how physical principles affect design outcomes (Atman et al. 2007). One of these principles is that energy must be transferred to any designed artifact by its power system, which is specified by the engineering designers, in order for the designed artifact to perform work.[1] Energy analysis is a part of the design process for every technology that requires electricity

[1]I recognize that some readers may interpret this sentence as treating the terms "energy" and "power system" as synonyms. It is not my intent to imply that energy and power are equivalent in engineering. Rather, the phrase "power system" is an engineering *term of art*. It refers to the portion of a technology that provides energy in the form and rate necessary for the technology to function.

K.B. Wendell (✉)
Department of Curriculum and Instruction, Center of Science and Mathematics in Context, University of Massachusetts Boston, Boston, MA, USA
e-mail: Kristen.Wendell@umb.edu

R.F. Chen et al. (eds.), *Teaching and Learning of Energy in K-12 Education*,
DOI 10.1007/978-3-319-05017-1_15, © Springer International Publishing Switzerland 2014

or other energy resource. Because of this, it is likely that there are opportunities within school engineering experiences for students to explore the applications of energy concepts. However, in many cases these opportunities have not been made explicit or have not been structured for elementary student learning. The purpose of this chapter is to identify opportunities for students to reason about energy during existing elementary school engineering activities.

To set the stage for this analysis, I begin the chapter by considering broadly how professional engineers view energy as they analyze requirements of and constraints on design. In the next section of the chapter I examine what national frameworks and standards have to say about what all students, not just those intending to become engineers, should learn by the end of elementary school about the role of energy in engineering design. This leads to the main inquiry of the chapter: what kinds of opportunities are there to reason about energy within existing elementary school engineering activities? I use as a test case the resource *Family Engineering* (Jackson et al. 2011), a guide for introducing children and their families to the world of engineering. It features 37 activities that draw upon a range of engineering disciplines and that are recommended for use with students aged 7–12. I close the chapter with a discussion of the difficulties students and educators might face in taking advantage of these opportunities to make energy explicit while conducting engineering design.

15.2 Energy in Professional Engineering Practice

A search of the extensive academic journal database Academic Search Complete for peer-reviewed manuscripts published between 1990 and 2012 with "energy" as a word in the title and "engineering" in the journal name resulted in 2,129 articles. This means that on average, the engineering research and development community produces about 100 journal articles strongly related to energy each year. This high publication rate suggests that energy concepts are central to engineering innovation (see Fig. 15.1). However, there is wide variety in how energy is treated in engineering design and analysis. Below are the topics of a representative sample of ten recently published engineering journal articles with key word "energy":

- Power generation using piezoelectric film (Tanaka et al. 2012)
- Impact of building shape on energy performance (Parasonis et al. 2012)
- Solar-based street lighting (Tsado and Ganiyu 2012)
- Energy balance of a low energy house (Džiugaitė-Tumėnienė et al. 2012)
- Vibration-driven energy harvesting (Akande et al. 2012)
- Energy audit of a brewery (Noah et al. 2012)
- Methods for monitoring pump conditions including energy consumption (Beebe and Jenkins 2012)
- Energy metabolism profile of certain kinds of cells under reduced gravity environment (Stoltz et al. 2012)
- Modeling and optimization of energy consumption in multi-robot systems (Vergnano et al. 2012)

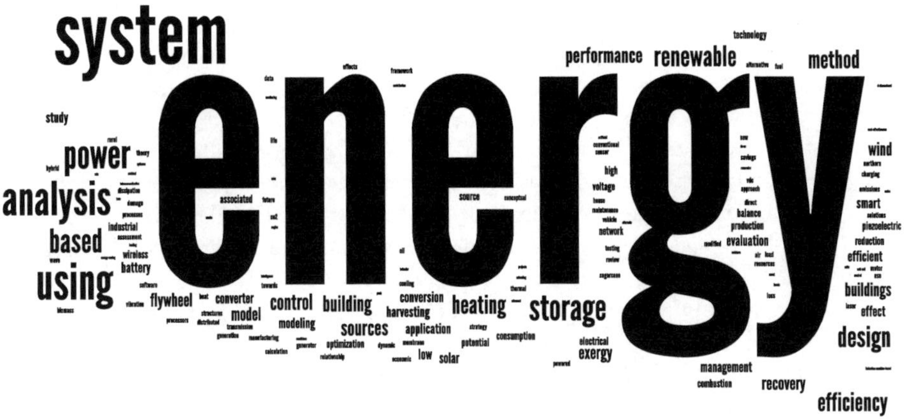

Fig. 15.1 Word splash created with the text of titles from 100 engineering journal articles strongly related to energy, published in 2012, and indexed by Academic Search Complete (word splash created with http://wordle.net)

- An energy consumption indicator for use during the design process (Ingeneer et al. 2012)

In this sampling of just ten reports, we see engineers using energy as a descriptor in six distinct ways: energy "performance," "consumption," "balance," "harvesting," "audit," and "metabolism." What this range of usage suggests is that energy concepts are viewed as useful across the engineering disciplines but that engineers may not use the term "energy" with the language precision used in science discourse. Some engineering endeavors treat energy as a quantity that can be produced (as in "harvested") and used up (as in "consumed"). This stands in contrast to scientists' and science educators' recent calls to avoid using language that suggests energy can be created or destroyed (National Research Council [NRC] 2012a). Other engineering endeavors consider energy as an essential ingredient of technology (as in energy "requirements"), but an ingredient that comes in limited supply (as in the need for an energy "audit"). The fact is that professional engineers use the term "energy" in several different and yet commonly accepted ways. Along a similar vein, discipline-based education research (NRC 2012b) into college engineering students' understandings of energy suggests that novice engineers hold a range of energy conceptions, some of which are scientifically accurate and others which might make physicists cringe (e.g., thinking of heat as energy rather than heat as a process of energy transfer) (NRC 2012b).

Despite this variety in the engineering research literature, energy is treated consistently in texts about engineering systems. In engineering, a "system" is a collection of interacting components that are designed collectively, whenever possible, to achieve a desired outcome (NRC 2009). Considerations of energy are ubiquitous in systems design and analysis, because along with matter and information, energy is one of the three main quantities that flow into and out of a system (Dandy et al. 2007). For example, in a system like a geothermal generator, thermal energy flows in and electrical energy flows out. Conversely, if we think of a kitchen refrigerator as a designed system, electrical energy is one of its inputs, and thermal energy transfer to the air near its coils is one of its outputs. There is also matter flow within the refrigerators' subsystems. For instance, a quantity of hot gas is input into the refrigerator's heat-exchanger coils during every compressor cycle. The matter output from these coils is condensed liquid refrigerant (which then flows through an expansion valve, vaporizes, and cools even more). Finally, we can view the refrigerator's thermostat setting as its information input, and its actual temperature reading as information output. The engineers who design and analyze systems such as these create models that relate inputs to outputs, and they keep careful track of whether the actual performance of the system matches the predicted outputs for a given set of inputs.

In designing, specifying, and supervising the construction of artifacts and processes, engineers rely on conceptual knowledge to make predictions about how changes in variables will affect how systems behave (Streveler et al. 2008). Because energy is treated as a system input and output for many technological systems, making predictions about relationships between variables requires knowledge about the conservation of energy and energy transfer. For example, as mechanical and chemical engineers work to specify equipment for exchanging energy within structures, manufacturing plants, or machines, they have to judge how changes in flowrates, temperature, or pressure will impact energy transfer rates (Streveler et al. 2008).

While analyzing energy inputs and outputs is a key aspect of engineering design across engineering disciplines, there are also engineering specialties whose main purpose is to transform and transfer energy for society's use. Mining engineering, petroleum engineering, and nuclear engineering – and new emerging fields like wind engineering and photovoltaic engineering – all focus on safely making natural resources available for transformation to energy that is usable by industrial and consumer technologies. These engineers talk about "providing energy" as something they design systems to do, as in the National Academy of Engineering "grand challenge" of providing energy from fusion. And a subspecialty of electrical engineering – power systems engineering – is devoted to transferring the electrical energy output by generators across vast distances with minimal losses. These energy-focused engineering fields do not work in isolation, of course. As we witnessed when millions in the northeastern United States were without electricity in the aftermath of Super Storm Sandy in late 2012, countless other technologies are dependent on the United States power generation system. The communication systems used by the New York Stock Exchange, for example, were slowed by the Super Storm power outages and thus struggled to keep up with the demands of global financial trading. Therefore even engineers in disciplines other than those listed above must take energy into account.

Another way in which energy commonly comes into play across different engineering disciplines is in the contemporary push for sustainable technologies

(Dandy et al. 2007). Designing for sustainability is now accepted by many engineers as part of their professional obligations (Dym 2010). To promote and work in a sustainable manner requires engineers to know about renewable energy resources and about how to minimize energy losses in mechanical and electrical systems. In the growing discipline of "green engineering," one guiding principle is to "ensure that all material and energy inputs and outputs are as inherently safe and benign as possible" (NRC 2004, p. 22). Here, again, we see energy treated as a quantity that flows through a system.

15.3 Elementary School Learning About Energy in Engineering Design

Clearly, future engineers need to know how the principle of energy conservation works as a constraint on every designed system, and they need to be able to analyze energy inputs and outputs of technological systems. And this involves correcting any naïve views they may still hold about concepts related to energy, such as heat, temperature, and electric current. But what about young students who do not yet know whether they will become engineers – what do they need to know about the role of energy in engineering design?

In describing energy flow as a "crosscutting concept" of science and engineering disciplines, the national *Framework for K-12 Science Education* emphasizes that the ability to analyze energy transfer is "a *tool* that students can use across virtually all areas of science and engineering," including the examination of engineering systems (NRC 2012a, p. 95, emphasis added). The *Framework* also recognizes that in engineering design, a major goal is "to maximize certain types of energy output while minimizing others, in order to minimize the energy inputs needed to achieve a desired task" (NRC 2012a, p. 95). Of course the *Framework* not only considers energy flow as a "crosscutting concept" that applies across disciplines but also identifies energy as a "disciplinary core idea" in the physical sciences.

15.3.1 Three Goals for Applied Knowledge of Energy

Synthesizing the *Framework*'s fifth-grade endpoints for energy as a physical science core idea with the role of energy in engineering, I have generated three "applied knowledge of energy" goals that students should reach by middle school. These goals represent knowledge that students should construct and apply during engineering design problem-solving.

First, students should be able to recognize when, to design a technological solution, it is necessary to transfer energy from place to place because a technology needs an energy input. This involves understanding that actions on the world require energy and that more energy cannot be spontaneously "created" or "produced." It also involves predicting *in relative terms* how much energy input is required by one technology compared to others. For example, an airplane requires more energy input than a bicycle; a pumpkin launcher needs more energy input than a cotton ball catapult. This energy needs goal is an application of the *Framework*'s Energy Definitions core idea PS3.A.

Second, once students identify the need for an energy input, they should be able to identify options for storing the required energy. This involves understanding that when we talk about "supplying energy" to a device, we are typically referring to the conversion of stored energy into a form that is useful for the technology. It also involves recognizing that some modes of energy storage are better (e.g., more concentrated, less dangerous, etc.) than others, and that energy is lost or dissipated whenever it is converted from a store. Both of these facts affect the sustainability of technologies. This energy storage goal applies the *Framework's* Energy in Chemical Processes and Everyday Life core idea PS3.D.

Finally, having considered modes of energy storage, students should be able to draw from a repertoire of ways to transfer energy from place to place (e.g., via moving objects, sound, electric current, and age-level accessible forms of electromagnetic radiation such as visible light and heat), including ways to convert stored energy into a desirable form for use in a specific technology. Like the energy storage goal, this energy transfer goal also entails recognizing that some energy is lost over the course of each energy transfer, and that engineers strive to design energy transfer strategies with minimal losses. This third goal applies the *Framework*'s Energy Transfer core idea PS3.B, as well as its Energy and Forces core idea PS3.C.

15.3.2 Alignment with Other Standards Documents

In summary, the three "applied knowledge of energy" goals are (1) *energy needs awareness* – recognizing when a technology needs energy input, (2) *energy storage options* – considering a variety of modes in which energy can be stored, and (3) *energy transfer strategies* – identifying several possible ways to transfer energy from place to place. These goals are consistent with the *Standards for Technological Literacy* of the International Technology and Engineering Education Association (ITEEA 2000/2002/2007), whose Standard 16 focuses on energy and power technologies and suggests that third through fifth graders know that "tools, machines, products, and systems use energy in order to do work" (aligned with goal 1 above) and that "a well-designed tool, machine, product, or system minimizes energy losses [and] should be designed to apply energy efficiently to do a useful task" (aligned with goals 2 and 3 above) (ITEEA 2000/2002/2007, p. 160). Likewise, the three applied knowledge of energy goals suggested here align with the Technology and Engineering Literacy Framework of the National Assessment for Educational

Progress (WestEd 2012). This framework, developed for an NAEP assessment first piloted in spring 2013, focuses on the three areas of technology and society, design and systems, and information and communication technology. Within the design and systems area, elementary students should know (as in goal 1 above) that energy is one of the resources needed to solve design challenges (along with tools, materials, people, capital, and time), and (as in goals 2 and 3 above) that the storage and flow of energy can be traced through a designed system.

15.4 Analyzing Elementary School Engineering for Energy Reasoning Opportunities

In 2009 the National Academy of Engineering produced a report on the status and prospects of K-12 engineering education (NRC 2009). It found that although energy is one of the most common science topics explicitly mentioned in K-12 engineering curricula, the concepts related to it (and most other science concepts) are "presented in the form of encyclopedia-like explanations that are subsequently reinforced in laboratory activities" (NRC 2009, p. 80). Indeed, in the report's appendix of analyses of 75 K-12 engineering curriculum units (from 15 different initiatives and projects), 22 have goals that explicitly include learning about or applying energy concepts. Seven of these 22 units are for elementary students. "Energy" is often listed in a curriculum's set of science learning objectives or teaching goals, as in "energy transfer," "concepts of energy," "sources of energy," "forms of energy," "potential energy," "kinetic energy," and "energy is the ability to do work." Therefore from the perspective of those who develop curriculum, there are certainly pathways to learning about energy within K-12 engineering experiences. However, the question remains of whether these pathways involve only direct instruction about the definition of energy and its different "forms," or whether there are true opportunities for students to reason critically about the role of energy in engineering design solutions.

My purpose in what follows is not to report on an exhaustive search for energy connections in all available elementary-level engineering resources. Instead, the goal is to create a beginning sense of the kinds of energy connections that might be found in any set of elementary school engineering experiences. Below I suggest (a) an approach to classifying school engineering activities for their possible connections to energy and (b) examples of the ways in which students might reason about energy within engineering design activities.

15.4.1 Procedures for the Analysis

Because this is not a systematic survey of instructional materials, but rather an example of how to approach any set of materials, I chose just one engineering activity guidebook as a test case. The book *Family Engineering* (Jackson et al. 2011) is a guide for introducing children (aged 7–12) and their families to the world of engineering. Schools and community organizations are encouraged to choose from the activities in the book to design an evening or weekend event at which families learn about engineering and engage in engineering design. Although this book is intended for informal education, its activities are quite similar to the engineering challenges posed to students in formal elementary education settings. I chose *Family Engineering* for this case study because it was supported by the National Science Foundation, carefully created by an interdisciplinary team of expert engineers and educators, and built upon past successes in children's engineering. Moreover, its 37 activity descriptions are brief and clear, and each of its activities can be completed in less than 60 min, which is a typical lesson length in elementary classrooms. Also, having served on its panel of reviewers, I was familiar with its content and knew that its goal was to expose students to a wide range of engineering fields (e.g., electrical engineering, mechanical engineering, chemical engineering) and concepts (e.g., the engineering design process, trade-offs, systems). This wide range makes it more likely to give us a sense of the range of energy connections that might be possible within elementary school engineering – more so than an engineering curriculum focused on a small number of design challenges.

With the goal of identifying opportunities for reasoning about energy, an analysis was conducted of the 37 activities in the *Family Engineering* guide. The first step in this analysis was to limit the pool of activities to those that feature hands-on work to design, test, or improve a physical device. This step resulted in a pool of 26 activities. The 11 activities that were excluded were intended for learning *about* what engineering is and what engineers do (e.g., matching pictures of products to the branch of engineering most closely associated) rather than for engaging in engineering itself. The 26 included activities are hands-on engineering challenges that involve the testing of physical devices. Some require participants to design a device "from scratch" and test it against requirements or metrics; others supply participants with initial designs and focus on testing and analysis.

The following procedure was used to analyze the 26 engineering challenges.

1. Review the one- to three-page activity description.
2. Write a brief summary specifying the design goal, the test to be conducted, and the available materials.
3. Note whether the activity involved designing and testing, or only the testing of existing designs.
4. Note any explicit mention of "energy" in the activity instructions, Engineering Connection summaries, or image captions.
5. Note any implicit energy connections to the "applied knowledge of energy" goals for elementary school students – i.e., implied need for energy input (goal 1), energy storage (goal 2), or energy transfer (goal 3).
6. Note any implicit energy connections most likely to be inaccessible for elementary school students – e.g., energy methods for structural analysis.
7. Order the challenges according to the strength of their energy connections.

15.4.2 Results of the Energy Opportunities Analysis

The results of this analysis are summarized in Tables 15.1, 15.2 and 15.3. Six of the 26 challenges make explicit reference to energy concepts (Table 15.1). These involve catapults, paper rockets, flashlight re-design, hot chocolate fountains, toy car windshields, and low flow showerheads. All but the flashlight re-design involve either making use of gravitational potential energy (showerhead showdown, hot chocolate fountain, cars) or transferring energy from elsewhere to increase gravitational potential (catapult, rocket).

Seven challenges (Table 15.2) do not explicitly mention "energy" but have clear implicit opportunities to reason about energy concepts, at a level within the reach of elementary school students. These seven challenges involve roller coasters, motorized artists, egg protectors, soundproof containers, grabber devices, animal puppets, and miniature drums. Energy comes into play through intentional energy dissipation (soundproofing, egg protection, drums), energy transfer with mechanical linkages (grabber device, animal puppet), transformation of gravitational potential (roller coaster), and transformation of electrical energy to energy of motion (artistic robot).

The 13 remaining challenges (Table 15.3) have less obvious energy connections or require more advanced knowledge and skill to apply energy concepts. They involve design of processes (mining for chocolate, ballpoint pen assembly line, traffic flow), stable structures (cantilever, pipe cleaner tower, aluminum boat, paper-towel-tube tower, bridges, building foundations, triangle frames, laminate beams), and materials selection or packaging design (waterproofing materials, equal-volume boxes).

From these results we can see that all of the challenges classified by *Family Engineering* as civil engineering activities are classified in the "less obvious energy connections" category. Of course, there is an entire field of civil engineering dedicated to using energy principles for structural analysis, but to make these approaches accessible to elementary students is not a straightforward task, nor is it necessarily productive. As mentioned above, energy concepts at the elementary school level typically involve directly observable energy transformations or transfers, as in energy transferred by motion, light, heat, or sound. The potential energy of stable, static structures is neither easily observed nor easily analyzed by elementary students.

Let's consider how students could make progress toward the three "applied knowledge of energy" goals as they work on two *Family Engineering* challenges: "Launchers," (Table 15.1) whose activity guide makes explicit reference to energy concepts, and "Artistic Robots," (Table 15.2) whose guide does not mention energy.

The problem posed in Launchers is to send a cotton ball flying through the air with only spoons, craft sticks, and rubber bands. Participants are instructed to measure the distance traveled by the cotton ball and adjust their design to launch it even farther. The *Family Engineering* guide recommends explaining the terms kinetic energy and potential energy to participants, but even without knowledge of

Table 15.1 *Family Engineering* challenges with explicit energy connections

Challenge	Explicit mention of energy	Additional energy opportunities[a]
Design and test		
Launcher (Aero): Build a launcher with craft sticks, rubber bands, and spoons. Measure distance traveled by a launched cotton ball.	"Launchers work by storing potential energy and then releasing it as kinetic energy (energy of motion) which is used to propel an object."	Energy transfer from motion of hands, to storage in stretched rubber band, to motion of launcher, to motion of cotton ball (Goals 1, 2, 3)
Blast Off! (Aero): Build a rocket with paper and cardstock. Measure distance traveled when launched with air from squeeze bottle.	"Propulsion is the energy needed to move the craft through the air."	Energy transfer from motion of hands, to storage in squeezed bottle of air, to motion of rocket (Goals 1, 2, 3)
Bright Ideas (Elec): Reconfigure a flashlight into a lightweight reading light using common craft materials. Test for hands-free reading capability.	"Electrical engineers are developing alternatives to incandescent lights, such as compact fluorescent light (CFL), which will help reduce energy consumption."	Electrical engineering as discipline that includes design of systems to deliver and transfer electrical energy (Goal 3)
Hot Chocolate Machine (Chem): Build a fountain with stacked paper cups. Test whether water, milk powder, and chocolate are mixed well.	"Chemical engineers ... try to use the least amount of energy and materials, and produce the least amount of waste."	Energy transfer from gravitational potential to motion of water; thermal energy of hot water impacts dissolving of milk and chocolate powders (Goals 2, 3)
Test of existing designs		
Against the Wind (Mech): Compare speeds of matchbox cars with different index-card "windshields" attached on top.	"How can engineers save energy through design?" "Engineers help us save energy by designing more aerodynamic cars and trucks."	Energy efficiency; implication that aerodynamic cars require less energy input because less energy of motion is dissipated to air resistance (Goal 1)
Showerhead Showdown (Env): Compare volume of water leaving two cups with different sizes and numbers of holes in the bottom.	"'Low flow' showerheads are engineered to use less water, save energy, and still provide a nice shower."	Implication that with low flow shower heads, less energy is needed to heat water (because the shower uses less hot water) (Goal 1)

[a]Goal 1 is energy needs awareness: recognizing when a technology needs energy input. Goal 2 is energy storage options: consider various modes of storing energy for a technology. Goal 3 is energy transfer strategies: identifying several possible ways to transfer energy from place to place

Table 15.2 *Family Engineering* challenges with implicit energy connections

Challenge	Implicit energy opportunity[a]
Design and test	
Thrill Seekers (Mech): Build a curvy track with clear plastic tubing. Test whether marble makes it all the way to the end.	Energy transfer from gravitational potential to motion of marble (Goal 3)
Artistic Robots (Mech): Build an artwork-creating device with a motor, battery, and common materials. Test whether marks are made on paper.	Energy storage in battery transferred to motor to motion of device (Goals 1, 2, 3)
Brain Saver (Biomed): Build a helmet for a raw egg using common materials. Test whether egg survives a drop.	Energy transfer from gravitational potential to motion of helmet and egg; energy transfer from helmet to cushioning materials instead of egg (Goal 3)
Soundproof Package (Acous): Build a sound-muffling package using a plastic storage container and common materials. Test whether noisemaker's volume is reduced.	Energy transfer (or not) from noisemaker to sound-muffling packaging (Goal 3)
Give Me a Hand (Biomed): Build a grabber device using common craft materials. Test whether it can pick up a cotton ball, eraser, pencil, and marble.	Energy transfer within human body from chemical storage to electric current to motion of hands to grabber device (Goals 1, 2, 3)
Create a Critter (Mech): Build movable 2D animal models with paperboard and brass fasteners. Test whether linkages/mechanisms work as planned.	Energy transfer from motion of hands to linkages to animal limbs (Goals 1, 3)
Test of existing designs	
Make It Loud! (Mat): Compare volume of paper-towel-tube drums with membranes made of different materials.	Energy transfer from motion of membrane to motion of air particles and tympanic membrane (Goal 3)

[a]Goal 1 is energy needs awareness: recognizing when a technology needs energy input. Goal 2 is energy storage options: consider various modes of storing energy for a technology. Goal 3 is energy transfer strategies: identifying several possible ways to transfer energy from place to place

those terms (which, according to the NRC 2012 *Framework*, can be misleading), students can use energy concepts as a tool for thinking through this design challenge. First, they could recognize that launching a cotton ball requires transfer of energy from somewhere else to a cotton ball at rest (goal 1, energy needs awareness). Because energy cannot be spontaneously produced at the site of the cotton ball, the only way that a cotton ball will move through the air is if the energy for that flight is input to the cotton ball. Second, students could review the available materials and consider ways to store the energy needed for cotton ball flight (goal 2, energy storage options). For example, a plastic spoon could be bent slightly backwards; a rubber band could be stretched and fixed at one end to the spoon and at the other end

Table 15.3 *Family Engineering* challenges with less accessible energy connections

Challenge	Implicit energy opportunity
Design and test	
Mining for Chocolate (Min): Devise a process that uses toothpicks, etc., to get the chocolate chips out of cookie. Calculate profit based on number of chips and state of cookie.	Energy stores and transfers involved in mining natural resources (Goals 1, 2, 3)
Assembly Line (Ind): Devise a process for quickly assembling deconstructed ballpoint pens. Measure time to assemble six pens.	Energy stores and transfers involved in industrial assembly lines (Goals 1, 2, 3)
Domino Diving Board (Civ): Build a cantilever with overlapping stacked dominoes. Measure distance from table to edge.	Energy methods for engineering structural analysis
Team Up! (Gen): Build a tall and stable tower with pipe cleaners, but do so as a team without talking and with only one hand per member.	Energy methods for engineering structural analysis
Learning from Failure (Gen): Build a boat with aluminum foil. Test how many pennies it can hold before sinking.	
Five Points Traffic Jam (Civ): Devise a plan for placing the signs and signals on a 2D map of a 5-point traffic intersection.	
Test of existing designs	
Tumbling Tower (Civ): Compare stability of various tube-and-cardboard tower configurations.	Energy methods for engineering structural analysis
Arches (Civ): Compare how many erasers are held up by a flat bridge and an arch bridge.	Energy methods for engineering structural analysis
Solid Ground (Geo): Compare stability of a block placed on rough gravel, sand, and smooth gravel.	Energy transfer from block to foundation (Goal 3)
Shifting Shapes (Civ): Compare stability (shape-shifting) of square and triangle shapes made with poster board and brass fasteners.	Energy methods for engineering structural analysis
Glue Is the Clue (Mat): Compare how many washers are held up by two glued index cards versus two unglued index cards.	Energy methods for engineering structural analysis
Boxing Beans (Pack/Mat): Compare how many beans fit inside four paper boxes of different shapes.	
Wrap It Up! (Mat): Compare water-proofing ability of five different materials wrapped around cotton swabs.	

to the table surface. Doing both of these things would store energy in the combined elasticity of the spoon and rubber band. Finally, students could identify a few ways that the stored energy could be transferred to the cotton ball (goal 3, energy transfer strategies). Heat or electric current could be applied to the rubber band, and perhaps this would cause it to fray and release its stored energy to the spoon. If the cotton ball were in contact with the spoon, then the spoon's energy of motion would launch the cotton ball into the air. Of course, generating heat or electricity to break the rubber band would require an additional energy input, and this would decrease the device's energy efficiency. A human hand could instead be used to cut the rubber band or to lift the tape fixing it to the table surface. With less energy loss than would occur through a heating device or electric circuit, this human input would release the band's energy to the spoon, which would then transfer energy of motion to the cotton ball.

Of course, through systematic trial and error, students can construct successful launchers without conducting this energy transfer analysis. In that case, a teacher could ask careful questions to prompt the students' reasoning about energy: "Why do you need a launcher device at all? Why does the rubber-band device launch the ball farther than the human hand can throw it?" These question provide an opportunity for students to see the rubber band as an energy storage device that allows more energy from the human to be transferred to the ball than is transferred by the act of throwing. To illustrate this point, compare the human energy output when flinging a cotton ball across the room versus the human energy output when stretching a strong rubber band.

Now, let's connect the "applied knowledge of energy" goals to a challenge with no explicit mention of energy but a clear opening to energy connections. In the "Artistic Robots" activity, participants are tasked with building an artwork-creating device with a motor, battery, felt-tip markers, and common household materials like plastic cups. They test whether their device – when turned on – makes marks on a piece of paper. Although the *Family Engineering* guide does not mention energy in the instructions for "Artistic Robots," it is another challenge where students could use energy concepts as a tool for design and analysis. First, with applied knowledge of energy, students could recognize that getting markers to move across a page without a human's touch requires a transfer of energy to the markers from a source other than a human's hand (goal 1, energy needs awareness). Because energy cannot be created spontaneously within the markers, the only way they will move is if energy is transferred to them from somewhere else. Where might the energy for the markers' motion be stored? In this challenge, students are provided with a battery, which they could identify as a portable source of stored energy (goal 2, energy storage options). Students could then ask, how can this store of chemical energy be transferred to the motion of the robot's markers? Teachers could support students in recognizing that a battery's energy is released through electric current when a complete circuit includes the battery. Students could then walk through several possibilities for transferring the electric current's energy to the markers (goal 3, energy transfer strategies). Electrical energy could be transferred to heat or light through the use of a resistive material, and this heat or light could be directed to the markers. But this would not cause them to move. Likewise, the electrical energy could be transferred to a noisemaker, but sound would have no effect on the markers' motion. The most desirable energy transfer from the battery would bring a moving object into contact with the markers. Realizing this, the students could see the motor as a device that transfers electrical energy to spinning motion. The motor's spinning axle can then transfer energy to the markers of the artistic robot. Finally, the markers' motion makes art on paper!

15.5 Learning Applied Knowledge of Energy: The Challenges

As we have seen in the *Family Engineering* activity analysis, not all elementary-level engineering challenges have accessible opportunities to apply energy knowledge. For example, bridge building is a common elementary school design activity. But to use knowledge about energy as a resource for predicting which bridge design is the strongest involves methods of engineering mechanics that are (because of the mathematics involved) beyond most high school students, let alone elementary students.

Even when students are engaged in an elementary-level engineering activities with clear opportunities to reason about energy, there are several challenges to that actually taking place. First, if students are given enough time and materials, they can complete many engineering design problems through trial and error methods alone. Engineering by systematic trial and error does not always require applying knowledge of science principles. Second, some design problems are solvable based only on prior knowledge of already existing artifacts. For example, students might design successful cotton ball launchers based only on catapults they have seen in photographs. They are basing their design ideas on everyday experience rather than on reasoning about energy or any other science concept. This sort of intuitive designing can enable very successful solutions without any associated science knowledge construction (Fortus 2003). To deal with these threats to science learning, students' intuitive experiences with design and their trial and error discoveries must be articulated and formalized (Fortus 2003). Teachers must plan thoughtfully for substantial reflection and discussion of *why* design artifacts succeed or fail. These activities must focus student's cognitive efforts on conceptual understanding rather than on task completion.

A third challenge to young students' learning "applied knowledge of energy" for engineering is the level of skill required of the teachers. Teaching scientific concepts within engineering design experiences places substantial demands on teachers. Many in-service elementary school teachers completed postsecondary education that required minimal coursework in science and typically no coursework in engineering or technological design (Tolman and Campbell 1991). These traditionally educated teachers may be accustomed to treating science as a body of knowledge that contains one correct answer for any given question (Lederman 1992; Tilgner 1990). Without extensive preparation and support, many teachers may not be able to adapt their science teaching methods to engineering activities that involve multiple acceptable solutions that cannot be anticipated ahead of time. To facilitate students' reasoning about energy during engineering design challenges, teachers need to have deep understanding of the scientific concepts of energy as well as great confidence in their ability to apply it to novel situations. During or after a design session in which each student creates a unique engineering solution, helping students learn about energy concepts requires the teacher to plan quickly how to apply her own knowledge of energy to the unpredictable creations of students, and to transition students from physically manipulating materials to cognitively operating on ideas. These demands on the teacher cannot be overlooked.

However, well-designed professional development opportunities can increase the chances that elementary school teachers will succeed in facilitating engineering design challenges. For example, researchers at the Center of Engineering Education at Outreach (CEEO) at Tufts University found that a 5-day summer workshop was adequate to prepare third- and fourth-grade teachers to enact two engineering design-based science curriculum units the following school year (Kendall and Wendell 2012). At the workshop, teachers worked through the two curriculum units as if they were students; they designed and tested mechanical devices to solve engineering problems, and they explored the underlying science concepts. Researchers at the Institute for P-12 Engineering Research and Learning (INSPIRE) at Purdue University found that first-grade through fifth-grade teachers who attended a 3-day summer engineering education academy and a 2-h follow-up session were able to design their own engineering challenges and help their students successfully complete them (Capobianco et al. 2011). During the summer academy, the teachers conducted hands-on activities (adapted from the Museum of Science Boston's Engineering Is Elementary materials) to explore how engineers design devices to capture wind energy and to clean water, and they learned how to integrate the engineering design process into standards-based science lessons.

15.6 Conclusion

This analysis of just one engineering curriculum resource suggests that many elementary school engineering challenges offer meaningful opportunities for students to reason about energy in ways that could contribute both to the success of their design solution and to their overall knowledge of energy concepts. By achieving three "applied knowledge of energy" goals – energy needs awareness (goal 1), energy storage options (goal 2), and energy transfer strategies (goal 3) – elementary school students would be equipped with an extremely important tool of engineering design and would be well prepared for more formal, complex learning about the science of energy in the middle and high school years.

References

Akande, A., Nechibvute, A., & Luhanga, P. (2012). Discussion on establishing a metrology standard for vibration-driven energy harvesting. *Journal of Energy & Power Engineering, 6*(5), 674–682.

Atman, C. J., Adams, R. S., Mosborg, S., Cardella, M. E., Turns, J., & Saleem, J. (2007). Engineering design processes: A comparison of students and expert practitioners. *Journal of Engineering Education, 96*(4), 359–379.

Beebe, R., & Jenkins, S. (2012). Condition monitoring methods for pumps. *Chemical Engineering, 119*(9), 34–39.

Capobianco, B. M., Diefes-Dux, H. A., & Mena, I. B. (2011, June 26–29). Elementary school teachers' attempts at integrating engineering design: Transformation or assimilation? In *Proceedings of the 118th annual conference of the American Society for Engineering Education*, Vancouver, BC.

Cross, N. (2004). Expertise in design: An overview. *Design Studies, 25*(5), 427–441.

Cunningham, C. (2009, Fall). Engineering is elementary. *The Bridge*, 11–17.

Dandy, G., Walker, D., Daniell, T., & Warner, R. (2007). *Planning and design of engineering systems* (2nd ed.). New York: Taylor & Francis.

Dyehouse, M., Diefes-Dux, H., & Capobianco, B. (2011, June 26–29). Measuring the effects of integrating engineering into the elementary school curriculum on students' science and engineering design content knowledge. In *Proceedings of the 118th annual conference of the American Society for Engineering Education*, Vancouver, BC.

Dym, C. L. (2010). Sustainability: Affirming engineering values. *International Journal of Engineering Education, 26*(2), 240–246.

Džiugaitė-Tumėnienė, R., Jankauskas, V., & Motuzienė, V. (2012). Energy balance of a low energy house. *Journal of Civil Engineering & Management, 18*(3), 369–377.

Fortus, D. (2003, March). *A comparison of design-based science & project-based science.* Unpublished preliminary examination paper. University of Michigan.

Ingeneer, L., Mathieux, F., & Brissaud, D. (2012). A new 'in-use energy consumption' indicator for the design of energy-efficient electr(on)ics. *Journal of Engineering Design, 23*(3), 217–235.

Jackson, M., Heil, D., Chadde, J., & Hutzler, N. (2011). *Family engineering: An activity and event planning guide.* Houghton, Michigan: Foundation for Family Science and Engineering and Michigan Technological University.

Kendall, A. L. M., & Wendell, K. B. (2012, June 10–13). Understanding the beliefs and perceptions of teachers who choose to implement engineering-based science instruction. In *Proceedings of the 119th annual conference of the American Society for Engineering Education*, San Antonio, TX.

Lederman, N. G. (1992). Students' and teachers' conceptions of the nature of science: A review of the research. *Journal of Research in Science Teaching, 29*(4), 331–359.

Museum of Science, Boston. (2012). *Catching the wind: Designing windmills.* Boston: Museum of Science, Boston. Retrieved from, http://legacy.mos.org/eie/20_unit.php

National Research Council. (2004). *The engineer of 2020: Visions of engineering in the new century.* Washington, DC: The National Academies Press.

National Research Council. (2009). *Engineering in K-12 education: Understanding the status and improving the prospects.* Washington, DC: The National Academies Press.

National Research Council. (2012a). *A framework for K-12 science education: Practices, crosscutting concepts, and core ideas.* Washington, DC: The National Academies Press.

National Research Council. (2012b). *Discipline-based education research: Understanding and improving learning in undergraduate science and engineering.* Washington, DC: The National Academies Press.

Noah, O., Obanor, A., & Audu, M. (2012). Energy audit of a brewery – A case study of Vitamalt. *Energy & Power Engineering, 4*(3), 137–143.

Parasonis, J., Keizikas, A., & Kalibatiene, D. (2012). The relationship between the shape of a building and its energy performance. *Architectural Engineering & Design Management, 8*(4), 246–256.

Stoltz, J.-F., Magdalou, J., Netter, P., Pinzano, A., Vandamme, K., Holy, X., & Petite, H. (2012). Impaired osteoblastogenesis potential of progenitor cells in skeletal unloading is associated with alterations in angiogenic and energy metabolism profile. *Bio-Medical Materials & Engineering, 22*(4), 219–226.

Streveler, R., Litzinger, T. A., Miller, R. L., & Steif, P. (2008). Learning conceptual knowledge in the engineering sciences: Overview and future research directions. *Journal of Engineering Education, 97*(3), 279–294.

Tanaka, Y., Matsumura, K., & Mutsuda, H. (2012). Study on flexible power generation device using piezoelectric film. *Journal of Energy & Power Engineering, 6*(3), 353–360.

Technology International Association Engineering Education. (2000/2002/2007). *Standards for technological literacy: Content for the study of technology* (3rd ed.). Reston: International Technology and Engineering Education Association.

Tilgner, P. (1990). Avoiding science in the elementary school. *Science Education, 74*(4), 421–431.

Tolman, M. N., & Campbell, M. K. (1991). Science preparation requirements of elementary school teachers in the United States. *Journal of Science Teacher Education, 2*(3), 72–76.

Tsado, J., & Ganiyu, M. A. (2012). Standalone solar-based street lighting. *IUP Journal of Electrical & Electronics Engineering, 5*(3), 23–33.

Vergnano, A., Thorstensson, C., Lennartson, B., Falkman, P., Pellicciari, M., Leali, F., & Biller, S. (2012). Modeling and optimization of energy consumption in cooperative multi-robot systems. *IEEE Transactions on Automation Science & Engineering, 9*(2), 423–428.

WestEd. (2012). *Technology and engineering literacy framework for the 2014 National Assessment of Educational Progress* (Pre-publication ed.). Retrieved from, http://nces.ed.gov/transfer.asp?location=www.nagb.org/publications/frameworks/prepub_naep_tel_framework_2014.pdf

Chapter 16
Launching the Space Shuttle by Making Water: The Chemist's View of Energy

Angelica M. Stacy, Karen Chang, Janice Coonrod, and Jennifer Claesgens

16.1 Introduction

In the Framework for K-12 Science Education (NRC 2011), energy is both a Crosscutting Concept and a Disciplinary Core Idea. This new delineation has raised the stakes. The implication is that scientific literacy is tied to a strong understanding of energy both within and across disciplines. Yet just as these documents recognize the importance of the concept of energy in science, energy is also recognized as one of the most difficult topics for students to learn. Benchmarks for Science Literacy (AAAS 1993) describes energy as a "mysterious concept," and both the National Science Education Standards (NRC 1996) and Framework (2011) caution about how the topic should be introduced to young students.

Energy is commonly described in terms of mechanical, chemical, and thermal energy, but the Framework cautions that this "is misleading, as it implies that the nature of the energy in each of these manifestations is distinct when in fact they all are ultimately some mixture of kinetic energy, stored energy, and radiation. Furthermore, what is meant by the first three terms above is seldom precisely defined." Recognizing the complexity of energy as a scientific construct, the Framework (2011) recommends that students may understand energy better through ideas of transformations, transfer, and conservation in physical science and flow in life science rather than as a concept unto itself.

Students confuse heat, work, and energy (Driver et al. 1994; Loverude et al. 2001). The distinctions between these energy ideas are not clear to students. For example, students confuse the ideas of heat and temperature (Lewis and Linn 1994). Students believe that it takes energy to break bonds (Boo 1998; Teichert and Stacy 2002). Driver et al. (1994) describes five main areas ofstudent misconceptions.

A.M. Stacy (✉) • K. Chang • J. Coonrod • J. Claesgens
Department of Chemistry, University of California Berkeley, Berkeley, CA, USA
e-mail: astacy@berkeley.edu

R.F. Chen et al. (eds.), *Teaching and Learning of Energy in K-12 Education*, 285
DOI 10.1007/978-3-319-05017-1_16, © Springer International Publishing Switzerland 2014

These are (1) energy is associated with animate objects; (2) energy is a causal agent stored in certain objects; (3) energy is linked with force and motion; (4) energy is fuel; and (5) energy is a fluid. The confusion among these concepts is observed in elementary school-aged children, and found to persist in college-aged students.

Traditionally the emphasis in energy instruction has been on types of energy (light, heat, sound, electrical) rather than processes associated with energy transformations as suggested by the Framework. Duit (1987) proposed that energy should be illustrated as something quasi-material, recognizing the natural tendency to think of energy as "stuff" before thinking of it as a transfer. Nordine et al. (2010) challenge the assumption of defining energy as stuff and suggest using the idea of "transformations." In this model, students track the flow of energy from trash to powering a stereo by tracking the transformations to types of energy. Hammer et al. (2012) suggest that understanding energy is more than a definition or a category. Implicit to understanding energy is how it is framed, or the "particular conversation," which affects how one thinks about it. In physics, energy is work while heat is often considered as inefficiency in moving an object, for example. In chemistry, energy is heat transfer (sometimes light); work is a byproduct of gases that expand because they are at a high temperature. In biology, energy is tracked through food chains. These are three very distinct ways of thinking about energy. The argument here is that context matters.

16.2 The Chemist's View: It Is All About Bonds

It is useful to frame the discussion of energy as it relates to chemistry with a specific example. Consider the reaction that occurs to launch the Space Shuttle: hydrogen gas, H_2, is mixed with oxygen gas, O_2. A spark ignites the mixture, and a change occurs explosively with enormous release of energy. All this happens because the hydrogen and oxygen atoms rearrange to produce water, H_2O.

$$2H_2\ (g) + O_2\ (g) \rightarrow 2H_2O\ \ (g)$$

The product water molecules are at a very high temperature. This means that the water molecules have a high kinetic energy and move very fast. The gaseous water molecules expand so rapidly that the Space Shuttle is thrust into orbit against the pull of gravity despite the large size and mass of the Shuttle. Thus, it is the *formation of water* that is associated with a release in energy. This release in energy that results from the *process* of making water is transferred by conduction as heat, by radiation as light, and by work in launching the Space Shuttle.

A key concept is that there is energy transfer anytime that a change occurs due to a rearrangement of the components in a small piece of the universe that is under observation. In chemistry, the change involves the rearrangement of atoms in reactant molecules to form new arrangements in product molecules. An understanding of such reactions and the link to the energy needs of society are critical components of science instruction for all citizens, and especially for students studying chemistry. We need scientists, leaders, and a voting public with sufficient understanding to make critical decisions regarding the use of such reactions and the selection of new technologies to build our energy future.

16.3 Challenges for Students and Reformulation of Instruction

Numerous studies point to the difficulties for students in learning about energy. We report here a portion of a study aimed at probing student conceptions of energy in the context of chemistry. Chang (2009) provides a more complete description of this research in her Ph.D. thesis in which she analyses data from semi-structured clinical interviews with seven students, 15–17 years old. These students were enrolled in a 6-week summer enrichment course in chemistry, with a focus on energy and thermochemistry. An early draft of Living by Chemistry Unit 5: Fire (Stacy et al. 2010b) was used for instruction. Each student was interviewed three or four times. The first interview was conducted pre-instruction, the next two probed student understanding after instruction on heat transfer associated with physical change and then on heat transfer associated with chemical change, and the fourth was conducted post-instruction. One premise of this study is based on the belief that information about student conceptions can be inferred from explanations students provided during interviews.

The various types of knowledge that students used to construct explanations during the interviews were sorted and classified as described in detail in Chang (2009). Topics that were found to be the most challenging for students included heat, heat capacity, bond energy, the terms exothermic and endothermic, and energies associated with dissolution. For this paper, we elaborate on student notions of (1) heat, (2) bond energy, and (3) the terms exothermic and endothermic. For each of these three topics, we suggest a reformulation of the curriculum to support student learning based on the results of the interviews.

16.3.1 Heat and Energy: Substance Versus Process

Our everyday language is laced with the metaphor of heat as being substance-like, or that objects contain heat. For example, it is common to speak about being out of energy, about getting energy from food, and about the calorie content of food. It is evident from this language that the conception of energy and heat is substance-like or fluid-like. Research on young children and middle school students confirms that it is a common belief that matter contains heat and that heat flows in and out of matter (Erickson 1979, 1980; Albert 1978; Tiberghien 1980; Lewis and Linn 1994). High school students and adults often employ similar thinking (Clough and Driver 1985; Bodner 1991; Lewis and Linn 1994; Jasien and Oberem 2002; Laburu and Niaz 2002; Chang 2009).

The interviews of students studying energy topics in chemistry revealed that even the highest performing students in the class speak of heat as if it were substance-like. In the following excerpt, one student named Warren (not his real name) implies that energy is in bonds:

> *... You have to look at the individual bonds, it's how much ... Like that's the bond energy in each bond. I'm not sure how to answer anymore than (mumbles)*

Warren further implies that heat is contained in his hand when explaining why evaporation of water from the skin results in a cooling sensation:

> *... Since heat is directly proportional to temperature using the equation* $q = mc\Delta T$, *um ... when the amount of heat energy in your skin decreases then so will the temperature. So, that means once heat leaves your skin, and transfers to the water to make it evaporate, then um, the temperature is going to go down, meaning you feel cooler.*

He also talks about heat as if it were a thing that can be added to substance to make a change occur:

> *Evaporation is the um phase change from liquid to gas. And the heat is just the, is the. Heat is ... like when heat is added to a substance it makes the molecules move faster and more randomly.*

The conception of heat as contained in objects is both intuitive and part of everyday language. It takes great effort to use the scientific language describing the process of *heat transfer to and from* substances rather than speak about adding and removing heat. Moreover, the heat-as-substance metaphor is helpful at times, for example when considering tactile sensations; the statement that "heat is added to a substance" is useful to Warren as he considers the physical change of evaporating water. However, it can be a deterrent for developing an understanding of bond energy and chemical change, as when Warren implies that energy is in a bond.

On the basis of data gathered during our study, we suggest that the goal of reformulating instruction of heat and energy in a high school chemistry class should not be to eradicate all thinking of heat as being substance-like. It is simply not possible as the metaphor is much too entrenched in the language used to describe heat transfer.

Instead, we propose that instruction should acknowledge that there are two models of heat, and assist students in linking the substance-oriented model (which is more intuitive) with the process-oriented model (which is more scientific). In the substance-oriented model, heat flows from an object at higher temperature to an object at lower temperature to reach thermal equilibrium. In the process-oriented model, heat transfer describes a kinetic-molecular view of energy transfer by the collisions of particles in random motion; the transfer of kinetic energy results when a collection of faster moving molecules collides with a collection of slower moving molecules such that both collections reach the same average speed at thermal equilibrium. Instruction might begin productively with physical changes such as raising the temperature or melting a substance. Students can consider both a substance-oriented model and a process-oriented model to explain their observations.

16.3.2 Bond Energy: Making and Breaking Bonds

One of the most important ideas in chemistry is the concept of bond formation. While the heat-as-substance metaphor provides an entry into thinking about physical change and heat transfer in a temperature gradient, it is a deterrent for developing an understanding of bond energy and chemical change. Students associate heat transfer with fires and digestion of food. In both cases, substances are broken apart giving the impression that breaking things apart releases energy. This is complicated by the fact that the products of these reactions are often gases that are not visible. Thus, rather than understanding that bond energy is the energy change associated with making or breaking a bond, students believe it describes the amount of energy contained in bonds. This can lead to confusing ideas about the energy changes associated with chemical reactions.

During the interviews, a student named Cassie was asked to explain the following observation:

Methane (CH_4) is the gas used in stoves and Bunsen burners. The following equation shows the burning of methane.

$$CH_4\ (g) + 2O_2\ (g) \rightarrow CO_2\ (g) + 2H_2O\ (g)$$

Why is heat transferred when methane burns?

Cassie argues that the "simpler" bonds in the product molecules contain less energy than the bonds of the reactants molecules (heat as substance-like). This leads her to the conclusion that when the atoms rearrange to form the bonds in the product molecules, energy is released by the reaction because the products contain less energy than the reactants.

Cause you're forming.... Like you're not just doing like a single replacement, like you're just... Like switching one or two. You're completely changing the whole thing, cause you're breaking every bond in the reactants and completely just match it up together in the products. So... I guess like making the, the simpler bonds in the products will release energy cause then the products, the bonds in the products hold in less energy than the bonds in the reactants.

The notion that bonds contain energy limits Cassie's understanding of how an exothermic reaction can release energy. From her perspective, breaking the bonds of the reactants, which contain lots of energy in her view, results in a release of energy since the atoms rearrange themselves into simpler bonds, which contain less energy. She completely misses the notion that energy is required to break bonds and energy is released when bonds form. Since bonding is one of the fundamental ideas in chemistry, the notion that leads students to conclude exactly the opposite of the normative view that energy is *required* to break bonds does limit understanding of molecules and chemical change.

On the basis of data gathered during our study, we suggest that the goal of reformulating instruction of bond energy and chemical change in a high school chemistry class is to encourage students to reevaluate the heat-as-substance model (a static description) and begin to consider the heat-as-process model (a description of change). If students can understand that energy describes a process, then it may become natural to think of bond energy as describing a process of breaking or making bonds as opposed to the final state of the bonds that are made. In other words, the energy experienced as transfer of heat during a chemical reaction is a result of the change.

We suggest that a key aspect of a reformulation of instruction on chemical change is to consider changes in kinetic and potential energy as underlying models for analyzing what happens in a chemical reaction. It is quite confusing to students if everything is referred to as simply energy. Moreover, the terminology needs to be made explicit with visualizations that make the bonding between atoms (potential energy), and the motion of molecules (kinetic energy) evident. The balanced chemical reaction that is typically offered to students in textbooks is insufficient, as students do not necessarily derive much meaning from such representations.

As an example of how to make these aspects of a chemical reaction more explicit, consider the images below representing a few molecules that participate in the reaction between hydrogen, H_2, and oxygen, O_2. The square box on the left defines the *system,* the collection of molecules being examined, before the reaction. The H_2 and O_2 molecules are moving randomly in the container as indicated by arrows; hence they have a positive kinetic energy. Nothing happens because energy is required to break the bonds. This observation that H_2 and O_2 co-exist in a container indicates that the pairs of H–H atoms and O=O atoms are strongly attracted to one another; they have a negative potential energy since attractive interactions are defined in this manner.

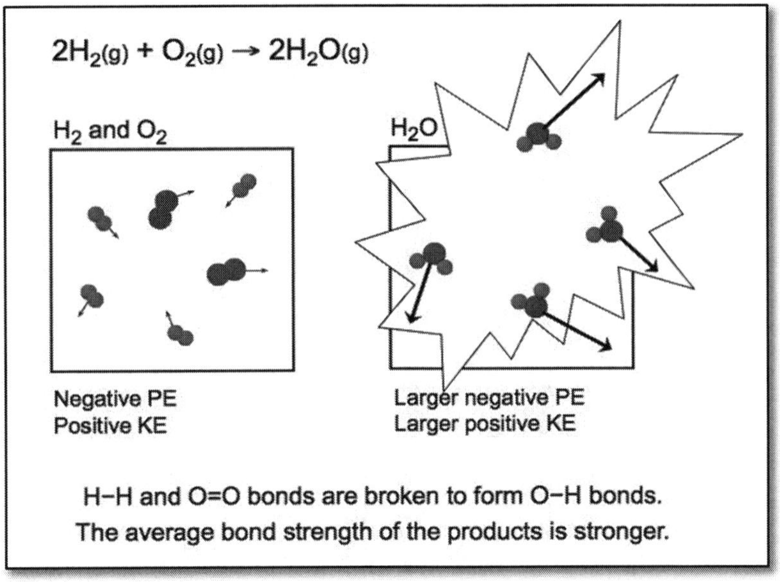

A spark is required to initiate the reaction. For example, a spark can cause H–H bonds to break. The H atoms produced are reactive, and cause a chain reaction as each makes a new bond. In a stepwise process, the atoms bonded as H–H and O=O in the reactant molecules rearrange to H_2O molecules. The temperature of the products is very high indicating that the H_2O molecules have a high average kinetic energy. The law of conservation of energy (total energy remains constant) suggests that if the kinetic energy increases, the potential energy of the water molecules is more negative than the potential energy of the reactants. Molecules at lower potential energy are more stable and have stronger bonds on average.

If this system is placed in contact with the surroundings, then thermal equilibrium is reached. Since the system is at a higher temperature due to the high kinetic energy of the water molecules, heat is transferred out of the system by conduction (or radiation): the water molecules with a high average kinetic energy collide with molecules in the surrounding air to raise the kinetic energy of the air molecules, while decreasing the kinetic energy of the water molecules. Ultimately, the atoms in the system have rearranged, thereby transferring energy to the surroundings and lowering the total energy in the system. The water molecules are more stable than the reactant molecules because the bonds are stronger (harder to break). It is the *making* of bonds that cause the transfer of heat.

There is a Molecular Workbench simulation showing the reaction to make fast-moving water molecules from hydrogen and oxygen (Molecular Workbench). There is also a demonstration showing what happens when a small piece of donut is reacted with liquid oxygen. This demonstration emphasizes that it is not the donut that releases the energy (although it is breaking apart), but the *reaction* between the donut and oxygen; the reaction is quite spectacular in the presence of so many oxygen molecules when delivered as a liquid. Matter is conserved as the atoms in the donut and the oxygen molecules rearrange to make the invisible gaseous products, carbon dioxide and water. The product gases are more stable with stronger bonds on average (lower potential energy). The difference in potential energy between the reactants and products is transferred to the surroundings as heat. [Do not try the donut and liquid oxygen reaction as it is quite dangerous.]

16.3.3 Exothermic and Endothermic: Physical Versus Chemical Change

Students also have difficulties with the terms exothermic and endothermic. For those who already understand, the terms may seem rather obvious. *Exothermic* indicates that the direction of heat transfer is *from* the system, while *endothermic* indicates that heat is transferred to the system. The system refers to the substances that are being studied, such as water in a pot that is heated on a stove, or the reaction between methane and oxygen. With a heat-as-substance model, students reason that heat is transferred *to* the system to raise the temperature and heat is transferred *from* the system, or "thrown away" as a student might say, to lower the temperature. While the heat-as-substance model might work for the physical process of raising the temperature of water, application of this model to chemical processes does not account for the observations. Although rarely stated explicitly, there are two parts to the process of burning methane in oxygen. The process produces carbon dioxide and water molecules with a high kinetic energy (high temperature), and then the kinetic energy is transferred to the surroundings as heat to reach thermal equilibrium. The temperature goes up and then it goes down. This is best explained with a process-oriented model of heat.

Students have great difficulty with these concepts as illustrated by Mary, one of the students interviewed. Mary is told that a white powder dissolves in water and the beaker that contains the solution feels hot. She is asked to describe the direction of heat transfer: into the solution, out from the solution, both or neither. Initially she says that heat must be transferred into the skin, which means that heat must be transferred away from the solution. Later on, the interviewer (I) asks Mary (M) to describe what happens to a thermometer when it is placed in the solution. Mary acknowledges she is not sure.

M: *Yeah. So, basically I think that if you're feeling heat that means that heat is being transferred into you, or onto your skin. So that must means it must be transferred away from the solution.*

I: *Okay.*

M: *Does that-?*

I: *Yeah, I understand what you're saying. Um, okay. Now say you put a thermometer into that solution. What's going to happen to the temperature?*

M: *Um, I think... (pause). I think that... it's good, I was thinking about that. Um... (pause). Because you would assume that it would go up cause you're feeling heat, but because heat is also being transferred out into the surroundings, that would mean that it would be cooler because heat's going away from it.*

I: *Wait. Can you explain that again?*

M: *Yeah, yeah. Well, I'm just going through my thinking process, but um. At first, I mean I was kind of like... you feel heat, so you'd assume that it was hot and so the thermometer, the temperature would rise. But also, I'm also thinking that if you're feeling heat, that means that heat is being transferred out of the solution into your hand or into the surroundings so transfer, heat transfer away from the solution would mean that the solution would be decreasing in temperature?*

Initially Mary says that heat is transferred from the solution to the hand on the basis of what she feels. She is correct. However, confusion arises when she is asked to describe what happens to a thermometer when it is placed in the solution. Mary evokes two arguments. Her first argument is based on the notion that a chemical change occurred causing the temperature of the solution to increase. Her second argument is that the temperature of the solution decreases, which is based on the notion that the temperature of substances will decrease to equilibrate with their surroundings; in her view, heat-as-substance is released to the environment.

Mary doesn't realize it, but both explanations are correct in the sense that the temperature increases and then decreases. For an exothermic chemical change, the products formed are at a higher temperature than the surroundings, which is observed as an increase in the temperature of the resulting solution. The products (or the resulting solution) then transfer energy to the surroundings to reach thermal equilibrium, which is observed as a decrease in the temperature of the solution. For exothermic physical changes, such as lowering the temperature of water, only thermal equilibrium occurs. Mary's confusion arises from the fact that she does not realize that thermal equilibration also occurs for chemical changes. But this is not her fault; rarely, if ever, does instruction address this explicitly.

On the basis of data gathered during our study, we suggest that a reformulation of instruction of exothermic and endothermic processes in a high school chemistry class needs to point out the distinction between energy changes in chemical and physical processes. The figure below makes explicit the use of these terms and how they relate to physical and chemical changes. For example, in order to raise the temperature of water to 100 °C and then boil the water, heat is transferred to the water. This physical change results in an *increase* in temperature; the process from the point of view of the system (the water) is *endothermic*. Hydrogen and oxygen react explosively to produce water. In this example, heat is transferred from the system (the hydrogen and oxygen) to the surroundings. This chemical change results in an initial *increase* in the temperature of the system, which then equilibrates with the surroundings; the process is labeled as *exothermic*. Notice that the temperature increases in both cases, but the labels on the two processes are different.

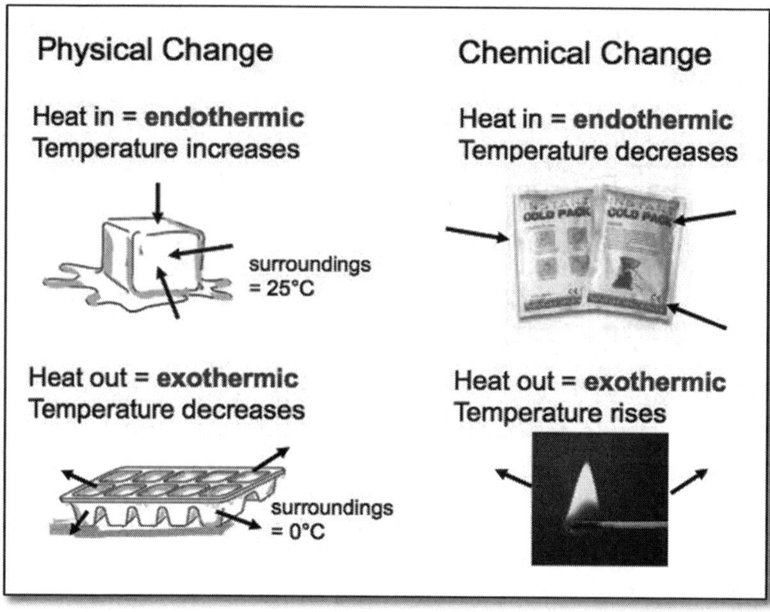

Finally, it is important to recognize that the example of dissolving posed to Mary during the interview needs clarification. In most high school textbooks, dissolving is labeled as a physical change. But is it? Consider the dissolution of ammonium nitrate in water. This endothermic process is used in cold packs for application to injuries because the temperature of the system (the cold pack) decreases. Thus, on the basis of energy considerations, dissolution fits best under chemical change.

16.4 Fire: Chemical Energy and Its Uses

Fire! This is a great theme to use to introduce high school chemistry students to ways of thinking about energy (*Living by Chemistry* 2010a, b), and situate the learning in a familiar context. Students have experience with fire and the heat that is transferred. They are immediately engaged when asked what burns and what does not burn. And when they test materials (small amounts, of course) to try to create flames, they find, much to their surprise, not everything burns.

The Fire Unit in the *Living by Chemistry* curriculum divides the concepts related to energy into four sections: (1) Observing Energy; (2) Measuring Energy; (3) Understanding Energy; and (4) Controlling Energy. The sequence and specific activities address the findings of Chang (2009) regarding student conceptions of energy as summarized above and include the recommended reformulations for instruction. The storyline of the Fire unit is provided below along with a description of the first three sections to show how a coherent understanding of energy might be built within a high school chemistry course.

16.4.1 Observing Energy

Before launching into mathematical and chemical representations, students need time to consider what the word "energy" means and how it is used. The best way to accomplish this is to provide students with opportunities to observe the heat transfer associated with change and discuss their observations. As a first step, students need to consider the difference between matter and energy. They determine patterns in the way the word energy is used in sentences, and ultimately try to provide a definition for the word. Example sentences with the word energy are given below.

- A plant needs energy from the Sun to grow.
- An athlete eats a snack bar for energy to continue running.
- Electrons move through a filament in a light bulb to cause it to glow.
- Pressure from steam provides the energy to move a locomotive.
- Water falling turns a water wheel.

Matter moves, falls, glows, melts, breaks apart, or burns. Generally, energy is seen as an ability to do work, as effecting change, and as a general abstract accounting quantity associated with change (Duit 2014).

In the first section of the Fire Unit, students explore systematically energy exchanges. They measure temperature changes when salts dissolve in water in order to distinguish between exothermic and endothermic processes. Then they examine heat transfer scenarios and experiences (such as alcohol evaporating) in order to consider that what they refer to as hot or cold is an exchange of energy in both cases, but in opposite directions.

After experiences with energy exchanges and considering the point of view of the system and surroundings, students are poised to consider the relationship between heat transfer and the temperature change of a substance. Rather than introduce the equation $q = mC_p\Delta T$ all at once, students are afforded the opportunity to do experiments in which the substances (the specific heat capacity) or the mass are held constant. They mix two samples of known quantities of water, each at a different temperature, and determine the final temperature. The data allow for conclusions regarding mass and thermal equilibrium. Next they put equal masses of two different substances at two different temperatures together, and determine the final temperature. It is a surprise to the students that when a hot metal is placed in cold water, the temperature of the water barely changes. This leads to a discussion of heat capacity and a kinetic view of heat transfer. Finally students explore the heat involved in phase changes such as melting ice and boiling water.

The concept of heat capacity is especially challenging as even the name implies an ability of the substance to hold heat. The kinetic model of heat as random motion of the atoms in a substance is introduced, and then used extensively to explain heat transfer and why one substance might require more heat in order to raise the temperature, i.e., getting the atoms in the substance to move faster. Students consider heating water in a metal pot on the stove. The metal feels hot while the water is still near room temperature. It is tempting to conclude that the metal heats up faster. Students are encourage to consider what it means that the amount of heat needed to raise the temperature of the metal is much smaller compared with that required to raise the temperature of water.

By the end of the first section of the Fire Unit, students have explored how heat transfer causes temperature changes associated with physical change, and they have discussed the difference between heat and temperature and the concept of heat capacity.

16.4.2 Measuring Energy

The second section of the Fire Unit in the *Living by Chemistry* curriculum offers students an opportunity to measure the amount of energy transferred when a piece of snack food (e.g., cheese puff, corn chip) burns. In an open inquiry exercise, students are provided with a tray of materials and asked to design their own experiment. This lesson is not about immediately doing things correctly. Rather, students need time to consider what it means to measure energy transfer. Several groups will ultimately design some variation of a setup with the burning cheese puff placed under a container of water. Other groups will attempt to measure the temperature of the flame, and others believe that food that burns the longest has the most energy. Some groups even place the cheese puff in water and boil the water. Groups might not weigh their cheese puff, or they might weigh the cheese puff at the beginning and not at the end of combustion. All of these actions promote engaging discussion. Students are then given an opportunity to refine their design, and ultimately they do a standard calorimetry experiment by burning alcohol below a known quantity of water.

Observations of students in the classroom suggest that this is a rich experiment to promote a conceptual understanding of energy (Chang 2009). Students learn that the unit for measuring energy is the calorie, which is the energy needed to raise the temperature of 1 g water by 1 °C. (The calorie is used to relate to food calories. Students convert to joules later.) This experiment makes it evident that energy is transferred from the burning of the food to heat a known quantity of water in order to determine the number of calories reported on the label of a snack food bag. Students learn that the calories refer to the heat transfer when the food is burned in air. Here again, the language of calorie content of the food is hard to avoid even though the students experience directly the transfer of energy to the water due to the reaction of the food with oxygen.

16.4.3 Understanding Energy

The next section of the Fire unit explores what is happening when the cheese puff burns. Since the cheese puff disappears in front of our eyes with only a small amount of black ash remaining, it is tempting for the student to conclude that the bonds break and energy is released. However, what they cannot see are the products that form (carbon dioxide and water) because they are gases. It is important to remind students that matter is conserved. The two product gases are very stable molecules with relatively strong bonds. The gases can be collected to show that mass is conserved, and they can be sensed as heat when they collide with our skin and transfer kinetic energy.

The third section of the Fire Unit in the *Living by Chemistry* curriculum explores bond breaking and bond making in order to understand the energy released from the burning cheese puff. Students consider the energy required to break the bonds in the reactants and the energy released when the bonds in the product form. For example, hydrogen, H_2, and oxygen, O_2, react to form water, H_2O. There is a net release in energy because the energy required to break the bonds of the reactants is less than the energy released when the bonds in the product form. Notice that the release in energy is due to the rearrangement of the atoms, and not to "high energy bonds" in the reactant molecules as is sometimes claimed in textbooks.

This analysis of bond breaking and bond making leads students directly into a core idea in our energy future: once energy is released from the reaction to form water, the water does not convert readily back to hydrogen and oxygen. Water is a more stable molecule that can only be changed with input of energy. Likewise, it is not possible to take the carbon dioxide and water and reform the cheese puff. Thus, when substances are burned, they are no longer available to us. The carbon dioxide and water will remain until there is an input of energy. Fuels get "used up".

Students are now positioned to consider the original source of the fuels. These are carbon-containing molecules produced by plants. Plants take in carbon dioxide and water to make fuels and food (fuel for our body) and oxygen. The reaction does not occur without input of energy. This is why plants need energy from the sun to grow. When we eat food and breathe in oxygen, we reverse this reaction, and breathe out the products carbon dioxide and water. The energy released by this process in our bodies, literally powers our body.

At the end of the third section in the Fire Unit, students are asked to make the link between combustion reactions and the energy required to move an object. The heat transferred from the combustion of fuel in an automobile engine causes the gaseous products to expand and do work to push the piston of the engine cylinder up. When coupled to a drive shaft, the car moves along the road. Energy is about the conversion of heat transferred from chemical reactions to the work needed to move an object. The source of the energy is a chemical reaction that results in the making of more stable bonds.

16.5 Learning in Context: The Fire Theme

The narrative describing the Fire Unit in the *Living by Chemistry* curriculum is meant to provide one example of a systematic path through lessons on energy that support students in developing a more coherent understanding of energy and the importance of energy considerations to our energy future. The specific activities in the curriculum draw on research regarding student conceptions of energy, and employ a student-centered pedagogy that offers a constructivist approach to support conceptual change. Important aspects of the lessons are that they draw on student experiences, provide opportunities for systematic observations with simple materials, measure and define concepts, and challenge students to apply their understanding to explain their experiences and to consider the energy challenges to our society. This effort to situate the learning in the simple but powerful context of fire offers students a rich experience with the topic of energy.

Acknowledgments This material is based upon work supported by the National Science Foundation under Grant No. 0352530.

References

Albert, E. (1978). Development of the concept of heat in children. *Science Education, 62*, 389–399.

Bodner, G. M. (1991). I have found you an argument. *Journal of Chemical Education, 68*(5), 385–388.

Boo, H. K. (1998). Students' understanding of chemical bonds and the energetics of chemical reactions. *Journal of Research in Science Teaching, 35*, 569–581.

Chang, K. (2009). *Learning thermochemistry: Understanding the challenges and promoting its success*. (Doctoral dissertation). Retrieved from ProQuest Dissertation and Theses Database.

Clough, E. E., & Driver, R. (1985). Secondary students' conceptions of the conduction of heat: Bringing together scientific and personal views. *Physics Education, 20*, 176–182.

Driver, R., Squires, A., Rushworth, P., & Wood-Robinson, V. (1994). *Making sense of secondary science: Research into children's ideas*. London/New York: Routledge.

Duit, R. (1987). Should energy be illustrated as something quasi-material? *International Journal of Science Education, 9*, 139–145.

Duit, R. (2014). Teaching and learning the physics energy concept. In R. Chen & A. Eisenkraft (Eds.), *Teaching and learning of energy in K-12 education* (pp. 67–85). Dordrecht: Springer.

Erickson, G. L. (1979). Children's conceptions of heat and temperature. *Science Education, 63*(2), 221–230.

Erickson, G. L. (1980). Children's viewpoints on heat: A second look. *Science Education, 64*, 323–336.

Hammer, D., Goldberg, F., & Fargason, S. (2012). Responsive teaching and the beginnings of energy in a third grade classroom. *Review of Science, Mathematics and ICT Education, 6*(1), 51–72.

Jasien, P. G., & Oberem, G. E. (2002). Understanding of elementary concepts in heat and temperature among college students and K-12 teachers. *Journal of Chemical Education, 79*(7), 889–895.

Laburu, C. E., & Niaz, M. (2002). A Lakatosian framework to analyze situations of cognitive conflict and controversy in students' understanding of heat energy and temperature. *Journal of Science Education and Technology, 11*(3), 211–219.

Lewis, E. L., & Linn, M. C. (1994). Heat energy and temperature concepts of adolescents, adults and experts: Implications for curricular improvements. *Journal of Research in Science Teaching, 31*(6), 657–677.

Loverude, M. E., Kautz, C. H., & Heron, P. (2001). Student understanding of the first law of thermodynamics: Relating work to the adiabatic compression of an ideal gas. *American Journal of Physics, 70*(2), 137–148.

National Research Council. (1996). *National Science Education Standards. National Committee for Science Education Standards and Assessment.* Washington, DC: National Academy Press.

National Research Council. (2011). *A framework for K-12 science education: Practices, cross-cutting concepts, and core ideas. National Committee for Science Education Standards and Assessment.* Washington, DC: National Academy Press.

Nordine, J., Krajcik, J., & Fortus, D. (2010). Transforming energy instruction in middle school to support integrated understanding and future learning. *Science Education, 95*(4), 670–699.

Stacy, A. M., Coonrod, J., & Claesgens, J. (2010a). High school chemistry student textbook. In L. Malek & J. Dowling (Eds.), *Living by chemistry* (1st ed.). New York: Bedford, Freeman & Worth.

Stacy, A. M., Coonrod, J., & Claesgens, J. (2010b). Teachers guide, unit 5: Fire. In L. Malek & J. Dowling (Eds.), *Living by chemistry* (1st ed.). New York: Bedford, Freeman & Worth.

Teichert, M., & Stacy, A. (2002). Promoting understanding of chemical bonding and spontaneity through student explanation and integration of ideas. *Journal of Research in Science Teaching, 39*(6), 464–496.

Tiberghien, A. (1980). Modes and conditions of learning an example: The learning of some aspects of the concept of heat. In W. F. Archenhold (Ed.), *Cognitive development research in science and mathematics* (pp. 288–309). Leeds: University of Leeds Printing Service.

Chapter 17
Energy in Chemical Systems: An Integrated Approach

Melanie M. Cooper, Michael W. Klymkowsky, and Nicole M. Becker

17.1 Introduction

Students are often told that chemistry is "the study of matter and the changes that it undergoes" (Chang and Goldsby 2012). What is less often emphasized is that understanding chemistry depends upon an understanding of the central role of energy in chemical systems. From the structure of individual atoms, to the folding of complex bio-molecules; from the simplest reactions, to the cascades of coupled reactions that have enabled living systems to remain organized and fight the tendency to disorder, understanding energy and energy changes are key. Unfortunately the central role of energy in the chemistry curriculum is often not made explicit, particularly in introductory college-level courses such as general and organic chemistry.

Based on a review of the literature related to students' understanding of energy ideas in chemistry contexts and on our experiences with introductory chemistry courses, we suggest that in introductory level college chemistry courses the concept of energy is often introduced from three different perspectives: the macroscopic; the atomic-molecular; and the quantum-mechanical perspectives. We discuss these three perspectives and the ways they may (or may not) be connected within the curriculum in the following Sections.

M.M. Cooper (✉) • N.M. Becker
Department of Chemistry, Michigan State University, East Lansing, MI, USA
e-mail: mmc@msu.edu

M.W. Klymkowsky
Department of Molecular, Cellular and Developmental Biology, University of Colorado, Boulder, CO, USA

R.F. Chen et al. (eds.), *Teaching and Learning of Energy in K-12 Education*,
DOI 10.1007/978-3-319-05017-1_17,

17.1.1 The Macroscopic Perspective

Although temperature change is a physical manifestation of the energy changes that take place on the atomic-molecular level, most instructional approaches to college-level chemistry do little to emphasize these origins: they do not explicitly connect the macroscopic (temperature) to the microscopic and molecular. For example, students might discuss energy changes of endothermic or exothermic reactions that are reflected by observable changes in temperature before ideas of molecular structure are introduced. Such energy changes are typically organized under the general headings of "Thermochemistry" or "Thermodynamics". Thermochemistry is concerned with the energy changes that take place when a macroscopic chemical system undergoes change, and usually these changes are observable as changes in temperature. For most students, learning thermodynamics topics in a general chemistry course begins with calculations using specific heats and temperatures, which are then related to enthalpy (H). Later (usually in the second semester general chemistry course), other thermodynamic functions such as entropy (S) and Gibbs energy (G) are also introduced, and while more abstract than observable changes in temperature changes, these functions also are related to changes in energy of macroscopic systems.

Change in Gibbs energy, for example, is a thermodynamic state function that is often represented by the equation $\Delta G = \Delta H - T\Delta S$ (where ΔH is the enthalpy change of a reaction, T is the temperature, and ΔS is the entropy of the system). This function is important in that is serves a proxy for the Second Law of Thermodynamics, allowing one to make predictions about the spontaneity of a process using only variables related to the system in question. Unlike enthalpy and internal energy, Gibbs energy is not conserved (because it includes an entropic term). Gibbs energy is one of the most important and useful thermodynamic functions because it allows predictions to be made about the direction of change. Biologists, for instance, might use the Gibbs function to determine the direction of change in biological systems and to understand how coupled reactions can drive thermodynamically unfavorable processes.

However, there is evidence that traditional approaches to teaching thermodynamics topics may leave students with fragmented or even incorrect understandings about what thermodynamic variables represent and how they relate to macroscopic changes in energy that correspond to changes in chemical systems. For instance, prior literature related to students' understanding of Gibbs energy suggests that students may not develop an understanding of Gibbs energy as a proxy for the Second Law of Thermodynamics. Instead, students may conflate change in Gibbs energy for a process, ΔG, with the amount of heat transferred in or out of a system (Thomas and Schwenz 1998) or they may believe that the magnitude of ΔG can be used to determine the rate of reaction (Sozbilir 2002). Factors such as entropy (S), which is often discussed in conjunction with energy changes in macroscopic systems, can be further sources of confusion. While entropy is best understood as related to the number of states or arrangements possible within a system (Lambert 2002), students may conflate entropy and energy (Carson and Watson 2002), believing that entropy represents a form of energy.

We believe that such difficulties may be related to the fact thermodynamic treatments of energy in introductory chemistry coursework seldom build on students' prior knowledge, but rather introduce a new set of ideas that are not linked to other knowledge and may appear to the student to be introduced solely for the purpose of doing calculations rather than as a foundation upon which to predict molecular behaviors.

17.1.2 Representational Difficulties Related to Macroscopic Perspectives on Energy

In addition to the fragmented nature of energy topics within chemistry curricula, the representational tools used to communicate energy ideas may contribute to student difficulties with energy topics. As Taber (2013) noted, "Learning chemistry involves both forming concepts that are sufficiently aligned with those of other members of the chemical community, *and* adopting the systems of symbols commonly used within the chemical community so as to be able to communicate with others about these concepts" (p. 4). Thus, students must develop fluency with representational resources commonly encountered in introductory chemistry classes, such as mathematical expressions and graphical representations. Mathematical representations are especially important; the use of mathematical resources to model and represent systems is a key scientific practice that has the potential to facilitate students' understanding of energy transfer and conservation in more complex systems (National Research Council 2012).

However, there is abundant evidence that students approach mathematical representations of thermodynamic functions, such as enthalpy or Gibbs Energy, in an algorithmic fashion and that even advanced chemistry students can fail to grasp what these mathematical representations of functions represent. For instance, a practicing chemist might be able to examine an equation such as $U = q + w$ and interpret it a representation of idea that the total energy of a system is the combination of heat and work done on the system, and as a representation of the conservation of energy. Students, however, may struggle to relate variables such as heat and work to the real-world phenomena or atomic-molecular level understandings of the system. For example, Hadfield and Wieman (2010) found that students enrolled in an upper-division physical chemistry course did not consider the expression $U = q + w$ to relate to the conservation of energy.

These difficulties may derive in part from the fact that students' everyday interpretations of energy-related terminology (for example "heat" or "work", both of which are commonly used in everyday speech) may be quite inconsistent with the way those terms are appropriately used in thermodynamic contexts (Jin and Anderson 2012; Kaper and Goedhart 2002; Lancor 2012). Thus, to successfully interpret expressions such as $U = q + w$, students must not only be able to interpret the mathematical expression, but hold understandings of terms like "energy", "heat", and "work" that are consistent with disciplinary understandings of those terms (Jewett 2008). This is particularly problematic since if students do not develop an appreciation of the concepts underlying thermodynamic functions, we feel it becomes nearly impossible for students to appreciate energy as a tool that they may use to predict and explain the outcomes of chemical processes.

17.1.3 The Atomic-Molecular Perspective

Energy concepts are critical to understanding how molecules form and behave. It is at the atomic-molecular level where the origins of the observable manifestations of energy change can be observed. As such, a second perspective on energy in the introductory chemistry curriculum often relates to energy at the atomic-molecular level in the context of the structure and interactions of matter.

Energy associated with bonding and intermolecular interactions are foundational parts of chemistry in that they enable predictions of molecular properties and energy transformations at the macroscopic level. It is possible to explain most of the properties and interactions of matter, ranging from the sizes of atoms to their interactions along the spectrum from London Dispersion Forces to covalent bonding, in terms of kinetic and potential energy. To understand bonding at a conceptual level in terms of energy, students must recognize that such interactions are based on attractive and repulsive forces, and that a stable interaction is formed when there is a balance between these forces, an "energy minimum" (Nahum et al. 2007). However, developing the ability to reason about energy at the atomic-molecular scale is not without difficulties. In reasoning about bond formation and stability, students may rely on heuristics such as the octet rule, rather than an understanding of how electrostatic forces contribute to bond formation (Taber 1998). Students may view ionic bonds, covalent bonds, and intermolecular interactions as distinct entities rather than understanding that all involve electrostatic interactions and energy minimization (Taber 1998). Furthermore, misconceptions related to the energetics of bonding interactions, such as the idea that bonds "store" energy and that energy is released when a covalent bond is broken are persistent sources of confusion. A number of researchers have found that even after instruction, typically over 50 % of students incorrectly believe that bonds release energy when they are broken (Barker and Millar 2000; Boo 1998).

While the construct of potential energy is often referenced when discussing intermolecular forces and bonding, chemistry curricula rarely address how electrostatic potential energy at the molecular level, which arises from electrostatic interactions, relates to the more familiar concepts such as gravitational potential energy. Unfortunately at present most students arrive at college having been exposed to kinetic and potential energy in macroscopic systems (for example a ball rolling down a hill), but in our experience, they have little understanding of how these ideas might translate to the molecular level. It is our hope that as students experience instruction that builds upon the NRC Framework for Science Education (NRC 2012) and the Next Generation Science Standards (NGSS) (Achieve 2013), they will develop a more coherent framework upon which to build some of these ideas. The NGSS emphasize the idea that energy is best treated in an interdisciplinary manner, with explicit connections between the macroscopic and the atomic molecular level. For instance, one connection that may be made is that electrostatic potential energy can be considered analogous to gravitational potential energy in that both depend on the distance between two interacting objects, both involve forces that mediate interactions between objects, and the equations relating the energy of both interactions take very similar forms. If students understand these similarities, they may be better equipped to differentiate gravitational potential energy from electrostatic potential energy in atomic-molecular systems. For example, electrostatic potential energy in the context of molecular systems differs from gravitational in that it is a far stronger interaction at the molecular level and there are two types of charges, meaning that there can be both attractive and repulsive forces, while only attractive forces are present within a gravitational field. Introductory chemistry texts often do not explicitly acknowledge the similarities, differences and difficulties in translating across scales from the macroscopic to the molecular and students are largely left to infer these for themselves.

We consider the origin of potential and kinetic energy changes at the atomic-molecular level critical to understanding the basis of thermodynamic ideas that are in common use. If students do not know how energy is transferred and stored at the atomic-molecular level, they will likely find it difficult to understand, for example, the origin of "chemical energy" – how or why chemical reactions can be used as a source of energy (from food to batteries). To this end, we must do more to ensure that students develop a robust understanding of core energy concepts (such as potential and kinetic energy) at the molecular level and to reinforce appropriate interpretations of energy as related to both macroscopic and atomic-molecular scales.

17.1.4 The Quantum-Mechanical Perspective

The third focus of energy instruction in university-level general chemistry courses centers on the idea that energy is quantized at the atomic-molecular scale. For most students (and for most people!) this idea is entirely counterintuitive as it has no counterpart in the macroscopic, observable world. Energy quantization is often taught in introductory chemistry courses primarily in connection with topics related to atomic structure. However, connections between quantization of energy from a quantum-mechanical perspective and atomic-molecular or macroscopic phenomena are seldom explicit. Rather than asking students to use the idea of energy quantization to explain phenomena such as why carbon, the building block of life, forms four bonds and not six, or why materials emit or absorb electromagnetic radiation of particular wavelengths, we typically emphasize more easily assessable ideas such as the recitation of electron configurations.

Again, there are a number of reports in the literature about students' understandings of the concepts of quantum chemistry (Taber 2002, 2004; Tsaparlis and Papaphotis 2009). Despite instruction, students may fail to grasp the relationship between energy quantization and orbital ideas in general chemistry contexts (Park and Light 2009; Taber 2002). Park and Light (2009) described the quantization of energy and the uncertainty principle as "threshold concepts". Clearly, if we want students to be able to cross this "threshold" and to relate energy ideas from the quantum-mechanical perspective to molecular-level structure as well as to macroscopic phenomena, more explicit attention is needed towards helping students connect energy ideas across the chemistry curriculum.

In summary, introductory college chemistry courses typically "cover" energy ideas from three perspectives: macroscopic, atomic-molecular, and quantum mechanical. However, in the context of most traditional courses, we feel such coverage may be fragmentary, not connected to students' earlier knowledge, and typically not set in a meaningful context.

17.2 How Should General Chemistry Students Learn About Energy Within an Introductory Chemistry Context?

It is clear from our analysis of traditional approaches to teaching energy concepts within the chemistry curriculum that the three dominant perspectives on energy, macroscopic, atomic-molecular, and quantum-mechanical, are rarely well-integrated. Indeed, there is ample evidence that students lack a coherent framework of energy concepts on which they can hang their understanding of energy changes associated with chemical change. While multiple perspectives on energy clearly have their place within the curricula, more must be done to help students connect energy ideas across the curriculum.

The problems inherent in traditional approaches to energy instruction are exacerbated by the fact that most assessments in traditional chemistry courses still emphasize rote problem solving and factual recall over conceptual understanding, and there may be little opportunity for students to synthesize and connect energy ideas across the curriculum. Many of the leading textbooks for general chemistry introduce energy topics in different orders (there are even editions of the same text with the topics juggled), so it is clear that there is no consensus on how to develop and connect energy concepts or even why they are important.

We believe that energy ideas must be developed in a scaffolded progression, which helps students to make sense of energy phenomena across macroscopic, atomic-molecular, and quantum mechanical levels. Our approach to designing an energy learning progression for general chemistry aims to reconcile these different perspectives and explicitly recognize places where energy is best treated by one or more of the perspectives. Our goal is to help students develop an integrated understanding of energy concepts that can help them make connections between the three perspectives discussed here.

17.3 The CLUE Approach to Energy in Chemical Systems

Our current work in this direction centers on developing, teaching and assessing the outcomes for a new general chemistry curriculum: *Chemistry, Life, the Universe and Everything* (NSF DUE # 0816692).[1] This approach is organized around three interconnected learning progressions for core ideas: structure, properties and energy, which are represented in Fig. 17.1 by three intertwined, interconnected strands. As each core idea is developed, its connections to the other core ideas are also emphasized. Our intent is to develop and connect structure, properties and energy throughout the course, rather than treating them separately. Figure 17.1 is intended to show how these strands are both intertwined and interconnected. The curriculum is structured so as to progress from simple systems, such as the atomic-level interactions of atoms and molecules, to more complex systems such as the networked reactions that drive thermodynamically unfavorable processes. At each stage in the curriculum, the three perspectives on energy are coordinated (Fig. 17.2) in order to give students access to a comprehensive view of the role of energy in chemical systems.

We have previously reported on assessments of learning outcomes for the structure-properties learning progression within the CLUE curriculum (Cooper et al.

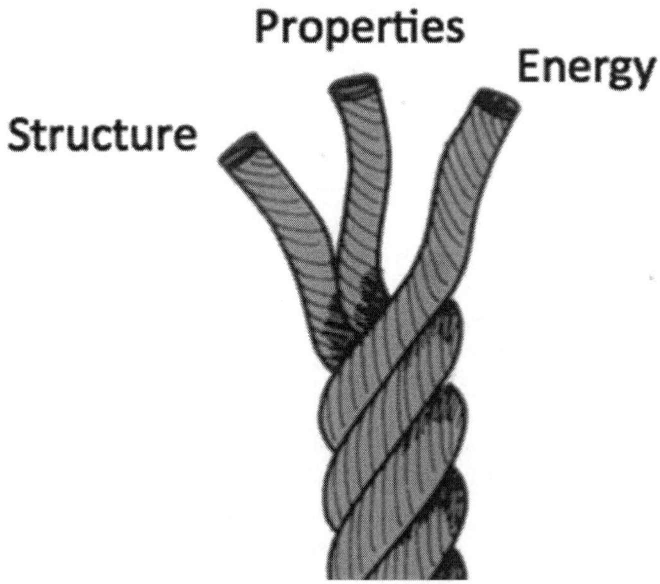

Fig. 17.1 The interconnected learning progressions of structure, properties and energy

[1]Selected course materials for the course available online at http://besocratic.colorado.edu/CLUE-Chemistry/index.html

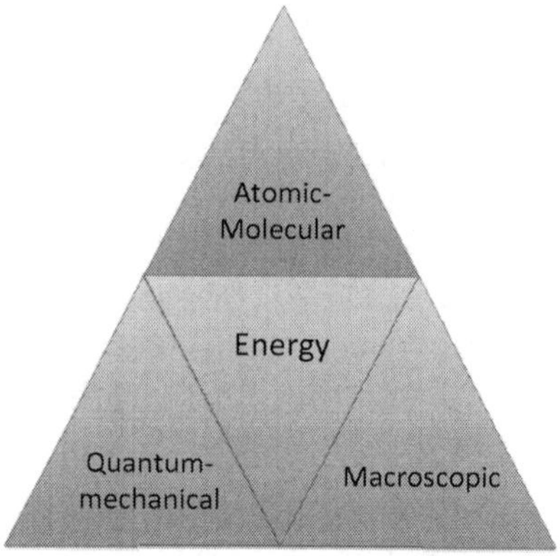

Fig. 17.2 Three perspectives on energy in the CLUE chemistry course

2012). Here, we provide a description of the design of the progression and the connections among ideas that support the development of a more integrated and robust conceptual framework for energy as it is needed to understand chemical systems. Research on student learning outcomes for the energy learning progression is underway and will be reported elsewhere.

While the initial development of the curricular materials pre-dated the release of the NRC Framework for STEM education (NRC 2012), the approach we describe here closely parallels many of the disciplinary core ideas related to energy in the physical sciences in the Framework. For example the Framework recommends a more coherent approach to the teaching of energy:

> Energy is best understood at the microscopic scale, at which it can be modeled as either motions of particles or as stored in force fields (electric, magnetic, gravitational) that mediate interactions between particles (p. 121).

The Framework also states that:

> The idea that there are different forms of energy, such as thermal energy, mechanical energy, and chemical energy, is misleading, as it implies that the nature of the energy in each of these manifestations is distinct when in fact they all are ultimately, at the atomic scale, some mixture of kinetic energy, stored energy, and radiation (p. 122).

While to date there is little empirical evidence that suggests that learning about molecular-level energy changes prior to learning about macroscopic energy changes in chemical systems is beneficial for students, we believe that this approach has the potential to addresses major impediments to student learning and to meet students where they are in terms of prior knowledge. Students who enroll in college to study STEM fields have already learned about the existence of atoms in K-12 coursework and it is highly likely that they have also been exposed to ideas about energy and energy changes at the macroscopic level. Ideally, students would come to college with a firm grasp of macroscopic energy ideas, including the relationships between different "kinds" of potential energy (gravitational, electrostatic, magnetic). Unfortunately at the moment this does not appear to be the case, and we must address energy in ways that are appropriate for understanding of foundational chemistry principles. While students' prior knowledge related to energy ideas at the atomic-molecular level may often be fragmented and incomplete, we believe that beginning with a discussion of energy at the atomic-molecular level with explicit connections to their prior understanding of energy ideas at the macroscopic level has the potential to serve as a foundation for better understanding discussions of macroscopic energy changes.

Thus, our current work centers the development of a learning progression that begins with discussion of energy ideas at the atomic level and connects to quantum-mechanical and macroscopic descriptions of energy. In the following sections, we present an overview of this progression and illustrate the ways in which energy ideas are integral within the course structure by using examples from the two-semester general chemistry sequence.

17.3.1 Chapter 1: Atoms

The CLUE approach to energy begins with an introduction to atomic structure and interactions in their simplest form. That is, quantization of energy levels is not the starting point of this introduction; rather we aim to link to students' prior knowledge about energy and interactions at the macroscopic-level by appealing to students' understanding of gravitational force and energy minimization at the macroscopic level. We begin by asking students what they know about energy both in the macroscopic "real world" and at the molecular level, and through class discussions begin to connect the two perspectives. We emphasize both the similarities between the gravitational forces and energy changes they have learned previously (for example a roller coaster ride), and the electromagnetic force that causes attractions and repulsions at the atomic-molecular level.

Throughout the CLUE curriculum, we use a number of activities in order to help students engage with the material. For instance the web-based simulation[2] shown in Fig. 17.3 uses the interactions of helium atoms as a simple model for reasoning about the potential energy changes that result from atomic-molecular interactions. The screenshots in Fig. 17.3 show a plot of potential energy versus distance between the nuclei and two bars representing the kinetic and the total energy of the system. As students interact with the simulation, for example by changing the distance between the helium atoms, they are prompted by the simulation to consider various energy changes that result from the changes they observe. The idea of energy conservation is introduced by drawing students' attention to the fact that total energy of the two-atom system remains constant as the atoms interact even though as the fluctuating dipoles in the electron clouds attract one another, the potential energy of the system decreases and the kinetic energy increases.

[2] Available online at http://besocratic.colorado.edu/CLUE-Chemistry/activities/LondonDispersionForce/1.2-interactions-1.html

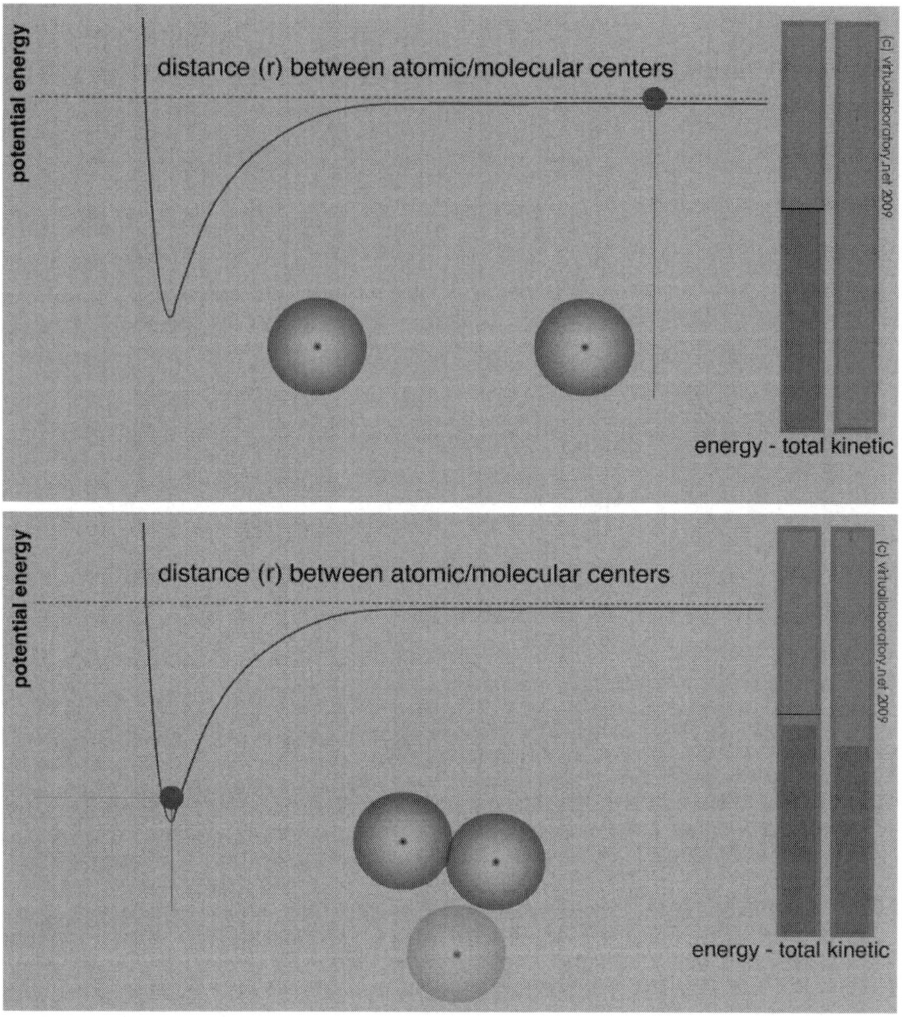

Fig. 17.3 Screenshots of the interactive simulations designed to provide students with atomic-molecular level visualizations of potential and kinetic energy and energy transfer. In the second screen shot, energy has been transferred to the "third" atom, allowing the initial pair to stabilize

The activity also introduces the idea of energy minimization in chemical systems. That is, a system will adopt the lowest potential energy configuration unless another source of energy is added. In this case, the lowest energy configuration occurs as the helium atoms interact through London Dispersion interactions. A third atom can be introduced that can either accept energy and stabilize the interaction or transfer energy to the two atom system.

We conclude Chapter 1 with a comparison of the energy changes associated with the formation of different types of intermolecular interactions. For example, we compare the energies associated with interactions between helium atoms caused by London Dispersion Forces (LDFs) and contrast these with the energy changes that occur as a covalent bond is formed between two hydrogen atoms. Throughout the CLUE curriculum, bonding and intermolecular forces are treated as a continuum of interactions that minimize potential energy in the system. As suggested by others, this approach may be beneficial since it exposes students to the commonalities in various types of interactions between atoms and molecules (Nahum et al. 2007).

17.3.2 Chapter 2: Electrons and Orbitals

In Chapter 2, we continue our discussion of bonding and intermolecular interactions by introducing the idea of quantization of energy levels at the atomic scale. However, rather than emphasizing memorization of electron configurations and shapes of electron orbitals, energy quantization is emphasized as an explanatory principle. Using different type of assignments, e.g. in class worksheets, homework assignments and on exams, students are asked to use the idea of quantized energy levels to explain phenomena such as atomic emission and absorption spectra and to help explain periodic trends such as effective nuclear charge (from which other periodic trends such at atomic radius, ionization energy, electronegativity and reactivity can be deduced). The introduction of quantized electronic energy levels also facilitates a discussion of the role of core and valence electrons, which can be used to reason about ideas such as why carbon has four valence electrons available for bonding (as opposed to six). These ideas provide a basis for discussion of bonding models in Chapter 3.

17.3.3 Chapter 3: Elements, Bonding, and Physical Properties

In Chapter 3, both valence bond and molecular orbital models of bonding are introduced. Again energy concepts are central to understanding the causes and effects of bond formation between atoms. Bonding models are explicitly compared in order to provide examples of different aspects of molecular structure for which each is appropriate. The CLUE curriculum emphasizes how a model of quantized molecular orbital bonding and anti-bonding orbital energies enables explanations related to physical properties. Using this model, students are asked to predict and explain observations such as why diamond is hard, translucent and has a very high melting point, while graphite is soft, shiny and conducts electricity, while also having a high melting point. Again in this chapter, the role of electrostatic interactions that lower the energy of the system, and the idea that these energies are quantized, are emphasized. A major aim of this chapter is to have students understand the idea that bond formation releases energy from the system, and bond breaking requires energy input.

17.3.4 Chapter 4: Heterogeneous Compounds

In Chapter 4, we emphasize the relationship between structure of molecules and their interactions. That is, we return again to ideas that were introduced in Chapter 1 about atomic interactions, and revisit them in the context of more complex systems. Much of this material is concerned with helping students understand the formalisms of depictions of molecular structures and to learn to decode the information they contain. It is not until students have had considerable practice working with depictions of molecular-level structure that we begin examining the impact of molecular-level structure on energy changes and physical properties at the macroscopic level. The role of energy minimization in determining the arrangement of atoms in molecules is highlighted as central to molecular-level interactions. We choose to emphasize this idea and language because it is what students will encounter in subsequent chemistry courses where energy minimization is an important concept.

17.3.5 Chapter 5: Systems Thinking

Chapter 5 introduces a thermodynamic perspective on energy changes and aims to connect the bulk properties of substances with molecular level interactions. We begin with the concept of phase changes, since only changes in intermolecular interactions are involved. At this point in the semester, students have already worked with the idea that to change from a solid to liquid or a liquid to gas, energy must be put into the system (and conversely that to change from gas to liquid, or liquid to solid a release of energy to the surroundings must take place). They have also encountered the idea that the stronger the attraction between the particles, the more energy is required to overcome the interactions between them. We aim to link this prior knowledge related to atomic-molecular and macroscopic energy ideas by introducing further detail about associated changes at the atomic-molecular and quantum mechanical perspectives. For instance, we discuss how adding thermal energy may result in an increase in temperature by increasing the kinetic energy of the particles, but that thermal energy may also causes increases in vibrational and rotational (quantized) energy levels which do not contribute to observed temperature changes.

At this point, the state function enthalpy (H) is introduced as a representation of the thermal energy of a system at constant pressure. Introduction to the enthalpies of phase changes and specific heats of substances allows for a quantitative discussion of ideas related to heat transfer and bond energetics. Systems at constant volume are mentioned but not emphasized at this point since most students will not encounter such systems in subsequent coursework. Using this approach, we are now able to address the idea of atomic-molecular interactions using the three perspectives on energy change as shown in Fig. 17.4.

Next, we introduce the term entropy using a probabilistic approach that is grounded in a discussion of changes at the molecular level (Lambert 2002). We also

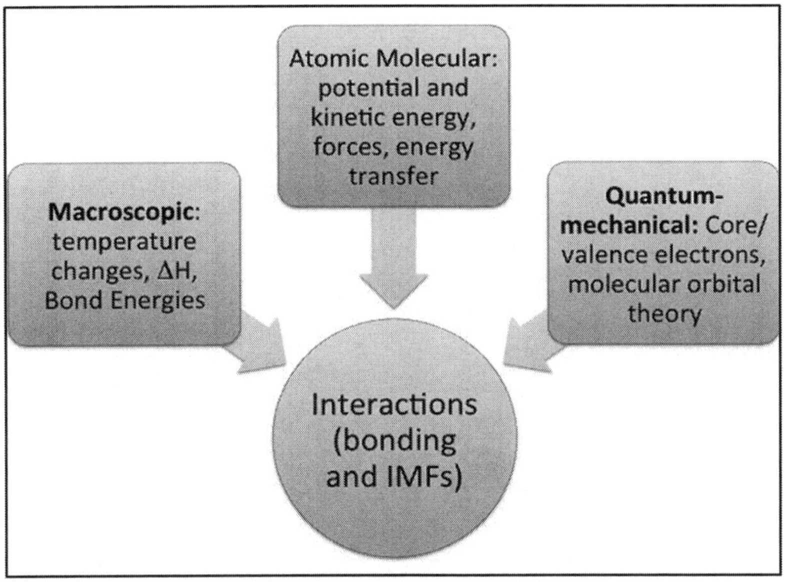

Fig. 17.4 Three perspectives on energy changes and interactions

introduce the state function entropy (S) and the Second Law of Thermodynamics, which states that the entropy of an isolated system always increases as the system evolves towards the more probable state in which energy is dispersed.

Gibbs energy (G) is then introduced as a proxy for the total entropy of the universe. We avoid the definition of Gibbs energy as the energy available to do work, since the term "work" at the molecular level is not a very useful concept, especially since it is often reserved for expansion work, for example during gas evolution. Instead, we emphasize the usefulness of the Gibbs energy function for predicting the direction of change in a chemical system and highlight the relationship between a negative Gibbs energy change and a positive total entropy change in order to illustrate the role of Gibbs energy as a proxy for total entropy.

Our objective in teaching thermodynamics topics in this way is to allow students to think about energy inputs and outputs as well as the molecular and macroscopic consequences of these energy changes. In concert with the concurrent development of an understanding of molecular structure, this approach is designed so that students may construct a coherent framework that allows them to predict and explain the direction of change in a chemical system. It is intended to provide a basis for understanding why some chemical processes require energy input and some produce energy, and how thermodynamically unfavorable processes can be driven by coupling them through common intermediates to more favorable processes. Figure 17.5 illustrates the ways in which molecular, quantum mechanical, and macroscopic energy ideas contribute to this framework.

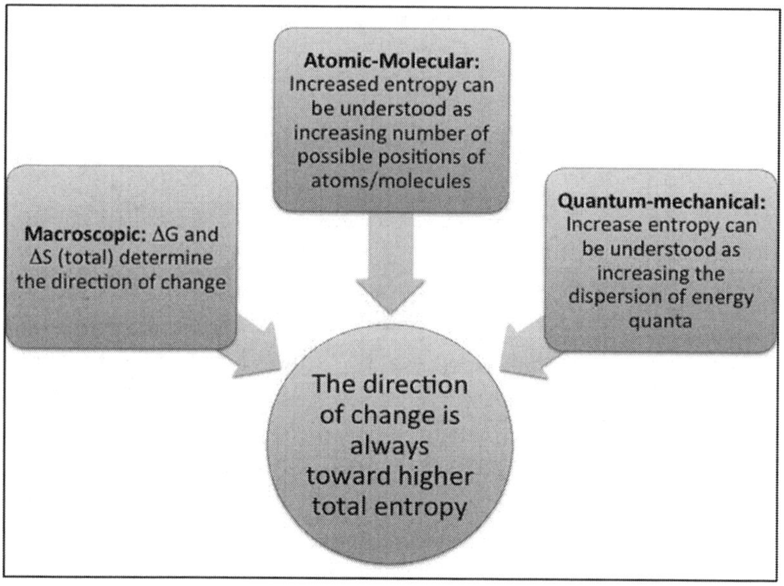

Fig. 17.5 Understanding the direction of change requires all three dimensions

The ideas developed in the first half of this curriculum are then used in the second semester to understand how energy (and structure and properties) affect the formation of solutions (Chapter 6), chemical reactions (Chapter 7), rate and extent of chemical reactions (Chapter 8), and how networked reactions can be used to drive thermodynamically unfavorable processes (Chapter 9).

17.4 Summary and Conclusions

We view energy as an integral component of introductory college level chemistry courses that has the potential to provide a framework for understanding both how and why chemical changes occur. As currently implemented within typical introductory chemistry courses, the discussion of energy is fragmented and rarely makes explicit connections across the various energy topics in the curriculum. This makes it exceedingly challenging for students to build on prior knowledge since most of the ideas that students have are based on macroscopic understandings of energy and energy changes.

The learning progression for energy that we have described aims to explicitly connect three commonly used perspectives related on energy (macroscopic, atomic-molecular, and quantum mechanical). Table 17.1 summarizes topics in the curriculum where the perspectives to energy instruction may be used.

Table 17.1 Three perspectives on energy and phenomena that they explain

Atomic-molecular perspective needed for:	Quantum-mechanical perspective needed for:	Macroscopic perspective needed for:
Interactions leading to potential energy minimization	Interactions of matter and electromagnetic radiation	Physical manifestations of molecular level energy changes
Energy transfer by collisions	Energy transfer by electromagnetic radiation	Thermochemistry and thermodynamics
Chemical and physical changes as systems	Periodic trends (effective nuclear charge etc.)	Temperature changes and chemical and physical processes
The origins of "chemical energy"	Valence and core electrons	Gibbs energy as a proxy for the second Law of Thermodynamics, and a predictor of change

By sequencing our discussion so as to begin with a discussion of atomic structure and by connecting new topics to prior understandings, we aim to scaffold students' ability to reason about the networked reactions that drive thermodynamically unfavorable processes. The development of this learning progression and assessment of student outcomes are ongoing, with data collected from student performances and interviews being used to refine and revise this approach.

What is clear is that in chemistry we cannot continue to treat energy concepts as if students already have a robust framework to build on. We must take time to reach back and reconstruct and re-develop energy ideas beginning at the molecular level and we must design and construct meaningful activities and assessments that encourage students to relate understandings of energy across the curriculum.

References

Achieve. (2013). *Next generation science standards*. Washington, DC: National Research Council. Retrieved from: http://nextgenscience.org/next-generation-science-standards

Barker, V., & Millar, R. (2000). Student's reasoning about basic chemical thermodynamics and chemical bonding: What changes occur during a context-based post-16 chemistry course? *International Journal of Science Education, 22*, 1171–1200.

Boo, H. K. (1998). Students' understandings of chemical bonds and the energetics of chemical reactions. *Journal of Research in Science Teaching, 35*, 569–581.

Carson, J., & Watson, E. M. (2002). Undergraduate students' understandings of entropy and Gibbs free energy. *University Chemistry Education, 4*, 4–12.

Chang, R., & Goldsby, K. (2012). *Chemistry* (11th ed.). New York: McGraw-Hill.

Cooper, M. M., Underwood, S. M., Hilley, C. Z., & Klymkowsky, M. W. (2012). Development and assessment of a molecular structure and properties learning progression. *Journal of Chemical Education, 89*, 1351–1357.

Hadfield, L. C., & Wieman, C. E. (2010). Student interpretations of equations related to the first law of thermodynamics. *Journal of Chemical Education, 87*, 750–755.

Jewett, J. W. (2008). Energy and the confused student III: Language. *The Physics Teacher, 46*, 149–153.

Jin, H., & Anderson, C. W. (2012). A learning progression for energy in socio-ecological systems. *Journal of Research in Science Teaching, 49*, 1149–1180.

Kaper, W., & Goedhart, M. (2002). 'Forms of energy', an intermediary language on the road to thermodynamics? Part I. *International Journal of Science Education, 24*, 81–95.

Lambert, F. L. (2002). Disorder – A cracked crutch for supporting entropy discussions. *Journal of Chemical Education, 79*, 187–192.

Lancor, R. A. (2012). Using student-generated analogies to investigate conceptions of energy: A multidisciplinary study. *International Journal of Science Education, 36*, 1–23.

Nahum, T. L., Mamlok-Naaman, R., Hofstein, A., & Krajcik, J. (2007). Developing a new teaching approach for the chemical bonding concept aligned with current scientific and pedagogical knowledge. *Science Education, 91*, 579–603.

National Research Council. (2012). *A framework for K-12 science education: Practices, crosscutting concepts, and core ideas* (Q. Helen, S. Heidi, & K. Thomas, Eds.). Washington, DC: The National Academies Press.

Park, E. J., & Light, G. (2009). Identifying atomic structure as a threshold concept: Student mental models and troublesomeness. *International Journal of Science Education, 31*, 233–258.

Sozbilir, M. (2002). Turkish chemistry undergraduate students' misunderstandings of Gibbs free energy. *University Chemistry Education, 6*, 73–82.

Taber, K. S. (1998). An alternative conceptual framework from chemistry education. *International Journal of Science Education, 20*, 597–608.

Taber, K. S. (2002). Conceptualizing quanta: Illuminating the ground state of student understanding of atomic orbitals. *Chemistry Education Research and Practice, 3*, 145–158.

Taber, K. S. (2004). Learning quanta: Barriers to stimulating transitions in student understanding of orbital ideas. *Science Education, 3*, 94–116.

Taber, K. S. (2013). Revisiting the chemistry triplet: Drawing upon the nature of chemical knowledge and psychology of learning to inform chemistry education. *Chemistry Education Research and Practice, 14*, 156–168.

Thomas, P. L., & Schwenz, R. W. (1998). College physical chemistry students' conceptions of equilibrium and fundamental thermodynamics. *Journal of Research in Science Teaching, 35*, 1151–1191.

Tsaparlis, G., & Papaphotis, G. (2009). High-school students' conceptual difficulties and attempts at conceptual change: The case of basic quantum chemical concepts. *International Journal of Science Education, 31*, 895–930.

Chapter 18
Energy Spreading or Disorder? Understanding Entropy from the Perspective of Energy

Rui Wei, William Reed, Jiuhua Hu, and Cong Xu

18.1 Introduction

"Why do some physical processes and chemical reactions happen spontaneously while others do not?" This is a fundamental question in the physical sciences (National Research Council [NRC] 2012). Answering it can explain everything from why an object falls to the Earth's surface to what determines chemical equilibrium (Lambert 2002b). The question also has deep significance for the life sciences, earth sciences, and engineering. For example, the conditions necessary for spontaneity explain the need for energetic coupling of spontaneous reactions with non-spontaneous reactions in living things, a fundamental characteristic of metabolism (Reece et al. 2011). Convection cycles, which play a key role in every major non-living Earth system, are perhaps best explained from the lens of spontaneous processes (Chen et al. 2010). Maximum theoretical efficiency in modern mechanical engineering is also inextricably linked, historically and presently, to this question (Dincer and Cengel 2001).

Given the significance of the question of spontaneity, one might conclude that after completing the 12th-grade, students should at least have familiarity with how this question is answered in contemporary science, recognizing that the answer is related to energy. Ideally, students should be able to go beyond that, and explain spontaneous processes in the context of the second law of thermodynamics (henceforth the second law). However, a comparison of K-12 science education standards from seven countries shows that only the standards from three countries involve entropy, the second law, and the direction of chemical reactions. In stark contrast, all

R. Wei (✉) • J. Hu • C. Xu
College of Chemistry, Beijing Normal University, No. 19 Xinjiekouwai Street, Beijing 100875, China
e-mail: weirui@bnu.edu.cn

W. Reed
Disney II Magnet School, 3900 N. Lawndale, Chicago, IL 60618, USA

R.F. Chen et al. (eds.), *Teaching and Learning of Energy in K-12 Education*, DOI 10.1007/978-3-319-05017-1_18, © Springer International Publishing Switzerland 2014

of the standards place strong emphasis on the first law of thermodynamics (energy conservation) and the forms of energy (Wang et al. 2014). Perhaps this is why even at the university level and among chemistry majors, research evidence suggests that students are by and large unable to adequately explain why some processes and reactions are spontaneous, whereas others are not (Sözbilir and Bennett 2007; Boo 1998).

Because some sort of energy change or transfer always accompanies spontaneous processes, energy is used for explaining in a generalized manner why such processes occur. Yet typical energy constructs, such as the forms of energy, the transfer and transformation of energy, and the conservation of energy are inadequate for making such an explanation. Still, perhaps because of the emphasis on these constructs in K-12 curricula, students often attempt to use them in explaining spontaneous processes. For example, many students believe that a spontaneous process occurs when the energy of a system decreases (Carson and Watson 2002). Of course, because of the conservation of energy, any decrease in the energy of a system is offset by an increase in the energy of its surroundings, and vice versa. There is no logical justification for prioritizing an energy decrease in the system under study over an energy decrease in its surroundings (Atkins 2006). Thus, spontaneous endothermic chemical reactions provide a counter example of the common misconception that these processes come about because of a decrease in energy of the system. Both the system under study and its surroundings must be taken into consideration when analyzing the cause of spontaneous processes in the system.

Other documented misconceptions for why spontaneous processes occur include the correct idea that the concept of entropy can be used to explain spontaneity. However, in these misconceptions, entropy is considered either without reference to energy at all or simply as another form of energy in a way similar to thermal or kinetic energy or enthalpy (Sözbilir and Bennett 2007). The latter belief is not surprising in light of the common understanding of spontaneous events in terms of negative changes in Gibbs free energy ($\triangle G$), a construct that can be used as a proxy for the total entropy change of the universe (the system plus its surroundings) but is often presented without this context made explicit. Another source of this misunderstanding is that $T\triangle S$, in $\triangle G = \triangle H - T\triangle S$, is often considered as energy that is not useful, or as dissipated heat, which is related to energy, while $\triangle H$ is often considered as the total energy change of the system. The common metaphor for entropy as a measure of disorder also contributes to this misunderstanding. This occurs when students, not unreasonably, consider the increased motion and collisions of particles that accompany increased thermal energy as an increase in disorder. Other students will persistently use, despite being instructed to do otherwise, vague or spatial understandings of disorder to describe entropy, without connecting the concept in any way to energy (Carson and Watson 2002). Of course, for many reasons, not least because that entropy is not conserved for real processes, it cannot be considered as yet another form of energy. However, as we argue in this chapter, entropy should be considered from an energy perspective.

Ultimately, it is entropy that is the key concept for explaining spontaneous processes. The second law, put in one way, states that systems spontaneously evolve toward the state of maximum entropy of the universe. Thus, it can be said that spontaneous processes are those that increase the total entropy of the universe. But what is entropy? And what is the best way to approach teaching the concept in K-12 science courses? Entropy is perceived as a difficult concept for students to grasp. We propose that a key reason for this is a lack of a widely accepted, accessible, and effective means of teaching entropy, or a specific set of learning progressions for understanding spontaneous processes across K-12 science education and multiple disciplines.

In this chapter, we suggest a greater emphasis in K-12 education on understanding spontaneous processes by analyzing entropy from an energy perspective. Our proposal for accomplishing this draws heavily on a widely discussed but underused metaphor—entropy as a measure of the extent of energy dispersal in space (Lambert 1999, 2002a, b, 2005, 2006a, b, 2007, 2011; Lambert and Leff 2009; Leff 1996, 2007, 2012a, b, c, d, e; Kozliak and Lambert 2005, 2008). With this metaphor, the second law can be restated as "energy spreads out spontaneously if not hindered from doing so." We also critically examine what remains a common method for introducing entropy in K-12 curricula, as represented by high school chemistry textbooks today—through the metaphor of entropy as "disorder" (e.g. Gao and Wang 2007; Song and He 2004; Wang 2007; Wilbraham et al. 2012). We analyze the advantages and disadvantages of the disorder and energy dispersal metaphors for entropy, with an emphasis on the metaphors' fidelity to key features of the entropy concept, relationship to the energy perspective, accessibility to younger students, and openness to updating in a learning progression. We conclude by reiterating what students should know about entropy and spontaneous processes as well as common student misconceptions, while describing challenges for widespread and successful adoption of teaching entropy from an energy perspective in K-12 education. Finally, we propose an outline of a program to develop a K-12 learning progression for understanding spontaneous processes from an energy perspective.

18.2 Understanding Entropy from the Energy Perspective

The two most widely used quantitative expressions for entropy, which can be shown to be equivalent, are the macroscopic Clausius formulation and the microscopic (and quantum) Boltzmann formulation. In both formulations, entropy connects intimately to energy. Here, we begin by discussing the Clausius formulation, which clearly shows that entropy is a function of energy. We then outline the major features of the metaphor for entropy as the dispersal of energy. In a subsequent section, we show how to update this metaphor to comply with Boltzmann's definition of entropy.

18.2.1 Clausius' Definition of Entropy and the Dispersal of Energy Metaphor

Clausius' mathematical definition of entropy change, $dS = dq_{rev}/T$, could be expressed verbally as "entropy change equals the amount of energy dispersed reversibly at a specific temperature *T*." This clearly does not imply that entropy is "disorder," and in fact Clausius himself used the term "transformation" to describe entropy. This "transformation" in the simplest sense is energy's dispersal from a source that is almost imperceptibly at a temperature above *T* to a receptor that is at *T*. More generally, Clausius' equation can be thought of as an index of the amount of energy dispersal at a specific temperature (q_{rev}/T). Though the specific temperature is clearly important, to a first approximation it follows that "entropy as the dispersal of energy" is a useful metaphor. When entropy increases, energy becomes more dispersed in space (Lambert 2002b).

18.2.2 Total Entropy and the Dispersal of Energy Between the System and Its Surroundings

It is crucial that useful entropy analysis takes into account both the system under study and its surroundings, because the second law holds that the *total* entropy of the universe never decreases. The second law tells us nothing about what can or cannot happen to the entropy of a system in the absence of its surroundings. Yet students often hold the misconception that spontaneous processes can be determined by changes in the system under study alone, without taking its surroundings into consideration (Sözbilir and Bennett 2007). This is not surprising, given that so many concepts in science concern themselves with the system alone. It is therefore useful to consider how our metaphor might be adapted when discussing the change of *total* entropy, such that the system and its surroundings are explicitly taken into consideration. We can simply say that when total entropy increases, energy becomes more dispersed between the system and its surroundings.

18.2.3 Connecting the Energy Dispersal Metaphor for Entropy to Spontaneous Processes

Total entropy is maximized in spontaneous processes. In other words, there will be more dispersal of energy between the system and its surroundings whenever a real spontaneous process occurs. This provides another formulation of the second law: "energy spreads out spontaneously if not hindered from doing so." For the purposes of this discussion, and for clarity in using these metaphors in K-12 courses, we propose that we consider energy "dispersal" to be a static function describing energy distribution—that is, entropy. Energy "spreading," on the other hand, is taken to be a dynamic process involving a change in entropy—in other words, the process of broadening the dispersal of energy between the system and its surroundings. A major advantage of framing an increase in entropy described by the second law in terms of energy spreading is that the surroundings are taken into consideration, which is often not explicit.

18.2.4 Using the Energy Dispersal Metaphor to Qualitatively Describe Spontaneous Processes

The spontaneous transfer of heat from higher temperature to lower temperature bodies follows the second law. In a more qualitative (and metaphorical) sense, even mathematically naïve students can understand that for energy to spread out maximally, the thermal energy of a hot object must be distributed to its surroundings, or the thermal energy of the hot surroundings must be distributed to a cool object. Similarly, particles carrying thermal energy will spread into a vacuum in order to maximize energy spreading (carrying their thermal energy over a larger space—a similar argument can be made for why mixed gases have more entropy than separated gases). The gravitational potential energy stored in the system of a suspended rock and the earth will, upon release of the rock, spread out as kinetic energy and eventually, upon impact on a surface, as heat to its surroundings. It should be noted that although this last example describes a mechanical rather than a thermodynamic system, the energy dispersal metaphor is still instructive in explaining why this spontaneous event occurs. In each of these contexts, even students that are still developing an understanding—of energy transfer, the conservation of energy, and energy transformation—should be able to use the idea of energy spreading to determine why certain spontaneous events occur. The dispersal of energy metaphor therefore provides accessibility to students for use in accurate entropy analysis and provides a natural avenue for taking both the system and its surroundings into account.

18.3 Movement to Replace the Disorder Metaphor with the Dispersal of Energy Metaphor

Lambert, an organic chemist, and Leff, a physicist, among others, have led a movement to shift the prevailing metaphor for entropy in introductory college chemistry and physics texts from the disorder metaphor to the dispersal of energy metaphor discussed above. Equivalent to this metaphor for entropy, and more relevant for K-12 educators, the *Framework* and *Next Generation Science Standards*

(*NGSS*) use language of "toward more uniform energy distribution" to describe the evolution of uncontrolled systems (or the second law) (NRC 2012). In this section, we begin by discussing the success of the disorder metaphor, and then outline some major problems with using the disorder metaphor in K-12 education. In this context, we suggest some criteria for a successful replacement metaphor, and finally describe how the dispersal of energy metaphor meets these criteria.

18.3.1 Disorder Metaphor Has Been Pervasive

Since the formulation of the macroscopic entropy concept by Clausius and later the microscopic entropy concept by Boltzmann in the nineteenth century, many metaphors have been proposed to describe entropy (Leff 2007). To be sure, the disorder metaphor must be considered one of the most pervasive; it has been both long lasting and widely used. There are many good reasons for this. First, increased spatial disorder is in fact observed in many spontaneous processes—from the mixing of two ideal fluids to the expansion of gas in a vacuum. Second, the idea of disorder on some level is readily accessible, and lends a workable way to approach teaching and discussing the Boltzmann's statistical mechanical formulation of entropy. Third, popular culture associations with ever-increasing disorder as a fundamental law of the universe are pervasive and appealing. Finally, the disorder metaphor has been so widely used for so long that it is difficult to quickly remove it from the discourse and curricula concerning entropy. The conception of entropy as disorder is also difficult to replace once incorporated by individuals (Sözbilir and Bennett 2007).

18.3.2 Entropy as Disorder: What's the Problem?

It is widely accepted that metaphors, by mapping abstract concepts to relatable everyday phenomena, can help students better understand and use science concepts (Duit 1991). If metaphors are considered as models of scientific phenomena, then it holds that all metaphors are somewhat limited in reflecting the phenomena they represent. Thus, for students to demonstrate adequate understanding of how a metaphor is used, they must explicitly consider the advantages and shortcomings of the metaphor in describing the phenomenon or concept it represents (Glynn and Takahashi 1998). This is an especially difficult task for a highly abstract concept such as entropy, and it is nearly impossible if the metaphor used for understanding that concept in the first place fails to approximate key features of the concept itself. Unfortunately, this is the case for the disorder metaphor for entropy. Here, we discuss three major problems with the disorder metaphor.

18.3.2.1 "Disorder" Is Vague

The first problem with the disorder metaphor is the vagueness of the term itself. Typical dictionary definitions for disorder are: "lack of order or regular arrangement," "confusion," and "(in medicine) a disturbance of normal functioning." The first of these definitions has a strong spatial connotation, which we discuss below. This is, to be fair, the definition that is typically emphasized in high school textbooks, though this does not guarantee that it will be the definition incorporated by students, and it does not reflect the more appropriate interpretation of entropy as it relates to available energy microstates. Using the "confusion" definition, a typical high school student might relate disorder to an inability to decide on a particular route when lost, for example. This, then, can be related to the disorder that some associate with higher temperatures, envisioned in terms of increased particle agitation. The variable definition of disorder makes the term itself confusing (Leff 2007).

There is also a problem that describing entropy as a measure of disorder does not in itself specify the level of analysis at which that disorder occurs. For example, ice cubes flying in space appear disordered macroscopically, though of course the ice cubes themselves are neatly ordered at the molecular level. Below the molecular level, the subatomic particles that make up the ice cube then have a higher degree of disorder. Which level of organization does "disorder" refer to (Donaldson 2011)?

18.3.2.2 Spatial Disorder Does Not Represent the Features of the Entropy Concept Well

The spatial disorder metaphor is related historically to the Boltzmann formulation of entropy. This particular formulation, known as the microscopic or statistical mechanical definition, holds that entropy increases logarithmically with the number of available energy microstates for a particular system (or, alternatively, with the number of ways of realizing the most probable microstates). In the modern quantum mechanical view of the Boltzmann formulation, microstates are possible ways of energy distribution, rather than spatial particle disorder (though particle disorder can be related to an increase in energy distribution). Clearly, entropy is an energy-related concept. Yet, the connection to energy using the spatial disorder metaphor of entropy is not explicit, and students may fail to recognize it (Granville 1985; cited in Cooper et al. 2014).

Another important and often ignored issue is that spatial disorder can just refer to a single "snapshot" of a particular system (i.e. one microstate), while according to the Boltzmann formulation entropy increases when there are more *available* microstates (i.e. the number of microstates), through which the system dynamically moves with an equal probability of being in each microstate at any particular instant. By simply considering snapshots, even experienced chemists can easily be fooled into naming what amounts to a lower entropy system as having higher entropy because one particular snapshot of that system is more likely to appear "disordered" (Styer 2000). So the disorder of one microstate and the number of available microstates must be differentiated in the teaching of entropy.

Avoiding a detailed discussion of the meaning of energy-microstate and the number of microstates, this brings us to the problem that we will draw wrong conclusions while analyzing entropy changes of some processes with the spatial disorder metaphor. In some situations, visual disorder actually decreases as entropy increases, such as spontaneous crystal formation in a supersaturated sodium sulfate solution. From a spatial viewpoint, there is more visual order after crystallization, so that the entropy of the system seems to decrease. But considering that the temperature actually decreases in this process, i.e. the system absorbs energy from its surroundings and the entropy of the surroundings decreases, the entropy of the system has to increase so as to make the total entropy increase. Thus the visual order of this case refers to higher entropy. There are other processes that increase entropy yet do not lead to more disorderly visual states. For example, within certain temperature bands, increasing the temperature of some liquid crystals leads to more alignment of the crystals, while entropy has increased (Leff 2007; Lambert 2002a).

Another issue is that the number of microstates available even for a relatively low entropy system (or considered as "more ordered" with this metaphor), such as a small ice cube relative to an equivalent amount of liquid water, is so staggeringly large that it cannot be called orderly in human terms (Kozliak and Lambert 2005).

Most of the above issues are beyond the normal realm of K-12 education. Still, the metaphors used in K-12 education should avoid perpetuating misconceptions in students who may pursue further study of the physical sciences at the university level. The final problem with the disorder metaphor, discussed below, reflects the perpetuation of misconceptions about entropy and spontaneous events that directly interfere with the goals of learning these concepts in K-12 education.

18.3.2.3 Disorder Metaphor Does Not in Itself Integrate the Entropy of the System and Its Surroundings

The disorder metaphor fails to approach entropy in such a way that considers the total entropy of the universe. While it is possible to consider the "disorder" of the system and the "disorder" of its surroundings, this must be done with additional effort, and is therefore often neglected, especially when giving qualitative explanations (Sözbilir and Bennett 2007). The formula for Gibbs free energy implicitly takes the entropy of the surroundings into consideration (through the enthalpy term), but students using this formula often miss a key feature of the second law—that the total entropy of the universe, not of a single system in the absence of its surroundings, never decreases (Carson and Watson 2002).

Typical life experiences which reflect common understandings of spatial disorder are, in fact, from the perspective of just the system under consideration, not indicative of entropy increases or spontaneous processes (Lambert 1999). A bedroom does not spontaneously, in the physical sense, become messy. In fact, the movement of clothes and other items around the room represents a non-spontaneous process only made possible by the agency of the individual (and, ultimately, associated spontaneous chemical reactions that increase the entropy of the universe). Buildings may rust spontaneously, but they do not typically fall apart spontaneously—instead this is brought on by weathering.

Yet, it is problematic that these connections are often made explicitly in textbooks. For example, the action of a pack of dogs running around after becoming unleashed is used to illustrate entropy in a recent high school textbook (Wilbraham et al. 2012). Two popular textbooks in mainland China (Song and He 2004; Wang 2007) use illustrations of scattered matches or a messy room as depictions of "increased entropy". These events only indirectly, and through complicated mechanisms that consider the system and its surroundings as a whole, follow the second law.

It might be argued that the metaphor should not be required to comply with all the constraints of the concept it represents. However, for the purposes of pedagogy, a metaphor should not directly contradict the meaning of the entropy concept where it can be used in meaningful contexts (such as living organisms or Earth systems), leading to more misconceptions than it prevents. This is especially true if the metaphor does not do a particularly good job of representing the salient features of the concept.

18.3.3 Criteria for a Successful Replacement Metaphor and How Dispersal of Energy Meets These Criteria

Replacement of the disorder metaphor by a new metaphor or set of metaphors in K-12 education is past due. Whatever good that the disorder metaphor does for helping students to understand spontaneous processes, it does a greater amount of harm in misleading students and preventing broader incorporation of the second law into the everyday thinking of lay people. A replacement metaphor must accomplish many of the things that entropy as disorder fails to do.

18.3.3.1 Criterion I: The Entropy Metaphor Should Comply with the Features of the Concept

The metaphor must be reasonably precise in meaning, must not contradict the principles of entropy analysis if it takes the form of everyday macroscopic experience, and must comply with as many as possible of the features, whether theoretical or empirical, of the contemporary formulations of entropy.

How does the energy dispersal or spreading metaphor for entropy improve the situation over the disorder metaphor? First, from the perspective of the second law, energy "spreading" is unambiguous. Whereas "dispersal" or "distribution" may be more challenging for younger students to grasp, those words also carry far less ambiguity than "disorder." Second, unlike "disorder," many of the everyday experiences that students are likely to associate with "energy spreading" (such as diffusion and heat transfer) are consistent with entropy increases from the perspective of the system used to illustrate the metaphor. Another major advantage of the "dispersal of energy at a given temperature" metaphor is that it fits Clausius' macroscopic formulation of entropy (from classical thermodynamics) quite well. The concept of "energy spreading" driving spontaneous processes also clearly relates the second law to energy. At the same time, the metaphor simplifies incorporation of many of the most relevant uses and consequences of the second law, among them the inclusion of both the system and its surroundings in analyzing event spontaneity, the determination of chemical equilibrium state, and the use of Gibbs free energy to analyze chemical reactions.

18.3.3.2 Criterion II: The Entropy Metaphor Should Prevent Potential Misconceptions

The metaphor should have a fundamental connection with other energy constructs, such that it makes the relationship between energy and entropy clear without furthering the misconception that entropy is another form of energy. Describing entropy explicitly as a measure of the *dispersal* of energy checks students' misunderstanding of entropy as another form of energy.

It should also work to prevent users of the metaphor from thinking of entropy from a limited perspective, such as that only the system or only its surroundings is considered. As students study chemical systems in more depth, the temptation for them to explain spontaneous reactions by ignoring the surroundings will be minimized because the energy spreading metaphor looks beyond the single system.

18.3.3.3 Criterion III: The Entropy Metaphor Should Be Accessible to Students

The metaphor used for understanding entropy should be accessible to young students, particularly as it relates to its use in explaining spontaneous processes. Many familiar instances of spreading, related to diffusion or heat transfer, for example, demonstrate entropy straightforwardly so that it is not difficult for students to grasp its meaning. In well-designed reading materials with the energy dispersal metaphor, entropy and the second law will be readily understood by chemistry teachers- even by beginners in chemistry- and be accessible to students not majoring in science (Lambert 2005, 2006a, 2011).

Though not suggested by the *Framework* or the *NGSS,* we propose that upon completing of the 8th-grade, students should have an understanding that spontaneous events are determined by the maximal possible spreading of energy. In this way, students will be prepared for the reinforcing and deepening of their understanding of these ideas in high school. For example, students can much more quickly develop a qualitative understanding of chemical equilibrium from the perspective of energy spreading. Many reactions go to equilibrium rather than completion. This is because the mixture of products and reactants distributes the energy still contained within the chemical system, which will contribute to higher entropy of the system, and thus also contribute to higher total entropy (Lambert 2002b). It should be noted here that the point of maximum mixing state does not exactly correspond to equilibrium state, because the total entropy will be affected not only by the change in entropy of mixing of system, but also by the change in non-mixing entropy of system and the entropy change of surroundings (Gary 2004).

Another example of the advantages of an earlier understanding of spontaneity through energy spreading is that an earlier and more persistent qualitative understanding of the second law will demystify Gibbs free energy if it is presented in advanced classes or university. Students are better off understanding Gibbs free energy as a proxy for the degree of energy spreading in a reaction than they are puzzling through how a constructed form of energy could determine the direction of chemical reactions.

18.3.3.4 Criterion IV: The Entropy Metaphor Should Be Amenable to Updating and Modification

Like any metaphor, entropy as the dispersal of energy has its limitations. However, in evaluating the relative merits of different metaphors for science concepts for K-12 learners, top criteria should include the extent to which a metaphor is amenable to updating for modification in light of new evidence or more complex formulations of the concept. This is especially true when considering how to build on the concept across grade levels, or in such a way that lays a solid foundation for advanced study. The spatial disorder metaphor is difficult to update for reflecting the connections between entropy, spontaneous reactions, and energy. In fact, because it appears that the disorder metaphor itself (but not a sophisticated understanding of how it relates to entropy) is fairly easily incorporated by students, it can be difficult to replace once it has been established (Sözbilir and Bennett 2007). The dispersal of energy metaphor, on the other hand, is amenable to updating for reflecting the statistical mechanical formulation of entropy.

In considering chemical systems, the energy dispersal metaphor is a good metaphor for new learners because it works well to relate entropy to energy and leads to a view of reactions that considers both the system and its surroundings. This makes understanding why either exothermic reactions or endothermic reactions can take place spontaneously simpler for students. But the energy dispersal metaphor can lead to confusion in certain cases, for instance, in considering a constant-volume spontaneous chemical reaction in an isolated system where the distribution of energy in space is maximized both before and after the reaction. Below, we illustrate three levels in the route of updating and modifying the understanding of entropy from the perspective of energy dispersal, which will extend the usefulness of the metaphor to account for all chemical systems.

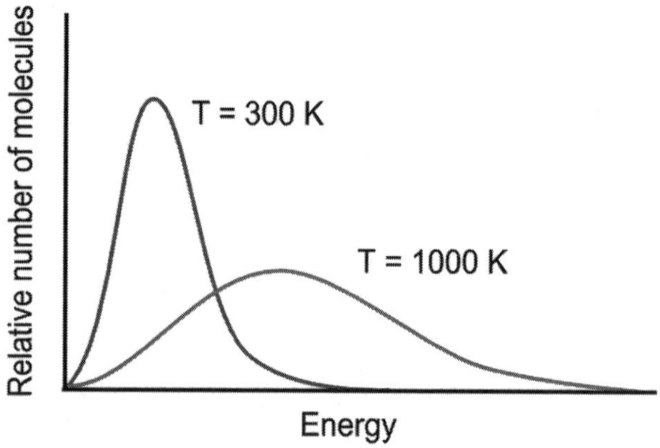

Fig. 18.1 Distribution of molecular energies at different temperatures

The first level understanding is the dispersal of energy between the system and its surroundings at the macroscale. This idea, which is discussed in Sect. 18.2 above and is elaborated on in Fig. 18.4 of Sect. 18.4.4, focuses on the energy redistribution in space when a reaction takes place.

The second level of understanding is the dispersal of energy across multiple energy levels at the microscale. To achieve this understanding, more advanced students can be led further to an understanding of the distribution of molecular energies across many discrete energy levels. On average, the distribution of particles across multiple energy levels, which can be expressed as in Fig. 18.1, does not change significantly for constant conditions (i.e.: pressure, volume, and temperature). Thus students can understand easily why higher entropy means that energy is more dispersed (energy distribution is in a broader band) from a molecular perspective, and vice versa. Because of the change in the possible energy levels which molecules in the system can occupy that accompanies a chemical reaction, as well as the change of the number and kinds of molecules, the distribution of energy across multiple energy levels also changes, even though the energy does not flow between the system and its surroundings (spreading in space).

The third level of understanding is that of entropy from the perspective of the number of energy microstates. The most fundamental contemporary view of entropy (S) is that it is logarithmically proportional to the number of available energy microstates of a system (W), such that $S = k_B \ln W$, where k_B is Boltzmann's constant. An energy microstate precisely describes one of many of these possible distributions of energy throughout all particles in a system. Because of collisions and other exchanges of energy between particles, the microstate of a system constantly changes. But as long as the number of *available* microstates stays the same, the entropy stays the same. According to the second law, the number of microstates of the universe (system + surroundings) can only remain the same or increase. A process will be spontaneous if it increases the total number of microstates of the universe.

If this statistical mechanical formulation is introduced to students, they could attempt to fit it within the context of energy dispersal and energy spreading metaphors. Students can make a connection between "energy spreading" and "energy having more ways of distributing itself" (that is, more energy microstates being available). It should be noted that in this case students must explicitly recognize that the energy is always in one microstate at a time, so it is incorrect to view energy as being "spread out" among microstates; energy can just be more dispersed among different energy levels in the energy distribution on average (see Fig. 18.1), which is caused by increasing the number of energy microstates. Extending the energy spreading metaphor in this way has the added benefit of reinforcing the probabilistic nature of entropy.

18.4 Conclusions

18.4.1 What Should Students Know About Entropy?

Students should be able to explain spontaneous processes qualitatively and to some extent be able to predict in which situations spontaneous processes will and will not occur. While the *NGSS* focuses primarily on heat transfer processes, we believe that students should be given more generalized tools that can apply to other physical processes as well as chemical reactions, along with applications in living organisms, Earth systems, and engineering. In order to make these explanations in a way that is consistent with the contemporary scientific view, students must at least implicitly involve entropy and the second law in their explanations, and should explicitly involve energy as well.

18.4.2 What Are the Challenges Teachers Face in Teaching Students This Knowledge?

There are both conceptual and structural problems in giving students the means to explain why spontaneous processes occur. Conceptually, entropy adds another layer of abstraction to already abstract energy concepts, such as transfer, transformation, and conservation. Both microscopic and macroscopic quantitative formulations of entropy are difficult, especially when applied to most meaningful situations to analyze entropy change. Entropy's application to "the universe" rather than a particular system under study sets it apart from the way that teachers approach many other scientific concepts, including energy concepts. Related to this, there are many ways in which physical and chemical systems interact such that entropy often decreases for particular systems, seemingly in violation of the general principle of the second law.

Beyond the conceptual problems, however, there are structural problems in teaching students about spontaneous processes. While the situation is slowly improving, for years teaching rules for analyzing spontaneous processes have been relegated by both standards and textbooks as optional or something reserved for the end of a high school chemistry course. While other ideas, including the first law of thermodynamics, are typically seen as important across grade levels, entropy and the second law in any form have perhaps been regarded as too difficult to even approach or approximate with younger students, with the possible exception of through heat transfer in the middle grades. This limited prioritization has led to a relatively limited work on student learning progressions and misconceptions, as well as fewer curricular resources than might otherwise be expected.

Ultimately, however, it is the pervasive yet inappropriate use of the disorder metaphor for entropy that has prevented more widespread incorporation of the second law into student thinking. Entropy had been described as disorder for a long time (American Association for Advancement of Science [AAAS] 1990, 1993/2009; National Academy of Sciences [NAS] 1996), although it is no longer described this way in the newest US national standards documents (NRC 2012), and has been eliminated from many college-level textbooks (a list of textbooks can be found on http://entropysite.oxy.edu/). This metaphor still persists, however, in recently published and widely used high school textbooks, both in China and the United States (e.g. Gao and Wang 2007; Song and He 2004; Wang 2007; Wilbraham et al. 2012). It is equally important that although the disadvantages of the disorder metaphor have been all but settled in certain academic/pedagogical debates, this message has not, by and large, reached teachers. In China (and we suspect we would find similar data in the United States), of 3,833 high school chemistry teachers that we surveyed, 61 % considered entropy as disorder, whereas only 17 % considered entropy as the dispersal of energy.

Because the metaphor of entropy as disorder has been so pervasive, most of students' misconceptions—that have been documented regarding entropy, the second law, and spontaneous processes—are directly or indirectly related to this metaphor. Common misconceptions include: entropy is another form of energy, related to thermal or kinetic energy; macroscale objects become spontaneously disordered; entropy increases whenever visible order decreases; disorder at the microscale refers to mixed-upness at any one instant, rather than an increase in available microstates; the physical imperative to increase entropy applies to a single system, rather than to the universe; entropy refers to instability; Gibbs free energy is not related to an increase in total entropy; and spontaneous processes are determined by a decrease in energy of the system under consideration (Boo 1998; Carson and Watson 2002; Sözbilir and Bennett 2007).

18.4.3 What Should Be Done to Meet These Challenges?

There have been several fruitful approaches to teaching entropy at the high school level in ways that go beyond the disorder metaphor (Bindel 2004; Hanson and Michalek 2006). However, we propose that the best way of addressing students' conceptual difficulties with explaining spontaneous processes is to develop a successful framework for teaching about energy that includes both entropy and the second law. This framework should include ways of teaching that are accessible to students both in terms of making concrete connections between the entropy concept and their everyday experience and in terms of making meaningful connections between entropy and other energy topics that students have studied. The framework should also avoid many of the misconceptions that are brought upon by using the disorder metaphor for entropy. Finally, it should be amenable to building student ideas about spontaneous processes, such that these foundations can begin in the elementary grades, progress through the middle grades, and conclude in high school with a strong basis for increasingly complex applications, including quantitative applications. We believe that the entropy component of such a framework can be built around the energy dispersal metaphor for entropy and the energy spreading metaphor for explaining what drives spontaneous processes.

In order to successfully enact such a framework that aids in teaching about spontaneous processes, we propose that standards, written curricula, and teachers need to incorporate the framework. Because most active teachers themselves have likely been taught that entropy is disorder, and because their disorder schema is so persistent, teacher education—from pre-service to professional development—must focus on actively discrediting the disorder metaphor, at least as the best or only way to teach entropy. An effective means of doing this will be to provide teachers with concrete counterexamples of the disorder metaphor. Fortunately, as discussed in Sect. 18.3.2 above, there are many specific counterexamples that illustrate how visual or microscopic spatial disorder can decrease while entropy increases (Lambert 2002a; Leff 2007). Of course, teachers must also be presented with a positive alternative metaphor meant to replace the disorder metaphor.

It is also fortunate that the *Framework* and *NGSS* adopted the view that "uncontrolled systems evolve toward more uniform energy distribution (NRC 2012, p 125)." Although this view is stated in clear language and is simply another way of stating that spontaneous processes occur when energy dispersal is maximized, one concern may be that because the view does not explicitly draw a link with entropy, many teachers will not make this connection. The discussion of "more uniform energy distribution" is only a single paragraph in the *Framework* (NRC 2012, p 125), with a single corresponding Performance Expectation (which is at the high school level) in the *NGSS*. However, we believe that this minimal inclusion of the second law in the *Framework* and *NGSS* provides an opportunity for educators and curriculum developers to build a new framework for teaching entropy.

Because entropy is so intimately connected with energy, we believe that the basis for learning about spontaneous processes must begin as soon as energy transfer and transformation (along with conservation) are explored in science education. We see the need for the integration of entropy into the energy concept system through modified energy learning progressions. This integration needs empirical backing for what students are capable of understanding at any particular age, and how some of the more complicated ideas relating to entropy can be supported by specific earlier learning experiences and understandings. Below, we provide an outline for developing a *potential* K-12 learning progression for understanding spontaneous processes, which we hope can form a starting point for a broader redesign of energy learning progressions. It is important to emphasize that this learning progression requires much further research, especially as it relates to students' prior ideas and the success of their incorporation of these ideas at different ages.

A few features of this learning progression are important to mention here. First, it prioritizes qualitative explanations over quantitative ones, because even extensive quantitative instructional practice with entropy-related concepts often does not necessarily have a meaningful effect on student understanding of spontaneous processes (Carson and Watson 2002). Second, the second law (though not by this name) is introduced before entropy. This is because the concept of "energy spreading" as a means of explaining spontaneous processes is actually conceptually less challenging than "the dispersal of energy at a specific temperature"—the metaphor for entropy itself. Third, technical vocabulary to describe any of these concepts, including "spontaneous," "entropy," and "the second law," is not introduced until these concepts have been thoroughly established otherwise in students' understanding.

18.4.4 A Proposed Outline of a K-12 Learning Progression for Explaining Spontaneous Processes

1. Some physical events happen naturally (i.e., heat transfer from hot to cold substances) while others apparently never happen naturally (i.e., heat transfer from cold to hot substances). [Early elementary]
2. Thermal energy can transfer into or out of a system (see Fig. 18.2). [Late elementary]
3. The part outside of a system can be thought of as its surroundings, or as another system. Thermal energy released from a system will transfer into its surroundings, and thermal energy released from the surroundings will transfer into the system. In the process of energy transfer between two systems, total energy is always conserved (see Fig. 18.3). [Middle school]
4. Thermal energy tends to spread from where it is concentrated to where it is less so. For example, if energy is concentrated in the system, it tends to spread into its surroundings. Otherwise, energy will transfer into the system. In this way it becomes maximally dispersed (see Fig. 18.4). [Middle school]
5. There can be barriers to prevent this energy spreading from happening rapidly (such as insulating material), but even with barriers slowing it down, thermal energy still tends to spread. [Middle school]
6. Forms of energy other than thermal energy also tend to spread from where they are concentrated to where they are less so. For example, in a chemical reaction, if the chemical energy stored in the system is more concentrated than

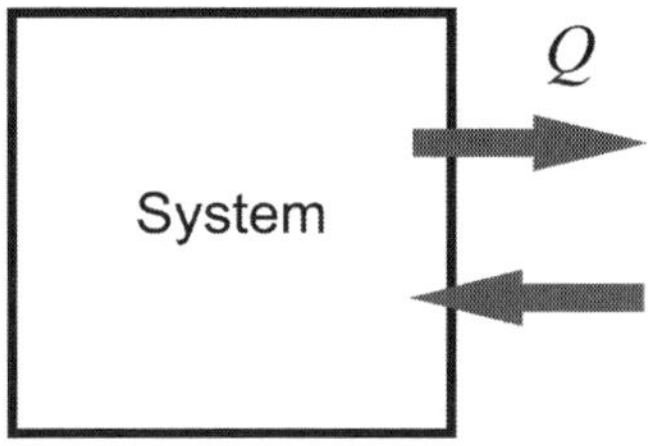

Fig. 18.2 A system can have thermal energy inputs and outputs. Q = thermal energy

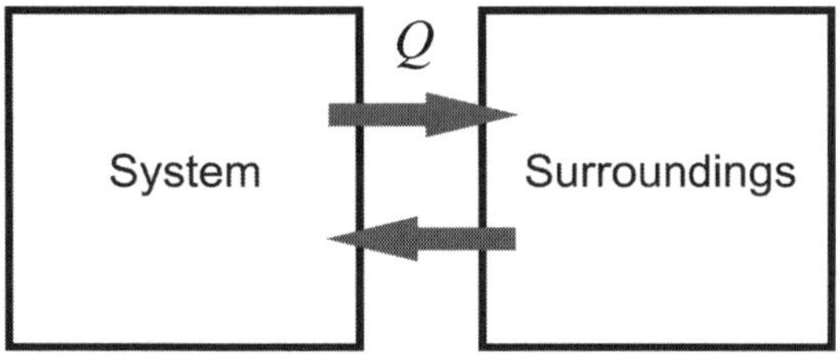

Fig. 18.3 Thermal energy can transfer between a system and its surroundings. Q = thermal energy

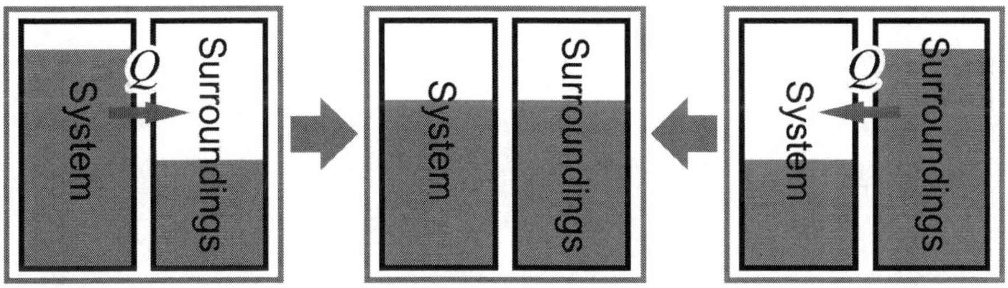

Fig. 18.4 Thermal energy tends to spread from where it is concentrated to where it is less so

its surroundings, the chemical energy will be transformed into thermal energy or other forms of energy to spread out of the system. Otherwise, other forms of energy from the surroundings will be transformed into chemical energy stored in the system. [High school]

7. Energy spreading is what determines whether or not simple events happen on their own (i.e., spontaneously). If energy is already dispersed as far as possible, events will not happen on their own. If an event causes energy to be less dispersed, it will not happen on its own. [High school]

Acknowledgments We would like to thank Jeff Nordine, Knut Neumann, Tobin Roger, and Hui Jin for their detailed reviews of the drafts of this chapter and their brilliant suggestions. Our thanks go to Frank L. Lambert for his correspondence discussing entropy with us. Thanks to Chi-Yan Tsui for his careful edit on the paper. The Chicago Transformation Teacher Institutes (NSF DRL 0928669) provided funding and support. We are also grateful to many other friends for kindly offering us their help.

References

American Association for the Advancement of Science. (1990). *Science for all Americans*. New York: Oxford University Press. Retrieved from, http://www.project2061.org/publications/sfaa/online/chap4.htm#26

American Association for the Advancement of Science. (1993). *Benchmarks for science literacy*. New York: Oxford University Press. Revised web version (2009) http://www.project2061.org/publications/bsl/online/index.php

Atkins, P. W. (2006). *Atkins' physical chemistry*. New York: Oxford University Press.

Bindel, T. H. (2004). Teaching entropy analysis in the first-year high school course and beyond. *Journal of Chemical Education, 81*, 1585–1594.

Boo, H. K. (1998). Students' understandings of chemical bonds and the energetics of chemical reactions. *Journal of Research in Science Teaching, 35*, 569–581.

Carson, J., & Watson, E. M. (2002). Undergraduate students' understandings of entropy and Gibbs free energy. *University Chemistry Education, 6*, 4–12.

Chen, S., Liu, Z., Bao, S., & Zheng, C. (2010). Natural convection and entropy generation in a vertically concentric annular space. *International Journal of Thermal Sciences, 49*, 2439–2452.

Cooper, M. M., Klymkowsky, M. W., & Becker, N. M. (2014). Energy in chemical systems: An integrated approach. In B. Chen, A. Eisenkraft, D. Fortus, J. Krajcik, K. Neumann, J. Nordine, & A. Scheff (Eds.), *Teaching and learning of energy in K – 12 education* (pp. 301–316). New York: Springer.

Dincer, I., & Cengel, Y. A. (2001). Energy, entropy and exergy concepts and their roles in thermal engineering. *Entropy, 3*, 116–149.

Donaldson, S. (2011). *Entropy is not disorder*. Retrieved from http://www.science20.com/train_thought/blog/entropy_not_disorder-75081

Duit, R. (1991). On the role of analogies and metaphors in learning science. *Science Education, 75*, 649–672.

Gao, P. L., & Wang, M. Z. (2007). *Huaxue fanying yuanli* [Principles of chemical reactions]. Jinan: Shandong Science and Technology Press.

Gary, R. K. (2004). The concentration dependence of the ΔS term in the Gibbs free energy function: Application to reversible reactions in biochemistry. *Journal of Chemical Education, 81*, 1599–1604.

Glynn, S. M., & Takahashi, T. (1998). Learning from analogy-enhanced science text. *Journal of Research in Science Teaching, 35*, 1129–1149.

Hanson, R. M., & Michalek, B. (2006). Give them money: The Boltzmann game, a classroom or laboratory activity modeling entropy changes and the distribution of energy in chemical systems. *Journal of Chemical Education, 83*, 581–588.

Kozliak, E. I., & Lambert, F. L. (2005). "Order-to-disorder" for entropy change? Consider the numbers! *The Chemical Educator, 10*, 24–25.

Kozliak, E. I., & Lambert, F. L. (2008). Residual entropy, the third law and latent heat. *Entropy, 10*, 274–284.

Lambert, F. L. (1999). Shuffled cards, messy desks, and disorderly dorm rooms – Examples of entropy increase? Nonsense! *Journal of Chemical Education, 76*, 1385–1387.

Lambert, F. L. (2002a). Disorder—A cracked crutch for supporting entropy discussions. *Journal of Chemical Education, 79*, 187–192. Revised web version http://www.entropysite.com/cracked_crutch.html

Lambert, F. L. (2002b). Entropy is simple, qualitatively. *Journal of Chemical Education, 79*, 1241–1246. Revised web version http://entropysite.oxy.edu/entropy_is_simple/index.html

Lambert, F. L. (2005). *Teaching entropy is simple—If you discard "disorder"*. Retrieved from, http://entropysite.oxy.edu/teaching_entropy.html

Lambert, F. L. (2006a). *The second law of thermodynamics*. Retrieved from, http://entropysite.oxy.edu/wiki_secondlaw.html

Lambert, F. L. (2006b). A modern view of entropy. *Khymia, The Bulgarian Journal of Chemistry, 15*, 13–21.

Lambert, F. L. (2007). Configurational entropy revisited. *Journal of Chemical Education, 84*, 1548–1550.

Lambert, F. L. (2011). *Entropy in general chemistry*. Retrieved from, http://entropysite.oxy.edu/wiki_entropy.html

Lambert, F. L., & Leff, H. S. (2009). The correlation of standard entropy with enthalpy supplied from 0 to 298.15 K. *Journal of Chemical Education, 86*(1), 94–98.

Leff, H. S. (1996). Thermodynamic entropy: The spreading and sharing of energy. *American Journal of Physics, 64*, 1261–1271.

Leff, H. S. (2007). Entropy, its language and interpretation. *Foundations of Physics, 37*, 1744–1766.

Leff, H. S. (2012a). Removing the mystery of entropy and thermodynamics – Part I. *The Physics Teacher, 50*(1), 28–31. Retrieved from, http://www.csupomona.edu/~hsleff/selpubs.html

Leff, H. S. (2012b). Removing the mystery of entropy and thermodynamics – Part II. *The Physics Teacher, 50*, 87–90. Retrieved from, http://www.csupomona.edu/~hsleff/selpubs.html

Leff, H. S. (2012c). Removing the mystery of entropy and thermodynamics – Part III. *The Physics Teacher, 50,* 170–172. Retrieved from, http://www.csupomona.edu/~hsleff/selpubs.html

Leff, H. S. (2012d). Removing the mystery of entropy and thermodynamics – Part IV. *The Physics Teacher, 50,* 215–217. Retrieved from, http://www.csupomona.edu/~hsleff/selpubs.html

Leff, H. S. (2012e). Removing the mystery of entropy and thermodynamics – Part V. *The Physics Teacher, 50*, 274–276. Retrieved from, http://www.csupomona.edu/~hsleff/selpubs.html

National Academy of Sciences. (1996). *National science education standards*. Washington, DC: The National Academy Press. Retrieved from, http://www.nap.edu/catalog/4962.html

National Research Council. (2012). *A framework for K-12 science education: Practices, crosscutting concepts, and core ideas*. Washington, DC: The National Academies Press. Retrieved from, http://www.nap.edu/catalog.php?record_id=13165

Reece, J. B., Urry, L., Cain, M., Wasserman, S., Minorsky, P., & Jackson, R. (2011). *Campbell biology* (9th ed.; International ed.). Harlow: Pearson Education.

Song, X. Q., & He, S. H. (2004). *Huaxue fanying yuanli* [Principles of chemical reactions]. Beijing: People's Education Press.

Sözbilir, M., & Bennett, J. M. (2007). A study of Turkish chemistry undergraduates' understanding of entropy. *Journal of Chemical Education, 84*, 1204–1208.

Styer, D. F. (2000). Insight into entropy. *American Journal of Physics, 68*, 1090–1096.

Wang, Z. H. (2007). *Huaxue fanying yuanli* [Principles of chemical reactions]. Hangzhou: Jiangsu Education Publishing House.

Wang, L., Wang, W., & Wei, R. (2014). What knowledge and ability should high school students have for understanding energy in chemical reactions? An analysis of chemistry curriculum standards in seven countries and regions. In B. Chen, A. Eisenkraft, D. Fortus, J. Krajcik, K. Neumann, J. Nordine, & A. Scheff (Eds.), *Teaching and learning of energy in K – 12 education* (pp. 87–102). New York: Springer.

Wilbraham, A., Staley, D., Matta, M., & Waterman, E. (2012). *Pearson chemistry*. Boston: Pearson Education.

Chapter 19
Constructing a Sustainable Foundation for Thinking and Learning About Energy in the Twenty-First Century

Lane Seeley, Stamatis Vokos, and Jim Minstrell

19.1 Introduction

Addressing the energy challenges of today and tomorrow will require energy experts in fields from municipal government to public health. These experts will need to draw from diverse, sophisticated, and nuanced understandings of energy in society that go far beyond static lists of energy facts. They will need to think and communicate using energy concepts that are rigorous, relevant, and fit known phenomena. Despite pervasive rhetoric (including in the Next Generation Science Standards) that energy is a unifying, crosscutting concept, historically energy instruction has been compartmentalized along disciplinary lines and appeared rigid. Students often associate the energy ideas they learn in school as a regimented program of taxonomy and bookkeeping. They understand their task as being to identify correctly forms and tabulate transfers and transformations. Students also learn a scientific concept of energy that is conserved; yet live in a world in which people are constantly 'using up' energy. We believe students can construct flexible, intuitive energy models that will empower them to make sense of phenomena, processes and resources that they care about in the real world by tracking energy transfers and transformations *locally* through detailed analysis of hypothesized mechanisms for such. We work with teachers to construct such models so that they can support similar energy engagement among their students.

L. Seeley (✉) • S. Vokos
Seattle Pacific University, 3307 Third Ave. W., Seattle, WA, USA
e-mail: seelel@spu.edu

J. Minstrell
Facet Innovations, Seattle, WA, USA

R.F. Chen et al. (eds.), *Teaching and Learning of Energy in K-12 Education*, DOI 10.1007/978-3-319-05017-1_19, © Springer International Publishing Switzerland 2014

19.2 The Energy Project at Seattle Pacific University

The Energy Project is a 5-year NSF-funded project with an overarching goal of increasing learner engagement with energy in K-12 classrooms. We work directly with elementary and secondary teachers to help them build their personal understanding and formative assessment practices in the context of energy. Our goals for energy learning include:

- Flexible application of the conservation principle and tracking of energy in 'real-world' processes
- Construction of a personally owned energy model that can be flexibly applied to novel scenarios.
- Application of energy models and representational strategies to socio-politically relevant energy questions.
- Recognition of the affordances and limitations of various energy representations.

Our progress toward some of these goals is reported elsewhere (Close et al. 2010, 2011; Close and Scherr 2011; Harrer et al. 2011; McKagan et al. 2011; Scherr et al. 2012a, b). In this chapter we describe how we have worked toward the preceding goals in workshops for K-12 teachers by:

- Providing representational strategies that recruit learner ideas about real situations, mandate energy tracking, encourage sense making and promote scientific questioning and reasoning
- Explicitly and implicitly reinforcing the idea that scientific language, representations, and classification strategies rest on a foundation of negotiated[1] understanding
- Scaffolding productive learner engagement with specific scenarios that foreground challenging aspects of the energy concept
- Supporting changes in learner engagement with energy concepts

We also share some preliminary research findings of significant changes in learner engagement with energy concepts. We conclude by discussing the critical, and in our minds unsolved challenge of developing a pedagogically accessible model for energy use, usefulness, and degradation that makes sense to learners and is widely applicable.

[1]By negotiation we intend a more stringent meaning than simple discussion. When a contract or an accord is negotiated, all signatory parties have the authority to hold up the negotiation until their concerns have been addressed. Therefore, negotiation within a learning community implies that each member is empowered to hold up the process until they feel that their questions or concerns have been addressed, or at least understood.

19.3 Providing Representational Strategies That Recruit Learner Ideas About Real Situations, Mandate Energy Tracking, Encourage Sense Making, and Promote Scientific Questioning

Energy is an inherently abstract concept. We don't see or touch or measure energy directly and the evidence for energy comes in a wide range of forms. Therefore, it is essential that learners construct meaningful representations of energy. Energy representations should be flexible enough that learners feel empowered to apply them to a wide range of scenarios and energy ideas. They must also be rigorous enough to problematize and refine learner thinking about energy. The Energy Project has promoted two dynamic energy representations, Energy Theater and Energy Cubes, which we find helpful for supporting constructive and creative thinking about energy. We have also encouraged learners to draw Energy Tracking Diagrams that capture dynamic energy processes in a static diagram. The dynamic energy representations provide a collaborative space in which learners can 'work out' the energy story associated with a specific physical scenario, while the static diagrams provide a medium for stabilizing their understanding of the dynamic energy story.

19.3.1 Energy Theater

Energy Theater is an activity that uses the body to represent the spatial and temporal evolution of "chunks" of energy in specially selected scenarios (Scherr et al. 2012a). Groups of 8–12 participants "become" chunks of energy and must "act out" the transfer(s) and/or conversion(s) of energy associated with specific scenarios. Energy Theater encourages learners to express their thinking about energy with their bodies and provides learners with a personal and bodily experience of energy conservation. The rules of Energy Theater are:

- Each person is a unit of energy
- Regions on the floor correspond to objects involved in the selected scenario
- Each person indicates his/her form of energy in some way (usually with a hand sign).
- People move from one region to another to represent energy transfer and change sign to represent energy conversion.
- The number of people in a region corresponds to the quantity of energy in a physical object.

Figure 19.1 shows a group of secondary science teachers who are using Energy Theater to represent the 'energy story' associated with a hand pushing a box across a floor at constant speed. The teachers on the right are representing chemical and motion/kinetic energy in the person/hand, the teachers at the left are representing motion and thermal energy in the box, and other teachers are leaving the box as thermal and sound energy into the floor and air.

Fig. 19.1 "Energy Theater" representation of a hand pushing a box across a floor at constant speed (Used with permission from Scherr et al. 2012a)

In Energy Theater every learner has an individual role to play so everyone has a vested interest in deciding on the essential energy transfers and conversions that are involved in the chosen scenario. We find that when we conduct this activity with teachers, a majority of the time is devoted to intense discussions about the underlying energy story and how the group can appropriately represent that story. For example, in the above scenario a person who will be transferring to the box as kinetic energy might be concerned that such a movement would cause the amount of kinetic energy in the box to increase even though the box is supposed to be moving at a constant speed. The group is then forced to negotiate a way to reconcile the dilemma. The instructor will intervene, when necessary, to help ensure that concerns are heard and considered by the entire group. Energy Theater also provides natural opportunities for formative assessment. Because learners represent energy processes in a public way, instructors and other learners have an opportunity to observe and respond to the reified ideas of others. We illustrate the nature of these conversations in a case study below.

Energy Theater foregrounds the questions of how, not whether, energy is conserved. Many secondary science students are familiar with the idea of energy conservation (Driver and Warrington 1985). They can recite the mantra that "energy is neither created nor destroyed." Therefore a central conceptual challenge in learning about energy is figuring out how energy is conserved in a wide array of dynamic physical processes. This involves answering questions like: Where does the energy start? Where does it go after that? What form does the energy take along the way? Energy Theater mandates energy conservation because learners—who represent units of energy—cannot spontaneously appear or disappear. The learners must individually and collectively decide where to begin, where and when to move, and what form to exhibit along the way. For example, when working through the energy story for a box pushed at constant speed, learners must decide where the energy comes from and they must show where the energy goes. They can't simply say that the energy goes "into friction" without locating and characterizing the form of the energy associated with that process.

We believe that effective scaffolding of Energy Theater and other energy tracking representations involves striking a balance between rigor and collective ownership. If the inherent rules of the representation are not attended to with sufficient rigor then the learners may avoid engagement with critical energy concepts. On the other hand, if the rules of the representation are too restrictive the learners may fail to recognize Energy Theater as a flexible resource for exploring energy processes. Therefore, it is important that instructors consider what aspects of the representation can be left open for discussion while retaining the rigor necessary in order to effectively problematize learner thinking about energy. To illustrate the importance of these strategic decisions we will present a case study of Energy Theater implementation in which the level of rigor was insufficient. We will then describe essential learner responsibilities for rigorous Energy Theater implementation as well as representational choices that can be left open to negotiation.

19.3.2 Case Study—Raising and Lowering a Ball at Constant Speed

This case study involves a group of students in a high school physics class who were using Energy Theater for the first time. They were acting out a sequence of scenarios consisting of lifting a ball at constant speed and then lowering the ball at constant speed.[2]

In the lifting scenario the students begin at the part of the room identified as the hand and are crouched down. They walk in single file into the center of the room where they spiral into a circle in the ball. As they walk into the circle they gradually rise up from a crouched walk until they are walking upright. Next the students act out the lowering of the ball at constant speed. Once again they begin in the hand and walk in single file into the ball. This time they gradually crouch down as they enter the ball.

The students display a surprising degree of confidence and appear to be representing these two physical scenarios effectively. They are clearly recognizing and differentiating the scenarios that they are representing and they have chosen regions of the room to represent the objects of interest. They appear to be working as a group and are coordinating their physical movements. In short, it is natural to be impressed with this group of Energy Theater novices. Upon closer inspection, however, there are crucial shortcomings in their adherence to the rules of Energy Theater. They are not clearly showing the type of energy that they represent. Further, it is unclear if they understand that they each represent a fixed quantity of energy.

[2]This pair of scenarios has been shown to be particularly challenging for many groups of learners, including physics faculty. In particular, most groups neglect the role of thermal energy conversion during the lifting stage. It is only after recognizing that energy cannot be conserved without thermal energy conversion during the lowering stage that they reconsider the lifting stage.

When they rise up/crouch down they may be representing an increase/decrease in the gravitational energy of the ball. If so, this is inconsistent with the rules of Energy Theater and, more importantly, eliminates the mandate that they show energy conservation. The students' insufficient adherence to the rules of Energy Theater is problematic because they avoid engagement with essential energy issues in these physical processes. Specifically, for the scenario in which the ball is being lowered at a constant speed the students do not engage with the question of where the energy goes and seem satisfied with a representation in which all of the participants start in the hand. Both of these characteristics of their representation suggest that they are not rigorously representing the energy story associated with this physical scenario.

Enforcing the basic rules of Energy Theater is a primary role of the instructor. After observing the student enactment of the lowering scenario the instructor might have asked the students, "When you crouch down are you still representing the same amount of energy?" The instructor can hold the students accountable to the rules of Energy Theater, which imply that each participant represents the same amount of energy throughout the representation. In doing so, the instructor is indirectly requiring students to devise an energy story that is consistent with the energy conservation principle. The instructor's role often involves asking students to clarify the form of energy that they are representing at all times. The instructor might have asked, "At the beginning of the scenario when you were all standing up in the hand, what form of energy were you representing?"

While it is critical that instructors help learners adhere rigorously to the rules of Energy Theater there are many representational choices that can be left open for the learning community to decide. Encouraging learners to decide about essential features of the energy story may help them recognize the representation as a flexible tool for constructing scientific understanding. It also may enhance learner feeling of ownership of the consensus ideas. For example, when analyzing the ball lowering scenario learners will sometimes discuss whether they should show energy that is transferred to the air as a result of air resistance. The instructor might ask them, "Do you think that air resistance has a significant effect on the motion of the bowling ball?" This question would provide learners with the opportunity to work toward recognizing that air resistance is not the primary reason that the ball speed does not increase. In addition, the question leaves the group to decide if they think that it is important to show the energy transfer associated with air resistance.

In a recent paper (Scherr et al. 2013) we describe teachers using Energy Theater to explore the energy processes associated with an incandescent light bulb that is providing constant illumination. In this case, Energy Theater supports learner efforts to distinguish between matter and energy. They are challenged to differentiate between electrons that flow around the circuit and the energy that those electrons carry from the wall outlet to the filament. We also describe how Energy Theater provides a shared representational space for learners to theorize mechanisms of energy transfer. In order to negotiate a sequence of energy steps for an incandescent light bulb, the teachers are compelled to decide whether the electrical energy is converted directly into light energy or if the filament glows because it is hot.

19.3.3 Energy Cubes

Energy Cubes is an activity in which learners use small cubes to represent "chunks" of energy. Designated regions on a whiteboard represent objects of interest. Groups of 3–5 learners move and turn over these cubes on the whiteboard to dynamically represent the transfer(s) and/or conversion(s) of energy associated with specific scenarios. For example, in Fig. 19.2, which depicts an Energy Cubes representation of a person lifting a box at constant speed, a learner might represent the

- energy conversion associated with the physiological effort of raising the hand by turning an energy cube in the hand from displaying a 'C' for chemical energy to showing a 'K' for kinetic energy
- mechanical transfer of energy from the hand to the box by moving an energy cube showing 'K' from the region depicting the hand to the region depicting the box
- increase in gravitational energy of the box by turning an energy cube in the box from showing 'K' to showing 'G' for gravitational energy
- mechanical transfer of energy from the box to the air by moving an energy cube showing 'K' from the region depicting the box to the region depicting the air
- dissipation of collective air motion by turning an energy cube in the air from showing 'K' to showing 'T' for thermal energy

The Energy Cubes representation provides some of the same pedagogical affordances as Energy Theater while allowing learners to work in smaller groups. Energy Cubes does not necessitate universal participation since participants can choose to let others handle the cubes. Learners can be challenged to coordinate their motions in order to show, for example, the constancy of the speed associated with the box lifting scenario. They would need to recognize that constant speed implies constant amounts of kinetic energy in the hand and box. Then, they would be challenged to choreograph their 'moves' so that the number of K's in both of these objects does not change. Additional energy scenarios for which we have used Energy Theater and Energy Cubes to engage and problematize learner thinking about energy are described below.

19.3.4 Energy Tracking Diagrams

Energy Tracking Diagrams are learner-invented representations of energy processes. Learners work in groups to draw a static representation (on a whiteboard; e.g. Fig. 19.3) of the dynamic representation they constructed using Energy Theater or Energy Cubes. In constructing these diagrams learners are challenged to show all the information that would be needed to recreate the dynamic representation in Energy Theater or Energy Cubes. This results in a diverse array of strategies for representing steps that have a complex distribution in space <u>and</u> time.

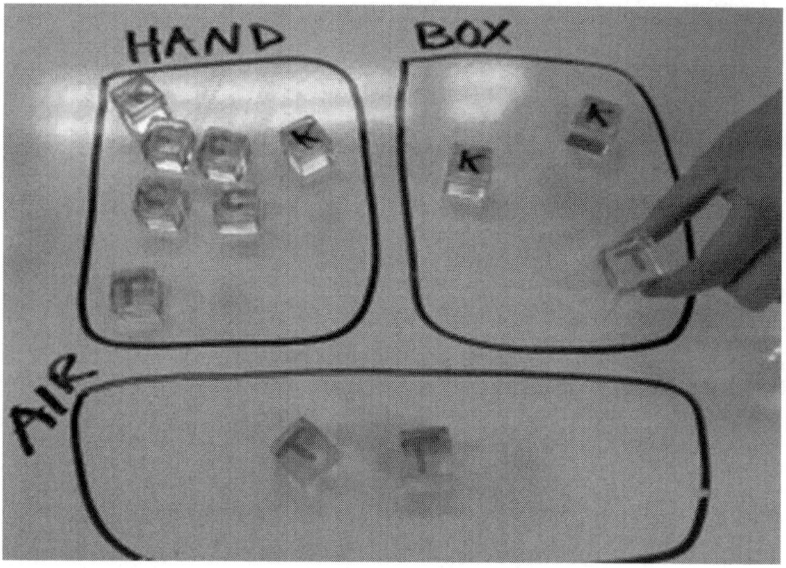

Fig. 19.2 "Energy Cubes" representation of a hand lifting a box vertically (Used with permission from Scherr et al. 2012a)

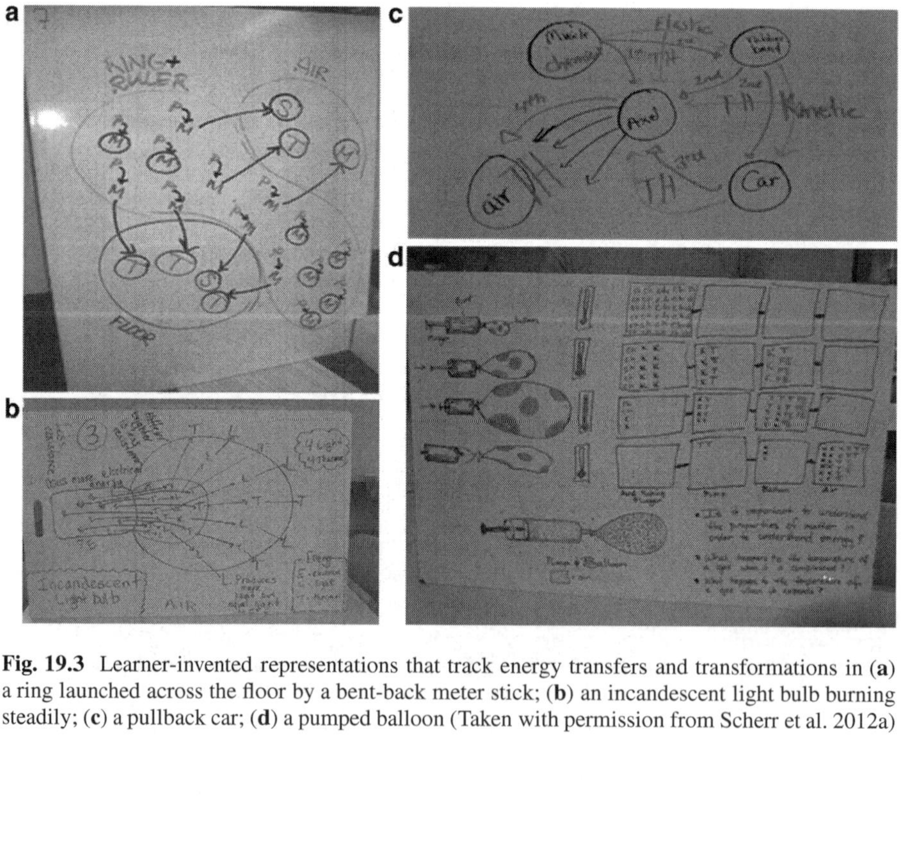

Fig. 19.3 Learner-invented representations that track energy transfers and transformations in (**a**) a ring launched across the floor by a bent-back meter stick; (**b**) an incandescent light bulb burning steadily; (**c**) a pullback car; (**d**) a pumped balloon (Taken with permission from Scherr et al. 2012a)

Scherr et al. (2012a) claim that "the variety in the diagrams' surface features is a testament to the learners' creativity and originality in producing the diagrams" (p. 6). Yet, the variety in the surface features of the diagrams does seem to present an instructional dilemma. How can an instructor or learner assess whether a diagram works or is sufficiently correct? We have adopted the approach of recruiting learners into the process of assessing their own work and the work of their peers. The utility and fidelity of an Energy Tracking Diagram can be evaluated based on features of the energy story that it coherently represents. Learners can ask general questions such as "how does the diagram show energy conservation?" They can also be guided to ask questions which are specific to the scenario of interest, such as, "If the light bulb has been on for a long time in steady operation then the temperature of the filament must be fairly constant. How is that feature represented in the diagram?" We have found that groups of teachers are generally willing to hold themselves accountable to construct diagrams that show the features of the energy story that they deem critical. In Sect. 19.4 we will show preliminary evidence that teachers who participate in our workshops become more likely to use diagrams constructively to track energy.

19.4 Explicitly and Implicitly Reinforcing the Idea That Scientific Language, Representations and Classification Strategies Rest on a Foundation of Negotiated Understanding

When learners work together to construct, evaluate and refine energy representations, as described in the preceding section they must actively question and critique their own work and the work of their peers. This will be possible within a community of discourse in which constructive discussion, questioning, and criticism are the mode rather than the exception (Brown and Campione 1994). Learners must be vulnerable in order to freely express their scientific ideas. Therefore, we front-load our workshops for teachers with an extended discussion of community rights and responsibilities. Throughout our workshops we continue to negotiate and re-negotiate the rights and responsibilities to which we all hold ourselves accountable.

Language lies at the heart of constructive scientific dialogue and language plays a particularly critical role in energy learning. Energy is a technical science word, but it is also a word that learners hear and use often outside of science class. As a consequence, learners bring many productive ideas about energy and they also use language about energy that they have acquired outside of the science classroom. In Energy Project workshops we have adapted an instructional approach to the regimentation of community discourse that was introduced by *The Algebra Project* (Moses and Cobb 2001). Close et al. (2010) have previously described the reasons for adopting the Algebra Project approach. "Through the Algebra Project we found an alternative instructional method that seeks less to direct the specific content of the learner's thinking and more to regiment the relationship between that thinking and its expression and communication through multiple representations." (p. 9)

The Algebra Project instructional approach foregrounds the distinction between *people talk* and *feature talk*. People talk is the speech we use in everyday situations. It is intuitive but often ambiguous and subjective. Feature talk is regimented and has a consensus meaning within a scientific community. Feature talk can take the form of language, and it can also take the form of regimented representational strategies. We challenge the participants to limit their use of feature talk that has not yet been negotiated within our learning community. In the process of negotiation every member of the community is encouraged to raise questions about the consensus understanding of a word's meaning. We focus on "idea first, name (for the idea) after" so that having the name does not cover lack of understanding of the idea. Initially it is the workshop instructors who challenge participants to explain the meaning behind their scientific language. The instructor may simply say, "I don't think we are all familiar with that term. Can you express your idea in more familiar language?" Eventually a classroom culture is established in which most participants are willing to demand a consensus understanding of newly introduced scientific terminology. Below are two specific examples that highlight the need to negotiate a shared understanding of energy language.

19.4.1 Potential Energy and the Potential to Have Energy

According to National Science Education Standards (National Academies 1996) *potential energy* is the energy "which depends on relative position." Many learners associate the phrase potential energy with the potential to have energy. For example, learners may claim that, "a meter stick has potential energy, (or potential elastic energy) because it can be bent." We have consistently found that some learners will still use the phrase *potential elastic energy* even after instructors have repeatedly referred to the energy as *elastic energy* or *elastic potential energy*. The phrase *potential elastic energy* is certainly more consistent with the idea of a potential to possess elastic energy. For related reasons learners may say that a bowling ball has potential energy or potential kinetic energy because it can be lifted up or because it can be rolled, etc. In our Energy Project workshops we foreground the difference between an intuitive use of *potential* as people talk and a regimented use of that word as feature talk. Participants discuss the difficulties that arise when some members of the learning community are using the word in a scientifically regimented way and others are understanding the word intuitively. This provides a motivation for negotiating language that is intuitive for everyone. For example, *elastic energy* is intuitively a type of energy, not a description of an objects elasticity or potential to have elastic energy.

19.4.2 Heat Energy, Thermal Energy and the Kinetic Energy of Particles

One reason for foregrounding and negotiating scientific language stems from the disparity between popular language and regimented scientific language. Disparities also exist between regimented scientific terms within different scientific communities. Kraus and Vokos (2011) did a scientific nomenclature study of energy concepts related to temperature in widely used college textbooks, pre-college curricula and various standards documents. They found a wide spectrum of terminology used to describe the energy contained by an object that is dependent on its temperature, including heat, heat energy, thermal energy, internal energy, average kinetic energy of the particles and translational part of the kinetic energy of the molecules. Kraus and Vokos suggest that teachers "qualify with the word 'energy' whatever terms they choose to use, as in 'the object contains heat energy' or 'there is heat energy transferred from the warmer object to the cooler object.'" They further recommend that teachers "begin first with the phenomena and observations, for which you want to build a scientific description. Next, as students begin to use new and different language to try to explain their observations, ask learners to qualify exactly what they are describing." (p. 7) Energy Project instructors discuss and attempt to model these recommendations in order that teachers can adopt them for their own teaching.

While we prioritize the negotiation of scientific language, we also recognize that this should not occur in isolation. All learners, and especially teachers, should be sensitive to the regimented scientific language of the broader scientific community as established through scientific articles, textbooks, published curricula, state and national standards. One might conclude that learners simply need to learn and adopt the regimented language of the broader scientific community. Unfortunately the regimented language itself is often not consistent from one scientific community to the next. This issue is particularly challenging in the case of energy language because energy is a central concept within all of the physical sciences. We hope that students will recognize energy as a *crosscutting* concept as they attempt to make sense of a system of scientific education that is characterized by artificial disciplinary divisions and parochial language. Furthermore, we hope that students will be empowered to apply their energy concepts to socio-politically relevant energy questions. This will require them to navigate energy language that is both inconsistent and is often manipulated to support a socio-political agenda. We would suggest that the best way to prepare our students for this challenge is help them become active and critical consumers of scientific language. Hopefully, when they hear about *sustainable* energy technologies or *efficient* appliances they will stop to consider the meaning behind the language.

19.5 Scaffolding Productive Learner Engagement with Specific Scenarios That Foreground Challenging Aspects of the Energy Concept

Many learners are familiar with the mantra that "energy is never created or destroyed" but lack the tools to make sense of this principle in everyday scenarios. We have chosen to use Energy Theater and Energy Cubes as primary strategies for representing energy scenarios. In doing so, we have chosen representations which mandate energy conservation. Since the people or cubes don't come into being or cease to exist, conservation of energy is required. Therefore, learners are not challenged to decide if energy is conserved but rather how energy is conserved. Thus, we begin by assuming "energy is conserved," a slogan with which most secondary students, and indeed the general public, are familiar. This challenge typically leads to two fundamental questions, where does the energy come from and where does the energy go? To unpack the meaning of "energy is conserved," we specifically choose to present physical scenarios for which these questions problematize learner thinking about energy. For example, the scenario involving lowering a bowling ball at constant speed was intentionally designed to raise the question of where the energy goes. Below we described another scenario that was designed as a context for learners to explore the question of where the energy comes from.

19.5.1 Rising Basketball in a Pool Scenario—Where Does the Energy Come from?

When considering a basketball floating upward from the bottom of a swimming pool, many learners readily identify several important aspects of the energy story associated with this physical scenario. The kinetic energy of the ball is increasing or leveling off as the ball moves upward. The gravitational energy of the ball is also increasing as the ball moves upward. In addition, many learners recognize that the thermal energy of the ball and water must also be increasing as the ball moves through the water. Tracking energy in this scenario leads naturally to the question of where all this energy is coming from. When thinking about the source of the energy most learners will recognize that buoyancy plays a central role. This connection then naturally leads to challenging questions. Is buoyancy a force or a type of energy? If buoyancy can be a type of energy, is it a new energy form or is it related to an existing energy form? These questions challenge learners to distinguish between force and energy and to consider the way in which energy forms should be categorized. Groups of teachers in our workshop typically recognize that buoyancy is more correctly described as an interaction between objects and, therefore, a force. They also will recognize that buoyancy does not seem to be a type of energy that is located in the ball. They might spontaneously, or after an instructor prompt, consider the change in location of the water as a result of rising ball. "Where does the that fills the space the ball leaves behind water come from?" In this way, they can recognize that, while there is additional energy associated with submersion of buoyant objects, the additional energy can logically relate to a form with which they are familiar, namely the gravitational energy of the water/Earth.

We typically scaffold learner engagement with a particularly challenging energy scenario by working through a series of related scenarios in which the energy story becomes progressively more challenging. Before learners analyze a basketball rising in a pool they might consider the energy story for an object falling through the air and then an object falling a similar distance through water. We are currently developing a set of web resources (www.energyprojectresources.org) in which we present a number of different physical scenarios that challenge learners to think carefully and creatively about various aspects of the energy concept. When well chosen, the selected scenarios open up critical issues in the content, while motivating learners to voice their ideas and to negotiate consensus understanding. These tend to be memorable for learners and become "benchmark" learning experiences to which subsequent experiences (in school or out) can be related and transferred (diSessa and Minstrell 1998).

19.6 Changes in Learner Engagement with Energy Concepts

In our workshop with teachers we have explicitly attempted to provide teachers with flexible tools for representing energy, build a culture of negotiated scientific language and present multiple scenarios which problematize the energy conservation principle. Two primary energy reasoning goals of our workshops are that:

- Teachers become more likely to rigorously attend to energy tracking when analyzing specific energy scenarios.
- Teachers become more likely to use diagrams constructively to track energy in specific scenarios.

We think these goals are also very relevant to all learners who need a model for engaging novel energy concepts that is both flexible and rigorous. By flexible we mean that the model can be applied in a wide range of energy scenarios and questions. By rigorous we mean that the model allows learners to rule out certain possibilities and refine their questions and ideas. In order to study teacher growth in these dimensions we have administered assessments before and after we work with them to develop representational tools and strategies for tracking energy. The following is an example question from one of these assessments (Fig. 19.4).

The ball lowering scenario was chosen based on the idea that learners who carefully attend to energy tracking will likely struggle with the question of where the energy goes. Gravitational energy is decreasing, chemical energy is presumably being "used up" and the kinetic energy is not changing. The idea that all of the lost gravitational energy and chemical energy could be transformed into thermal energy is counterintuitive for many learners as we will show below.

***Lowering a Bowling Ball** - A person carefully lowers a bowling ball from eye level to waist level. During this motion the bowling ball moves downward at a slow, constant speed.*

(a) Describe what is happening with energy during this process. If you aren't completely sure what is happening with energy, describe what you know and feel free to speculate when you are uncertain. Please feel free to include diagrams.

(As you go, write down questions that you ask yourself and need to answer in order to provide a reasonably complete description of the energy processes involved. Please write these questions in the box at the bottom of this page.)

Fig. 19.4 An item used in an Energy Project workshop for teachers to assess growth in engagement with energy questions

The pretest of this question was administered at the beginning of a 2-week workshop for secondary science teachers in the summer of 2012. The post-test was administered at the beginning of the second week of the workshop. During the intervening week participants had been introduced to Energy Theater, Energy Cubes, and Energy Tracking Diagrams. They had worked through several scenarios including a scenario involving raising a bowling ball at constant speed. We had not yet considered the lowering scenario as a part of class instruction. We wanted to offer teachers the option of drawing diagrams but not to imply that diagrams were required. A total of 22 teachers in our workshop completed both the pre and post-assessments.

19.6.1 Results—Attending to Energy Tracking

The question asked the participants to "describe what is happening with energy" as an effort to encourage energy tracking. Nevertheless, on the pretest, only 4 of 22 participants provided answers that demonstrated an effort to identify the ending form and location of the energy. Of these four, three cited that energy was transforming into thermal energy but did not clarify whether this increase of thermal energy was incidental or critical to the energy story. Only one participant articulated a concern over where the energy was going. She asked "Is kinetic energy increasing if it isn't accelerating?" Several participants cited work being done on the bowler but did not track the energy associated with that work to the bowler.

On the post-test, 18 of 22 participants explicitly focused on where and into what form the energy went in their response. Of these, five gave a clear answer that the energy was transformed into thermal and the remainder expressed their inability to figure out where the energy was going. The transition in the participants' inclination to track energy can be most clearly seen by following individual participants. One participant summed up her energy analysis in her pretest by writing,

> The energy ... must have been transferred to the bowler as he lowered the ball. Also, the energy was transferred from potential energy to kinetic energy while moving.

While she is clearly cognizant of energy forms and transfers she does not follow the energy when it is transferred to the bowler. One week later the same participant writes a lengthy inquiry into the energy process that includes an Energy Tracking Diagram. She circles the gravitational energy that is originally in the ball and remarks, "converted, but I don't know where or to what?"

She describes the increase in thermal energy in the air but also apparently decides that this increase in thermal energy cannot be sufficient to account for the energy decreases in her analysis. "I still have questions about gravitational energy units in the ball. I can't track them?"

Many other participants articulate an inability to account for where the energy goes. Another participant writes,

> If a ball is being lowered and decreasing the gravitational energy, where is that energy going if it is moving at a constant rate? Can't go back to chemical, so is it lost to the environment as thermal? Or does it become "stored"???? IDK! ☺

And another participant writes,

> In this case, the potential energy becomes....? Kinetic energy in the hands? But the hands don't speed up. Thermal energy? Certainly not all of it.... Maybe as it is converted into kinetic energy, it is then moved into the arms as elastic energy at a constant rate so there is only one K present in the ball at all times. The increasing elastic energy represents the effort of to hold the ball by muscle increasing over time. But is that force?

Even the participant whose pretest response most completely addressed the question of where the energy goes demonstrated an increase in their scientific questioning and efforts at sense making. On his pre-test he correctly identified that, "KE was turned into (thermal? elastic?) energy in the muscles." On the post-test, the question raised a more elaborate and refined set of questions for this participant.

> GPE <u>must</u> go somewhere -> into arm is only choice <u>but</u> KE of arm does not increase because arm speed is constant... Definitely <u>does not</u> get reclaimed in stored chem. PE in muscles (like a hybrid with regenerative braking...) How do arm muscles <u>receive</u> energy from an external source (<u>not</u> through digestion, ATP, etc...)? Go up, muscle PE to ball gravitational PE make some sense but going <u>down</u>, loss of GPE becomes ....? Don't know.

This participant's original response seems satisfactory to us and to the participant. They apparently recognize that the 'arm muscles <u>receive</u> energy' yet express uncertainty about how to describe or account for this accumulated energy in the muscles. Nonetheless, on their post-test they raise new questions about their analysis of the energy transfers and transformations. They articulate a reclaimed energy model and intuitively rule it out. They make a scientific comparison with the lifting scenario and apparently decide that while the motions are simply reversed the energy story cannot simply be reversed. If it could, then muscles would be acting like a car with regenerative braking. We infer that they are making use of the intuition that we cannot 're-charge' our muscles by lowering bowling balls.

Fig. 19.5 Examples of diagrams that are primarily used to illustrate an idea

19.6.2 Results—Using Diagrams as Reasoning Tools

We also observed an increase in both the prevalence of diagrams in participant responses and the apparent use of diagrams as reasoning tools when analyzing this scenario. On the pre-test only 5 of 22 participants included a diagram in their answer. Of these 5 diagrams, we classified three as being primarily used to illustrate an idea (Fig. 19.5).

A week later we see a dramatic increase both in the prevalence of diagrams and in the degree to which diagrams were used as tools for tracking energy. Sixteen out of twenty-two participants included diagrams in their analysis and of these, twelve were clearly using these diagrams as tools for tracking the energy in this scenario. Figures 19.6 and 19.7 show examples of two such diagrams.

We think that the complexity and evidence of progressive refinement in these diagrams suggests that they are being used constructively by these participants in their efforts to figure out what is happening with the energy in this scenario.

19.6.3 Summary of Preliminary Findings

In this preliminary study we saw a consistent increase in the degree to which participant responses raise ideas and questions about where the energy goes. We also observed an increase in the prevalence and constructive use of diagrams. There are a number of possible explanations for these changes:

- The in-class analysis of a similar scenario involving raising a bowling ball may have primed participants for engagement with this scenario.
- Participants may have become acculturated to the kinds of questions and representations that were more highly valued by the Energy Project instructors.
- Participants may have progressed in their ability and/or inclination to track energy.

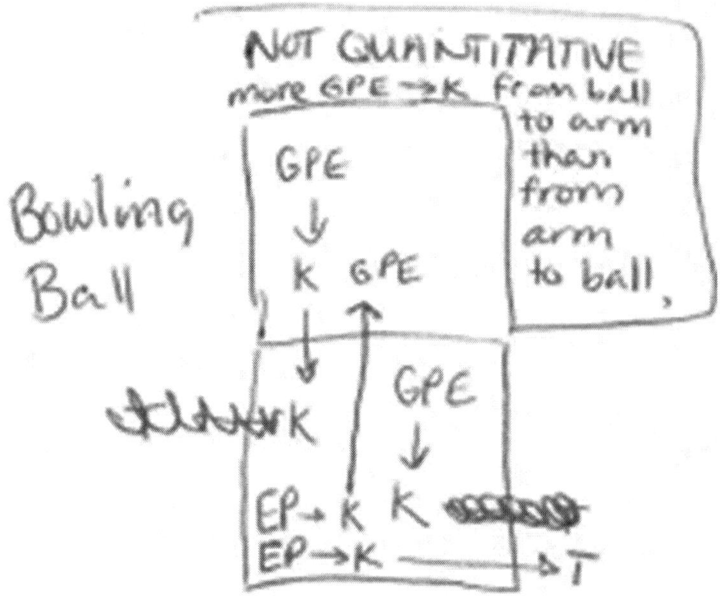

Fig. 19.6 An energy diagram that was used to arrive at a self-consistent analysis of where the energy went

- Participants may have progressed in their ability to use energy tracking diagrams constructively.

We suspect that all of these factors influenced the changes that we observed in participant responses. Nevertheless, this preliminary study demonstrates that the way in which participants document their analysis of a challenging energy scenario changed significantly as a result of participating in a single week of professional development. Furthermore, we believe this change corresponds to increased learner engagement with energy concepts.

19.7 Developing a Pedagogically Accessible Model for Energy Use, Usefulness and Degradation That Makes Sense to Precollege Students and Is Widely Applicable

Thus far, we have described an approach to learning about energy that prioritizes local energy tracking through the detailed analysis of mechanisms that cause energy transfers and transformations. Energy tracking serves to reinforce the principle of conservation of energy. However, in the real world, energy is understood as needing our stewardship for the purposes of its conservation. In the popular press, citizens

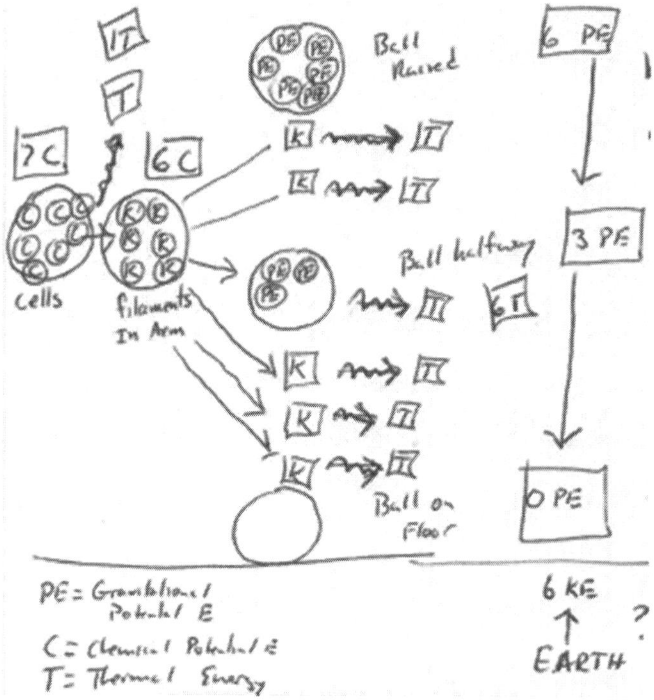

Fig. 19.7 An energy diagram that was successfully used by a participant to refine his questions about where the energy went

encounter a resource model in which energy is bought and sold, used and wasted, and can be conserved only through human efforts. If learners merely adopt a science classroom definition for energy conservation, which they cannot connect with their understanding of energy that can be used well or wasted, then they will be less likely to apply energy models from the science classroom to the energy issues that they care about. They will have no ownership of the ideas.

The challenge of constructing an accessible model that fully incorporates energy usefulness remains an unanswered question for us. This model must include the ways in which energy degrades but also integrate naturally with the conservation model. In addition to the 'standard' model of irreversibly increasing entropy, the literature suggests models for energy dissipation, which include energy spreading (Leff 2012) and entropy as freedom (Amin et al. 2012). It seems that an appropriate model may need to include both objective and subjective components. Consider the way in which a light bulb transforms electrical energy into thermal energy in a lighted room. There is an objective sense in which the energy is dissipated because there is no way to reverse this process and transform the thermal energy completely back into electrical energy. There is also a subjective sense in which the resulting thermal energy in the room is more useful if the occupant wants the room warmer and less useful if the occupant wants the room cooler. As one of the teachers in our workshops pointed out, "The heat from a light bulb isn't wasted if you are a chick in an incubator." We expect that a model for energy usefulness that can empower learners to address socio-politically challenging energy issues will integrate objective scientific principles with subjective normative priorities. We also hope to identify specific scenarios that catalyze learner engagement with concepts relating to energy usefulness (Daane et al. 2012). We expect that by supporting learners in the construction of models that incorporate energy usefulness, we will empower them to identify and engage with the energy questions that they care about. We are actively pursuing this goal.

19.8 Conclusions

We are working with teachers to build a model for energy that is precisely conserved while it is often degraded both objectively and subjectively. We hope to empower teachers to constructively engage with energy questions using flexible representational strategies within classroom learning communities that are characterized by negotiation, consensus building, and sense-making. In the preceding pages we have described instructional strategies which we have found to be effective in summer Energy Project workshops for teachers. In collaboration with Facet Innovations (www.diagnoser.com) we are also developing web resources to help teachers adapt Energy Project instructional strategies in their classrooms and to support formative assessment practices in the context of energy. We have found that this approach encourages teachers to represent, negotiate, and refine their energy understanding through engagement with conceptually challenging energy scenarios. We have shown preliminary evidence that teachers become more likely to rigorously attend to energy tracking and to use diagrams constructively to track energy in specific scenarios. We anticipate that through empowering teachers to constructively engage with their own energy questions we will also empower them to facilitate similar engagement on the part of their own students. We believe that these strategies can play a central role in preparing the next generation of global citizens to engage with the energy challenges of today and tomorrow.

References

Amin, T. G., Jeppsson, F., Haglund, J., & Strömdahl, H. (2012). Arrow of time: Metaphorical construals of entropy and the second law of thermodynamics. *Science Education, 96*(5), 818–848.

Brown, A. L., & Campione, J. C. (1994). Guided discovery in a community of learners. In K. McGilly (Ed.), *Classroom lessons: Integrating cognitive theory and classroom practice* (pp. 229–270). Cambridge, MA: MIT Press.

Close, H. G., & Scherr, R. E. (2011). *Differentiation of energy concepts through speech and gesture in interaction* (pp. 151–154). Physics Education Research Conference Proceedings, Omaha, NE.

Close, H. G., DeWater, L. S., Close, E. W., Scherr, R. E., & McKagan, S. B. (2010). *Using the Algebra Project method to regiment discourse in an energy course for teachers*. Physics Education Research Conference Proceedings, Portland, OR, 9–12.

Close, E. W., Scherr, R. E., Close, H. G., & McKagan, S. B. (2011). *Development of proximal formative assessment skills in video-based teacher professional development* (pp. 19–22). *Physics Education Research Conference Proceedings*, Omaha, NE.

Daane, A. R., Vokos, S., & Scherr, R. E. (2012). *Conserving energy in physics and society: Creating an integrated model of energy and the second law of thermodynamics* (pp. 114–117). Physics Education Research Conference Proceedings, Philadelphia, PA.

diSessa, A. A., & Minstrell, J. A. (1998). Cultivating conceptual change with benchmark lessons. In J. Greeno & S. Goldman (Eds.), *Thinking practices in mathematics and science learning*. Mahwah: Lawrence Erlbaum Associates.

Driver, R., & Warrington, L. (1985). Students' use of the principle of energy conservation in problem situations. *Physics Education, 20*(4), 171–176.

Harrer, B. W., Scherr, R. E., Wittmann, M. C., Close, H. G., & Frank, B. W. (2011). *Elements of formative assessment in learners' discourse about energy*. Paper presented at the 2011 Physics Education Research conference, Omaha, NE.

Kraus, P. A., & Vokos, S. (2011). The role of language in the teaching of energy: The case of 'heat energy'. *Washington State Teachers' Association Journal.*

Leff, H. S. (2012). Removing the mystery of entropy and thermodynamics—Part I. *Physics Teacher, 50*(1), 28–31.

McKagan, S. B., Scherr, R. E., Close, E. W., & Close, H. G. (2011). *Criteria for creating and categorizing forms of energy*. Physics Education Research conference proceedings, Omaha, NE, 279–282.

Moses, R. P., & Cobb, C. E. (2001). *Radical equations: Civil rights from Mississippi to the algebra project*. Boston, MA: Beacon.

National Academies. (1996). *National science education standards* (p. 180). Washington, DC: The National Academies Press.

Scherr, R. E., Close, H. G., Close, E. W., & Vokos, S. (2012a). Representing energy. II. Energy tracking representations. *Physical Review—Special Topics: Physics Education Research, 8*(8)020114, 1–11.

Scherr, R. E., Close, H. G., McKagan, S. B., & Vokos, S. (2012b). Representing energy. I. Representing a substance ontology for energy. *Physical Review—Special Topics: Physics Education Research, 8*(8)020115, 1–11.

Scherr, R. E., Close, H. G., Close, E. W., Flood, V. J., McKagan, S., Robertson, A. D., Seeley, L., Wittmann, M. C., & Vokos, S. (2013). Negotiating energy dynamics through embodied action in a materially structured environment. *Physical Review—Special Topics: Physics Education Research, 9*(9)020105, 1–18.

Chapter 20
Conclusion and Summary Comments: Teaching Energy and Associated Research Efforts

Joseph Krajcik, Robert F. Chen, Arthur Eisenkraft, David Fortus, Knut Neumann, Jeffrey Nordine, and Allison Scheff

20.1 Introduction

The Energy Summit and the chapters in this book started with the premise that energy is both a critical disciplinary idea as well as a crosscutting concept, as elaborated in the Framework for K-12 Science Education (National Research

J. Krajcik (✉)
College of Education, Michigan State University, East Lansing, MI, USA
e-mail: krajcik@msu.edu

R.F. Chen
Environmental, Earth and Ocean Sciences, University of Massachusetts Boston, Boston, MA, USA
e-mail: Bob.Chen@umb.edu

A. Eisenkraft
Education and Human Development, University of Massachusetts Boston, Boston, MA, USA
e-mail: arthur.eisenkraft@umb.edu

D. Fortus
Weizmann Institute of Science, Rehovot, Israel
e-mail: David.fortus@weizmann.ac.il

K. Neumann
Leibniz-Institute for Science and Mathematics Education (IPN), Kiel, Germany
e-mail: neumann@ipn.uni-kiel.de

J. Nordine
San Antonio Children's Museum, San Antonio, TX, USA

Leibniz-Institute for Science and Mathematics Education (IPN), Kiel, Germany
e-mail: jeffn@sakids.org; nordine@ipn.uni-kiel.de

A. Scheff
Massachusetts Department of Higher Education, Boston, MA, USA
e-mail: ascheff@bhe.mass.edu

R.F. Chen et al. (eds.), *Teaching and Learning of Energy in K-12 Education*, DOI 10.1007/978-3-319-05017-1_20,

Council 2012). Energy serves a central role in our everyday lives, as well as in all science disciplines. We were influenced by the argument presented in Framework for K-12 Science Education that energy is a critical concept that cuts across the disciplines and as such all learners need a solid understanding of this idea. However, the general population and many professionals, including K-12 science teachers, many science graduate students and scientists, lack a solid understanding of energy across all disciplines. Many of the challenges learners face in understanding the energy concept result not only because energy is a challenging concept but also because energy is seldom taught as a unifying idea; it is more likely taught using different language in different disciplines. For example, most learners never develop a rich conceptual understanding of what is meant by "energy is stored in chemical bonds." This problematic situation most likely arises because there are substantive differences in how the energy concept is used across disciplines that result from shorthand usage of language. Although many scientists can translate between the various shorthand ways of using energy, this language is never clearly explained to students and practitioners, including teachers and curriculum developers. In fact, many graduate students do not fully understand the idea of energy. This has led to many misunderstandings of energy including "energy being stored in chemical bonds" as meaning "energy is released when bonds break." As such, throughout the globe, we face challenges in teaching the energy concept, both because energy is such a challenging, misunderstood concept and different language is used to express different manifestations of it.

One of the advantages of treating energy as a crosscutting concept is that learners can develop connections among the various ideas and in various contexts. Research has demonstrated (Bransford et al. 2000) that when students make connections among ideas in multiple contexts they are better able to use that understanding to solve problems, explain phenomena and learn more. We refer to this type of understanding as integrated understanding (Fortus and Krajcik 2011). It is this integrated understanding of the energy concept that we hope students will develop.

Our goal in holding the Energy Summit – as well as producing this book – grew out of an effort to bring direction to the teaching of energy and its associated research efforts with hope of developing future directions on how to support learners in developing integrated understanding of the energy concept. In this chapter we summarize the major findings from our work at the energy summit and discuss research steps forward.

20.2 Summary Majors Findings

The chapters in this book show that energy is a complex concept and a difficult idea to teach across the grades. Yet to develop a rich understanding of energy, energy must be taught across the grades and in multiple contexts. The various points listed below expand on these ideas.

- A fully coherent understanding of energy can only be obtained when considering energy at the nanoscopic level, i.e., the energy of motion and position associated with molecules, atoms, and their interactions. Unfortunately, many students never develop this depth of understanding because of current instructional approaches. We need to learn more about how to support students in explaining phenomena by using energy at the nanoscopic level.

- In considering how to develop student understanding of energy across time, several key ideas about energy play a central role: (i) energy forms, (ii) energy transfer, (iii) energy transformation, (iv) energy dissipation, and (v) energy conservation. However, we need to learn more about how to introduce these ideas and apply these ideas at various grade levels and in all the disciplines.
- The contemporary use of the word energy has many connotations and different meanings, particularly in everyday contexts. Many of these uses support learners in developing misunderstandings of the ideas regarding energy, particularly that energy is a "something" and that it is stored in chemicals. As such, researchers and educators should help students bridge the discourse between everyday and scientific language.
- The teaching of energy can be used to connect the four major areas of science and engineering, helping learners see the relationships among other science ideas. Additionally, the teaching of energy should span to other subject areas, such as history and economics. By connecting energy to four major areas of science, engineering and other disciplines, students are more likely to develop integrated understanding that will allow them to use the energy construct across contexts.
- Students' prior understandings and teaching of energy at earlier ages significantly impacts the learning of more complex energy ideas. However, we need to learn more about what key experiences and ideas at early ages promote student understanding of more complex ideas. Teaching energy ideas at early grades is not about teaching the same ideas but in an easier manner; it is figuring out what ideas are essential to understanding energy and if students at a particularly grade level can understand those ideas with appropriate instruction. For instance, what ideas might be precursor ideas to students developing the idea of energy transfer?
- Many graduate students and other professionals, including K-12 teachers, have conceptual difficulties that originate from inadequate K-12 and undergraduate instruction on energy. This statement certainly calls for more appropriate professional development throughout the K-12 system. However, this statement also entails that we need to improve the way energy is taught in undergraduate science courses responsible for preparing future teachers.
- While large-scale investigations can provide valuable findings about students' general progression in understanding energy, in-depth analysis of curricula and their effect on students learning are required to add to the current research base. What we need are fine grained studies to determine what phenomena and instructional approaches can help students move from one level to the next level in a learning progression.
- The literature demonstrates that promising approaches to K-12 energy instruction help students bridge between everyday and scientific notions of energy. Unfortunately, the field has not been very good about communicating these ideas, and these ideas have not be readily picked up by commercial publishers. Furthermore,

there is little effort in the media to distinguish between the scientific use of the term 'energy' and the everyday usage.

- K-12 students need to learn a concrete and broadly applicable approach to the idea of energy that serves as conceptual tools that are useful for reasoning about the role of energy in a variety of contexts and can be used to solve problems, explain phenomena and learn more. Students in K-12 need to see the concept of energy as a conceptual tool that they can use when confronted with a problem or explain a phenomenon.
- In order to support students in reasoning about energy in new and meaningful ways, teachers and instructors of introductory college science courses need professional development that help them learn, use, and adapt new conceptual representations of energy.
- College students hear about the concept of energy in each discipline, but few references link the energy content ideas in chemistry with those in physics, biology, earth science, or engineering. Discipline-based content courses must encourage students to see these connections. For that to happen, college professors must work across disciplines to better understand how energy is used in different content areas and how the energy concept is introduced in the introductory and advanced courses. They must then find ways to coordinate their instruction to support understanding of energy as a crosscutting concept in each of their course assignments.

20.3 Future Research Directions

Because energy is such an important idea we need systematic research on how to support students' learning of energy from the elementary grades through college. We need to learn more about which are the initial ideas, contexts and phenomena that young children can learn that can be used to build a more sophisticated notion of energy. For instance, perhaps helping young children learn that objects in motion can cause something to happen and what happens depends on the speed and mass of the object might serve as an important precursor to understanding kinetic energy. As such, we need to know how to promote more sophisticated ideas in students. What learning experiences and at what grade levels help support students to reach the next level? Typically what happens in school is that an idea is presented at one grade level and the same idea is presented at a higher grade level without building on the previous idea. Often the idea is present at the same level of complexity. A classic example is that matter is anything that has mass and occupies volume. This type a revisiting of ideas does not support richer and more sophisticated development. We need to learn what ideas will promote the next level of understanding of an idea. Often the idea might not be related to energy, but it is an essential concept. For instance, students cannot track energy unless they know how to count and know that parts can make a whole. Below we summarize the major research opportunities that confront the field.

- Further research is required to learn novel instructional approaches to teach about energy in elementary, middle and high school, and to determine if these approaches have a positive effect on the construction of an integrated understanding of energy. We need to learn more about how students develop understanding of the various key ideas about energy, as for example, energy transfers and transformations. We also need to learn what are the precursor ideas to energy transfer and transformations.
- Attainable and meaningful levels of understanding about energy for different stages in the kindergarten thru postsecondary continuum need to be determined. In particular, we need to learn more about how to support learners in developing more sophisticated understandings of energy that cut across the disciplines and determine what prohibits learners from reaching these more advanced levels. We also need to learn more about how students' everyday conceptions influence their progression in developing an integrated understanding about energy. Research studies are needed to describe how students should be able to use energy concepts at the end of elementary, middle, and high school.
- Comprehensive synthesis of existing research on what levels of understanding students can develop and how those levels of understanding were obtained needs to occur. While the research that does exist is not extensive, the field would still benefit from a synthesis. These synthesis studies are critical as we need to build from what we already know.
- More systematic studies examining the design of innovative curricula that build understanding across time and are based on what is known about teaching and learning should be intensified. These studies in particular should include research on the effect of different approaches in teaching about the key ideas of energy like, for example, how energy should be introduced. These studies can help us learn more about how the various contextual factors, such as different energy curricula, play in how students progress in understanding the key ideas about energy.
- While curricular studies are taking place, we need to develop reliable and valid measures of assessing student understanding where students are applying the energy idea across multiple contexts to explain phenomena or solve problems. These assessments are crucial to studying the curricular interventions.
- More research should be done to articulate a learning progression for energy. For this learning progression to be useful, it needs to encompass all relevant disciplinary areas of science, as well as other areas such as history, culture, economics, and technology. While challenging, this learning progression will provide a common framework for all in education to follow. With the existence of a common learning progression about energy, we can teach and learn about energy efficiently and effectively. Yet, challenges remain. Developing a common learning progression for energy will take interdisciplinary teams and perhaps will span generations. This type of collaborative research is not seen often in educational research and seldom in the hard sciences, yet the benefits and this interdisciplinary and generational collaboration could result in significant changes to education. While cross-sectional approaches will help us learn some

about how students' ideas develop across time under particular conditions, more longitudinal studies need to occur. While these longitudinal studies are challenging and resource-intensive, they will give us the most insightful findings with respect to how best to support student learning across years.

- Without the existence of a common learning progression about energy, how can we teach and learn about energy efficiently and effectively? To overcome these challenges, research-based strategies should be promoted and shared within the science education community to help educators teach energy efficiently and effectively.
- More research needs to be done on how language affects the teaching and learning about the integrated understanding of energy. We know now that many students and citizens are confused by the multiple uses of the term energy.
- More research is needed to identify key phenomena for teaching energy as a crosscutting concept and the conceptual tools that help students to reason about these phenomena. In particularly, we need to know what phenomena and instructional aspects can help students move from one level of the learning progression to the next.
- Studies into students' ideas about energy and their ability to use it to make sense of phenomena should be connected to curricular interventions that incorporate a defined set of conceptual tools. While students' difficulties in using the energy concept are relatively well-documented, much more should be done to understand the efficacy of particular instructional approaches and their impact on subsequent student learning.
- Research studies are also needed to plan, design, implement and study various professional development experiences for teachers of K-12 systems and instructors of introductory college courses. We need to find answers to questions such as what types of experiences can support teachers and instructors at different grade bands learn about energy as a crosscutting concept. Curriculum materials also need to be designed to be educative in nature, supporting teacher learning of the energy idea.

20.4 Concluding Thought

The research recommendations we suggest point to a multi-year and multi-disciplinary effort, one that will require educational researchers and scientists from nations throughout the globe if we hope to be successful. Identifying teaching methods based on empirical research is a first step forward. But most importantly we cannot get caught in the trap and just speak to those in one discipline. As such it is important to engage scientists across various disciplines to agree on a common language with respect to teaching energy and to improve instruction at the undergraduate level so that learners see connections with use of energy among the disciplines. There has been some good research published (Nordine et al. 2011) that illustrates that students can learn sophisticated ideas about energy when appropriate curriculum materials and instructional approaches are used, but the impact of this research has not had a great impact on the teaching of energy, perhaps because of siloed nature of the disciplines and siloed nature in which science is taught in K-12 education. The effort we envision is one in which a K-12 and possibly a K-14 learning progression is developed that includes phenomena and teaching nuggets that can push students to the next level of sophistication as well as the assessment instruments needed to measure students' abilities to use the energy concepts in various contexts. Such a learning progression can guide the development of full curriculum and instructional strategies.

CNN (Hennen et al. 2013) reported that the United Nations released a document that stated that they are 95 % confident that 50 % of global warming has been contributed by human activity. More than 800 authors and 50 editors from countries around the globe took part in development of the report. The report clearly points to humans as being responsible for at least half of the global increase in temperature since 1950. For future generations, this is not a pretty scenario as increased temperatures will result in more severe weather and ocean rising thereby destroying habitats. While it is clear that we have failed our children in controlling global warming, we can't fail in educating them about energy as a crosscutting concept; they will need to understand this idea in the hope of reverting this climatic effect for future generations.

Our hope is that the ideas in this publication will spur a similar international effort to understand how people can develop an integrated understanding of energy as a crosscutting concept and, as a result, allow learners to solve problems, explain phenomena and learn more when needed.

References

Bransford, J. D., Brown, A. L., & Cocking, R. R. (2000). *How people learn: Brain, mind, experience and school*. Washington, DC: National Academy Press.

Fortus, D., & Krajcik, J. (2011). Curriculum coherence and learning progressions. In B. J. Fraser, K. G. Tobin, & C. J. McRobbie (Eds.), *The international handbook of research in science education* (2nd ed.). Dordrecht: Springer.

Hennen, D., Miller, B., & McLaughlin, E. C. (2013) CNN, September 28th, 2013.

National Research Council. (2012). *A framework for K – 12 science education: Practices, crosscutting concepts, and core ideas*. Washington, DC: National Academies Press.

Nordine, J., Krajcik, J., & Fortus, D. (2011). Transforming energy instruction in middle school to support integrated understanding and future learning. *Science Education, 95*, 670–699.

Appendices

Appendix A: Agenda

Saturday, December 15, 2012

9:00 am to 10:30 am Welcome and Introductions

11:00 am to 12:30 pm First Paper Session

Group A

Opportunities for Reasoning about Energy within Elementary School Engineering Experiences
Kristen Wendell
Critical Friend: Ramon Lopez
Additional Discussant: Nicos Papadouris
Readers: Nicole Becker, Sungyoun Choi, Melanie Cooper, arthur.eisenkraft@umb.edu, David Fortus, Cari Herrmann-Abell, Tom Kim, Sara Lacy, Ramon Lopez, Ann Novak, Sebastian Opitz, Nicos Papadouris, Roger Tobin, Lei Wang, Kristen Wendell

Energy in Chemical Systems
Nicole M. Becker, Michael W. Klymkowsky, and Melanie M. Cooper
Critical Friend: Sara Lacy
Additional Discussant: Cari Herrmann-Abell
Readers: Nicole Becker, Sungyoun Choi, Melanie Cooper, arthur.eisenkraft@umb.edu, David Fortus, Bob Geier, Cari Herrmann-Abell, Tom Kim, Sara Lacy, Ramon Lopez, Ann Novak, Sebastian Opitz, Nicos Papadouris, Roger Tobin, Lei Wang, Kristen Wendell

R.F. Chen et al. (eds.), *Teaching and Learning of Energy in K-12 Education*,
DOI 10.1007/978-3-319-05017-1,

Group B

A physicists musing on teaching about energy
Helen Quinn
Critical Friend: Xiufeng Liu
Additional Discussant: Allison Scheff
Readers: Reinders Duit, Hui Jin, Joe Krajcik, Jane Lee, yarlehavi@gmail.com, Xiufeng Liu, Alycia Meriweather, Kongju Mun, Knut Neumann, ppelletier@boston.k12.ma.us, Helen Quinn, Wei Rui, Allison Scheff, Angelica Stacy, Xin Wei

Teaching and Learning the Physics Energy Concept
Reinders Duit
Critical Friend: Hui Jin
Additional Discussant: Angelica Stacy
Readers: Reinders Duit, Hui Jin, Joe Krajcik, Jane Lee, yarlehavi@gmail.com, Xiufeng Liu, Alycia Meriweather, Kongju Mun, Knut Neumann, ppelletier@boston.k12.ma.us, Helen Quinn, Wei Rui, Allison Scheff, Angelica Stacy, Xin Wei

Group C

Repairing Engineering Students' Misconceptions about Energy and Thermodynamics
Margot Vigeant, Michael Prince, Katharyn Nottis, and Ronald Miller
Critical Friend: George DeBoer
Readers: Charles W. (Andy) Anderson, Bob Chen, Costas Constantinou, Jenny Dauer, George DeBoer, Orna Fallik, Sarah Jardeleza, Robin Millar, Jeff Nordine, Mihwa Park, Lane Seeley, Sonia Underwood, Margot Vigeant, Holger Wendlandt

The Energy Project: Constructing a sustainable foundation for thinking and learning about energy in the twenty-first century
Lane Seeley, Stamatis Vokos, and Jim Minstrell
Critical Friend: Costas Constantinou
Additional Discussant: Charles W. (Andy) Anderson
Readers: Charles W. (Andy) Anderson, Bob Chen, Costas Constantinou, Jenny Dauer, George DeBoer, Orna Fallik, Sarah Jardeleza, Robin Millar, Jeff Nordine, Mihwa Park, Lane Seeley, Sonia Underwood, Margot Vigeant, Holger Wendlandt

12:30 pm Lunch

1:30 pm to 3:00 pm First Report-Out and Discussion

Group A

Participants: Nicole Becker, Costas Constantinou, George DeBoer, Sarah Jardeleza, Hui Jin, Tom Kim, Joe Krajcik, yarlehavi@gmail.com, Jeff Nordine, Ann Novak, Nicos Papadouris, Mihwa Park, ppelletier@boston.k12.ma.us, Helen Quinn, Lei Wang

Group B

Participants: Bob Chen, Sungyoun Choi, Jenny Dauer, Reinders Duit, Cari Herrmann-Abell, Ramon Lopez, Alycia Meriweather, Robin Millar, Hannah Miller, Kongju Mun, Knut Neumann, Sebastian Opitz, Allison Scheff, Lane Seeley, Roger Tobin, Sonia Underwood

Group C

Participants: Charles W. (Andy) Anderson, Melanie Cooper, arthur.eisen kraft@umb.edu, Orna Fallik, David Fortus, Sara Lacy, Xiufeng Liu, Wei Rui, Angelica Stacy, Margot Vigeant, Xin Wei, Kristen Wendell, Holger Wendlandt

3:30 pm to 5:00 pm Second Paper Session

Group A

Looking Through the Energy Lens: A Proposed Learning Progression for Energy in Grades 3-5
Sara Lacy, Roger Tobin, Marianne Wiser, and Sally Crissman
Critical Friend: Lane Seeley
Additional Discussant: Melanie Cooper
Readers: Nicole Becker, Melanie Cooper, Jenny Dauer, Reinders Duit, Hui Jin, Sara Lacy, Jane Lee, Alycia Meriweather, Knut Neumann, Jeff Nordine, Sebastian Opitz, Mihwa Park, Lane Seeley, Margot Vigeant, Lei Wang

A Tale of Two Energies: What Do History, Language, and Research Tell Us about Students' Understanding of Energy?
Hui Jin, Xin Wei
Critical Friend: Reinders Duit
Additional Discussant: Margot Vigeant

Readers: Nicole Becker, Melanie Cooper, Jenny Dauer, Reinders Duit, Hui Jin, Sara Lacy, Jane Lee, Alycia Meriweather, Knut Neumann, Jeff Nordine, Sebastian Opitz, Mihwa Park, Lane Seeley, Margot Vigeant, Lei Wang

Group B

Developing and Using Distractor-Driven Multiple-Choice Assessments Aligned to Ideas about Energy Forms, Transformation, Transfer, and Conservation
Cari F. Herrmann-Abell and George E. DeBoer
Critical Friend: Angelica Stacy
Additional Discussant: Costas Constantinou
Readers: Bob Chen, Costas Constantinou, arthur.eisenkraft@umb.edu, Orna Fallik, Bob Geier, Cari Herrmann-Abell, Tom Kim, Robin Millar, Kongju Mun, Ann Novak, Helen Quinn, Wei Rui, Angelica Stacy, Sonia Underwood, Xin Wei, Kristen Wendell

Towards a research-informed teaching sequence for energy
Robin Millar
Critical Friend: Kristen Wendell
Additional Discussant: Helen Quinn
Readers: Bob Chen, Costas Constantinou, arthur.eisenkraft@umb.edu, Orna Fallik, Bob Geier, Cari Herrmann-Abell, Tom Kim, Robin Millar, Kongju Mun, Ann Novak, Helen Quinn, Wei Rui, Angelica Stacy, Sonia Underwood, Xin Wei, Kristen Wendell

Group C

Contextual Dimensions of the Energy Concept and Implications for Energy Teaching and Learning
Xiufeng Liu, Mihwa Park
Critical Friend: Charles W. (Andy) Anderson
Additional Discussant: Ramon Lopez
Readers: Charles W. (Andy) Anderson, Sungyoun Choi, George DeBoer, David Fortus, Sarah Jardeleza, Joe Krajcik, yarlehavi@gmail.com, Xiufeng Liu, Ramon Lopez, Nicos Papadouris, ppelletier@boston.k12.ma.us, Allison Scheff, Roger Tobin, Holger Wendlandt

Distinctive features and underlying rationale of a philosophically-informed approach for energy teaching
Nicos Papadouris & Constantinos P. Constantinou
Critical Friend: George DeBoer
Additional Discussant: Roger Tobin

Readers: Charles W. (Andy) Anderson, Sungyoun Choi, George DeBoer, David Fortus, Sarah Jardeleza, Joe Krajcik, yarlehavi@gmail.com, Xiufeng Liu, Ramon Lopez, Nicos Papadouris, ppelletier@boston.k12.ma.us, Allison Scheff, Roger Tobin, Holger Wendlandt

Sunday, December 16, 2012

9:00 am to 10:30 am Second Report-Out and Discussion

Group A

Participants: George DeBoer, arthur.eisenkraft@umb.edu, Joe Krajcik, Sara Lacy, Jane Lee, Ramon Lopez, Alycia Meriweather, Robin Millar, Mihwa Park, Wei Rui, Allison Scheff, Angelica Stacy, Margot Vigeant, Xin Wei, Holger Wendlandt

Group B

Participants: Charles W. (Andy) Anderson, Bob Chen, Melanie Cooper, Jenny Dauer, Reinders Duit, Orna Fallik, David Fortus, Cari Herrmann-Abell, Sarah Jardeleza, yarlehavi@gmail.com, Kongju Mun, Sebastian Opitz, Nicos Papadouris, Helen Quinn, Lei Wang

Group C

Participants: Nicole Becker, Sungyoun Choi, Costas Constantinou, Hui Jin, Tom Kim, Xiufeng Liu, Knut Neumann, Jeff Nordine, Ann Novak, ppelletier@boston.k12.ma.us, Lane Seeley, Roger Tobin, Sonia Underwood, Kristen Wendell

11:00 am to 12:30 pm Third Paper Session

Group A

Launching the Space Shuttle by Making Water: The Chemists View of Energy
Angelica Stacy, Karen Chang, Janice Coonrod, and Jennifer Claesgens
Critical Friend: Cari Herrmann-Abell
Additional Discussant: Reinders Duit

Readers: Nicole Becker, Melanie Cooper, George DeBoer, Reinders Duit, arthur.eisenkraft@umb.edu, Cari Herrmann-Abell, Sarah Jardeleza, Joe Krajcik, Sara Lacy, yarlehavi@gmail.com, Alycia Meriweather, Sebastian Opitz, Angelica Stacy, Lei Wang, Holger Wendlandt

Developing Energy Conception of Students in Chemistry Class: Research on the learning and teaching of energy by BNU Group
Lei Wang, Weizhen Wang, Rui Wei, Zhiying Jing, Tao Jiang, Qingrui Meng
Critical Friend: Melanie Cooper
Additional Discussant: Nicole Becker
Readers: Nicole Becker, Melanie Cooper, George DeBoer, Reinders Duit, arthur.eisenkraft@umb.edu, Bob Geier, Cari Herrmann-Abell, Sarah Jardeleza, Joe Krajcik, Sara Lacy, yarlehavi@gmail.com, Alycia Meriweather, Sebastian Opitz, Angelica Stacy, Lei Wang, Holger Wendlandt

Group B

Conservation of energy: An analytical tool for student accounts of carbon-transforming processes
Jenny Dauer, Hannah Miller, and Charles W. (Andy) Anderson
Critical Friend: Margot Vigeant
Additional Discussant: Kristen Wendell
Readers: Charles W. (Andy) Anderson, Jenny Dauer, David Fortus, Tom Kim, Ramon Lopez, Robin Millar, Hannah Miller, Kongju Mun, Jeff Nordine, Mihwa Park, ppelletier@boston.k12.ma.us, Helen Quinn, Margot Vigeant, Xin Wei, Kristen Wendell

A space physicist's perspective on energy transformations and some implications for teaching about energy at all levels
Ramon E. Lopez
Critical Friend: Helen Quinn
Additional Discussant: Robin Millar
Readers: Charles W. (Andy) Anderson, Jenny Dauer, David Fortus, Tom Kim, Ramon Lopez, Robin Millar, Hannah Miller, Kongju Mun, Jeff Nordine, Mihwa Park, ppelletier@boston.k12.ma.us, Helen Quinn, Margot Vigeant, Xin Wei, Kristen Wendell

Group C

Mapping Energy In the Boston Public School Curriculum
Robert F. Chen, Allison Scheff, Erica Fields, Pam Pelletier, and Russ Faux
Critical Friend: Nicos Papadouris

Additional Discussant: Xiufeng Liu
Readers: Bob Chen, Sungyoun Choi, Costas Constantinou, Orna Fallik, Hui Jin, Xiufeng Liu, Knut Neumann, Ann Novak, Nicos Papadouris, Wei Rui, Allison Scheff, Lane Seeley, Roger Tobin, Sonia Underwood

Several Often-Neglected Perspectives of Energy in Chemical Education
Rui Wei and Lei Wang
Critical Friend: Roger Tobin
Additional Discussant: Hui Jin
Readers: Bob Chen, Sungyoun Choi, Costas Constantinou, Orna Fallik, Hui Jin, Xiufeng Liu, Knut Neumann, Ann Novak, Nicos Papadouris, Wei Rui, Allison Scheff, Lane Seeley, Roger Tobin, Sonia Underwood

12:30 pm Lunch

1:30 pm to 3:00 pm Third Report-Out and Discussion

Group A

Participants: Charles W. (Andy) Anderson, Nicole Becker, Sungyoun Choi, Jenny Dauer, David Fortus, Sarah Jardeleza, Xiufeng Liu, Alycia Meriweather, ppelletier@boston.k12.ma.us, Helen Quinn, Allison Scheff, Lane Seeley, Angelica Stacy, Roger Tobin, Holger Wendlandt

Group B

Participants: Costas Constantinou, Melanie Cooper, Reinders Duit, arthur.eisen kraft@umb.edu, Orna Fallik, yarlehavi@gmail.com, Ramon Lopez, Kongju Mun, Knut Neumann, Jeff Nordine, Sebastian Opitz, Nicos Papadouris, Wei Rui, Xin Wei, Kristen Wendell

Group C

Participants: Bob Chen, George DeBoer, Cari Herrmann-Abell, Hui Jin, Tom Kim, Joe Krajcik, Sara Lacy, Robin Millar, Hannah Miller, Ann Novak, Mihwa Park, Sonia Underwood, Margot Vigeant, Lei Wang

3:30 pm to 5:00 pm K-12 Educator's Perspectives

Participants: Orna Fallik, Alycia Meriweather, Ann Novak, ppelletier@boston.k12.ma.us, Holger Wendlandt, Moderator: Jeff Nordine

Monday, December 17, 2012

9:00 am to 10:30 am	Energy Round Tables
11:00 am to 12:30 pm	Future Steps
12:30 pm to 1:30 pm	Closing Thoughts

Appendix B: Biosketches of Authors and Editors

Robert F. Chen is a professor in the School For the Environment and the Director of the Center for Coastal Environmental Sensing Networks (CESN) at the University of Massachusetts Boston. He received his A.B. from Harvard University in Chemistry and Physics and his Ph.D. in Oceanography from Scripps Institution of Oceanography. After a postdoctoral fellowship at the Woods Hole Oceanographic Institution, he has remained at UMass Boston since 1993. His research interests include the cycling of chromophoric dissolved organic carbon (CDOM), carbon biogeochemistry in coastal systems, and the development of smart sensor networks in shallow water systems. He is also dedicated to ocean and environmental science education and outreach at the local, national and international levels. He was the Principal Investigator of the Watershed-Integrated Sciences Partnership (WISP; wisp.umb.edu), COSEE OCEAN (coseeocean.net), and the Boston Energy in Science Teaching (BEST) (bostonscience.net) projects and has been involved in Ocean Literacy and Energy Literacy efforts. He has published over 50 peer-reviewed articles and is an active researcher in the area of coastal observations, carbon cycling, and contaminant distribution and fate.

Melanie M. Cooper is the Lappan-Phillips Professor of Science Education and Professor of Chemistry at Michigan State University. She received her B.S. M.S. and Ph.D. in chemistry from the University of Manchester, England. Her research has focused on improving teaching and learning in large enrollment general and organic chemistry courses at the college level, and she is a proponent of evidence-based curriculum reform for example the NSF supported "Chemistry, Life, the Universe & Everything". She has also developed technological approaches to formative assessment that can recognize and respond to students free-form drawings such as the beSocratic system. She is a Fellow of the American Chemical Society and the American Association for the Advancement of Science, a member of the Leadership team for the Next Generation Science Standards (NGSS), and the National Research Council advisory Board on Science Education (BOSE). She has received a number of awards including the ACS award for research on teaching and learning 2014, the Norris award for Outstanding Achievement in teaching of chemistry in 2013, and the 2010–2011 Outstanding Undergraduate Science Teacher Award from the Society for College Science Teaching.

Jenny Dauer received a B.S. (2000) in Secondary Education and Biology at Penn State University. She worked in informal education designing inquiry experiences for elementary school children and their parents before returning to Penn State University for an M.S. (2005) in Ecology. Her graduate work involved researching nutrient cycling and biogeochemistry in forests, which she continued during her Ph.D. (2012) at Oregon State University in Forest Science. Jenny took a position at Michigan State University 2011–2013 working with Andy Anderson on student learning about matter and energy in carbon transforming processes. Her work involved directing the Carbon TIME (Transformations in Matter and Energy) Project, an NSF program to pilot and disseminate middle and high school curriculum, and research on student practices such as tracing matter and energy, and inquiry investigations about matter and energy. Currently Jenny is an Assistant Professor of Practice at the University of Nebraska-Lincoln in the School of Natural Resources where she is teaching introductory biology classes and researching student systems-thinking in agroecosystems and student understanding of climate change.

Reinders Duit is an emeritus Professor of Physics Education at the Leibniz-Institute for Science Education and Mathematics Education (IPN) in Kiel, Germany. He studied Physics and Mathematics, as well as Pedagogy, Educational Psychology, and Philosophy. His Ph.D. (1972) and Habilitation "On the role of Energy in Physics Instruction" (a kind of second Ph.D., 1985) are in Physics Education. Main concern of his work has been to closely link analytical research on science content for instruction and empirical research on teaching and learning. The theoretical framework for this work has been a Model of Education Reconstruction that is based, on the one hand, on "classical" ideas of the German pedagogical tradition of instructional planning and includes, on the other hand, more recent theoretical positions of teaching and learning as provided by constructivist epistemological views.

Arthur Eisenkraft is the Distinguished Professor of Science Education, Professor of Physics and Director of the Center of Science and Math in Context (COSMIC) at the University of Massachusetts Boston. For 25 years, he taught high school physics and was a science coordinator. He is past president of the National Science Teachers Association and has served on NRC committees resulting in the reports How People Learn, Tech Tally, America's Lab Report, and Exploring the Intersection of Science Education and twenty-first century Skills. He is currently chair of the Science Academic Advisory Committee of the College Board. He is project director of the NSF-supported Active Physics Curriculum Project that is introducing physics instruction for the first time to all students and leading a similar effort with Active Chemistry. He is chair and co-creator of the Toshiba/NSTA ExploraVision Awards, involving 15,000 students annually. His current research projects include investigating the efficacy of a second generation model of distance learning for professional development; ascertaining the effect of professional development on mandated curriculum changes; and exploring how to infuse engineering into high school curriculum.

David Fortus has a Ph.D. in Science Education from the University of Michigan, a M.Sc. in theoretical physics from the Technion in Israel, and a B.Sc. in aeronautical engineering, also from the Technion in Israel. He began his career as a science education researcher by developing learning environments that foster the transfer of scientific knowledge to real-world situations. For this work he was given awards by the National Association for Research in Science Teaching (NARST) and by the American Psychological Association (APA). Since moving to Israel his focus has shifted to the study of the environmental factors that lead to declining motivation to engage with science, in and out of schools. His publications range from science education to theoretical physics to legal economics. He is an associate editor of the Journal of Research in Science Teaching (JRST). Before joining the Weizmann Institute of Science in Israel, he was an assistant professor at Michigan State University, a high school physics teacher, and a project director in the aerospace industry.

Cari F. Herrmann-Abell, Ph.D., joined AAAS Project 2061 in 2005 as a postdoctoral fellow for the Center for Curriculum Materials in Science. Herrmann Abell is currently a senior research associate contributing to the development and evaluation of curriculum and assessment resources aligned to K-12 science learning goals and the application of Rasch modeling and other statistical methods for analyzing the psychometric properties of assessments. She is principal investigator on a project funded by the U.S. Dept. of Education's Institute of Education Sciences (IES) to develop three vertically-equated instruments to measure students' understanding of energy from elementary to high school. In addition to presenting her work at professional conferences, Herrmann Abell also leads workshops on the Project 2061 item development process for researchers and classroom teachers. She received her Ph.D. in chemistry from the University of North Carolina at Chapel Hill and her B.S. in chemistry and mathematics from Muhlenberg College in Allentown, PA.

Dr. Hui Jin is an assistant professor at School of Teaching and Learning, The Ohio State University. Her research interests include: learning progressions, conceptual development, environmental education, and secondary science teaching and teacher education. Her JRST article on a learning progression for energy was chosen as one of 'Top Five JRST Articles Recommended for Science Teachers' in 2012. Her current research explores students' argumentation of social-ecological issues and teachers' pedagogical content knowledge as it relates to matter and energy.

Joseph Krajcik serves as the director of the CREATE for STEM Institute, a joint effort of the College of Natural Science and the College of Education at Michigan State University to improve the K – 16 teaching and learning of science and mathematics through innovation and research. He is currently the principal investigator and co-principal investigator for two National Science Foundation grants to design assessments and curriculum materials aligned with the Next Generation of Science Standards (NGSS). He served as lead writer for the Physical Science Design team for the Framework for K – 12 Science Education and for the Physical Science Standards for the NGSS. Joe serves as co-editor of the Journal of Research in Science Teaching. He has authored and co-authored curriculum materials, books, software and over 100 manuscripts. He served as president of the National Association for Research in Science Teaching, from which he received the Distinguished Contributions to Science Education Through Research Award. Joe is a former chemistry and physical science teacher.

Sara Lacy is senior scientist at TERC in Cambridge, Massachusetts where she has developed science curriculum for K-12 students and courses designed to boost the physical science and pedagogical content knowledge of K-8 teachers. Her current research and development work focuses on energy. How can elementary school students develop a conceptual understanding of energy that they can use to describe phenomena they encounter in school and in their everyday lives? How can their teachers acquire the science and pedagogical content knowledge they need to teach about energy? Sara received a Ph.D. in Civil Engineering from Princeton University. She has taught engineering students at Rutgers University and at Tufts University and developed an engineering course for K-8 teachers enrolled in an online Masters in Science Education program at Lesley University.

Xiufeng Liu is Professor of Science Education and Associate Dean for Interdisciplinary Research in the Graduate School of Education, University at Buffalo, State University of New York. He obtained Ph.D. from University of British Columbia (Canada). Before his current position, he taught high school chemistry in China, and was a faculty member at St. Francis Xavier University and University of Prince Edward Island, both in Canada. Xiufeng conducts research in closely related areas of technology-enhanced science assessment, applications of Rasch measurement in science education, and public understanding of science. Among the books he has published is Using and developing measurement instruments in science education: A Rasch modeling approach (2010, Information Age Publishing). In recent years, he has been actively involved in bridging science education research in China and in the Western world.

Ramon E. Lopez is a professor of physics, specializing in space plasma physics and space weather. He also has an extensive background in physics education, with a particular interest in spatial intelligence and visual cognition, and has graduated one Ph.D. student in physics education research and has another one working on his research. In addition, Ramon has a broad background in school district reform and standards development. At the upper-division undergraduate and graduate level, Ramon has applied active learning techniques commonly used in lower division courses (such as peer-instruction).

Robin Millar taught physics and general science for 8 years in secondary schools before moving to the University of York, where he is now Salters' Professor of Science Education. He teaches on the undergraduate and postgraduate programmes in education and on the pre-service teacher training programme for secondary science teachers, and directs the Centre for Innovation and Research in Science Education. His main research interests are teaching and learning in science (especially physics), science curriculum design and development, and the assessment of science learning. He has directed several large research projects in science education, and was co-ordinator of the ESRC-funded Evidence-based Practice in Science Education (EPSE) project. He also played a leading role in several major curriculum development projects, including Salters' GCSE Science, Science for Public Understanding and Twenty First Century Science. He is a member of the Science Expert Group for the OECD PISA 2015 study. He was President of the European Science Education Research Association from 1999 to 2003, and of the Association for Science Education in 2012.

Knut Neumann is Director of the Department of Physics Education at the Leibniz-Institute for Science and Mathematics Education (IPN) and Professor of Physics Education at the University of Kiel, Germany. After graduating from the University of Düsseldorf, he did a Ph.D. in Physics Education at the University of Teacher Education Heidelberg. He worked as a post-doctoral research associate in the Research Group and Graduate School "Teaching and Learning of Science" at the University Duisburg-Essen, before he was appointed Deputy Director of Physics Education at the IPN and Associate Professor of Physics Education at the University of Kiel. In 2013 he was promoted to Full Professor and Director of the Department of Physics Education. During his career Knut developed a particular interest in how to assess students understanding of core physics concepts and skills. He currently supervises several projects on the teaching and learning about core physics concepts (e.g. energy and matter) and skills (e.g. carrying out experiments). In addition to these activities he is interested in the investigation and improvement of instructional quality in physics.

Jeffrey Nordine is the Chief Scientist at the San Antonio Children's Museum in San Antonio, Texas, and a visiting professor at the Leibniz-Institute for Science and Mathematics Education (IPN) in Kiel, Germany. At the museum, he develops science exhibits and learning experiences for elementary-aged children and provides science teacher professional development for elementary teachers. At IPN, he studies how to support students in developing scientific understandings that can form the foundation for productive future learning in both in-school and out-of-school contexts. Jeff has been an assistant professor of science education at Trinity University and a high school physics teacher in San Antonio, Texas. He has published articles in both research and practitioner journals for science teachers at all levels, kindergarten through college. Jeff earned his B.A. in Physics and MAT in Teaching from Trinity University and his M.A. in School Administration and Ph.D. in Science Education from the University of Michigan.

Nicos Papadouris is a post-doctoral research associate with the Learning in Science Group at the University of Cyprus. He has a Ph.D. in Science Education from the University of Cyprus. His research interests concentrate on the development of decision making skills and student understanding of energy related concepts as well as the design of curriculum materials for promoting epistemic awareness.

Helen Quinn is Professor Emeritus of Particle Physics and Astrophysics at SLAC National Accelerator Laboratory. She received her Ph.D. in physics at Stanford in 1967. She has taught physics at both Harvard and Stanford. Helen is an internationally recognized theoretical physicist who holds the Dirac Medal (from Italy), the Klein Medal (from Sweden) and the Sakurai Prize (American Physical Society). She is a member of the American Academy of Arts and Sciences, the National Academy of Science and the American Philosophical Society. She is a Fellow and former president of the American Physical Society. She is an Honorary Officer of the Order of Australia. Helen is Chair of the National Academy's Board on Science Education (BOSE). She served as a member of the BOSE study that developed the report "Taking Science to School" and led the committee for the "Framework for K-12 Science Education", the basis of the Next Generation Science Standards (NGSS). She is now serving on an NRC committee to develop recommendations for assessments in the light of the Framework and NGSS.

Allison Scheff is the Executive Director of Science, Technology, Engineering, and Math (STEM) at the Massachusetts Department of Higher Education and the Executive Director of the Governor's STEM Advisory Council. In this role, she strategically works with private and public stakeholders, including P-20 educators, government officials, and employers, to ensure that students are interested in STEM and well prepared to enter post-secondary STEM fields. Previously, Allison served as the Associate Director of the Center of Science and Math in Context at the University of Massachusetts Boston. While at UMass Boston, she served as the Project Director of the Boston Energy in Science Teaching project, Associate Project Director of the Boston Science Partnership, and co-designed the Science Education Fellowship program. Allison began her career in education as a 7th grade science in New Orleans, Louisiana, as part of the Teach for America program. She has a Bachelors degree in Economics from the University of Virginia and a Masters degree in Education Policy from Teachers College, Columbia University.

Lane Seeley earned his Ph.D. in experimental condensed matter physics at the University of Washington. His doctoral work focused on testing microscopic and mesoscopic models for phase changes in the nucleation of ice from liquid water. Since joining the faculty at Seattle Pacific University in 2001 he has worked closely with colleagues to build a close-knit physics department that is primarily focused on student learning. Lane has worked with departmental colleagues on several grant funded projects aimed at supporting K-12 physics and physical science teachers. He has played an active role in the development of web based diagnostic tools for physical science teachers. Most recently, Lane has been a lead researcher on the SPU Energy Project, a research effort aimed at studying and supporting energy learning among K-12 teachers. Lane's current research interests include; building bridges between the energy we learn about and the energy we care about, studying growth in learner's ability and disposition to use a rigorous energy model creatively and flexibly, understanding some of the real and perceived obstacles to student centered science instruction.

Angelica M. Stacy is Professor of Chemistry and Associate Vice Provost for the Faculty at the University of California, Berkeley. She received her Ph.D. from Cornell University in 1981, did postdoctoral work at Northwestern, and then she began in her current position in 1983. Her scholarly work includes the development of nano-scale materials for energy applications, and studies of student learning of chemistry at the high school and college levels. She has developed a high school chemistry curriculum called Living by Chemistry, a research-based set of classroom and resource materials for students and teachers. Angy served on the AP Chemistry Redesign Team, and is assisting with test development. She also served on the Physical Science Design team for the Next Generation Science Standards. In recognition of her accomplishments, she has received a number of awards, including the National Science Foundation Teacher Scholar Award, the UC Berkeley Chancellor's Award for Advancing Institutional Excellence, and the UC Berkeley Distinguished Teaching Award.

Margot Vigeant is a professor of chemical engineering and an associate dean of engineering at Bucknell University. She earned her B.S. in chemical engineering from Cornell University, and her M.S. and Ph.D., also in chemical engineering, from the University of Virginia. Her primary research focus is on engineering pedagogy, primarily at the undergraduate level. She is particularly interested in the teaching and learning of concepts related to thermodynamics, which is where her interest in the cross-cutting concept of energy springs from. She is also interested in active, collaborative, and problem-based learning, and in the ways technology in general and games in particular can be used to improve student engagement.

Professor Lei Wang is the Chair of the Institute of Chemical Education and the Vice-Chair of the Institute of Curriculum and Instruction at Beijing Normal University, Beijing, China. She is also serving as the Vice-Director of Chemical Education Committee of Chinese Chemical Society, Executive Director of Chemistry Teaching Committee of Chinese Education Society, Associate Editor of Chinese Journal of Chemical Education, member of "Expert Committee on National Basic Education Curriculum and Textbook" and "Expert Committee on National Teacher Education Teaching Resources" of Ministry of Education of the P.R. China. As the Co-PI, professor Wang directed the development of Middle School Chemical Education Standards of China and High School Chemical Education Standards of China. She was also the Chief Editor of a set of high school chemistry textbooks (eight books, Shandong Science and Technology Press), which is one set of the most widely used chemistry textbooks in mainland China.

Rui Wei is an assistant professor at Institute of Chemical Education, Beijing Normal University (China). He received his Ph.D. (2008) and M.Ed. (2005) in Chemical Education, B.S. (2003) in Chemistry, from Beijing Normal University. He teaches various courses in secondary chemical education and supervises teaching interns. His research interests include history and philosophy of chemistry, secondary school chemistry education and teacher education, metaphor and analogy in science education, vocational education in science curricula, and museum science education.

Kristen Bethke Wendell is Assistant Professor of Elementary Science Education at the University of Massachusetts Boston. Her teaching and research interests include pre-service teacher education in engineering and the integration of engineering design into children's science, reading, and writing experiences. She was graduate policy fellow at the National Academy of Engineering and received her B.S.E. from Princeton, her M.S. in Aeronautics and Astronautics from MIT, and her Ph.D. in Science Education from Tufts University.

Printed by Publishers' Graphics LLC